Robotics, AI, and Humanity

Joachim von Braun • Margaret S. Archer
Gregory M. Reichberg • Marcelo Sánchez Sorondo
Editors

Robotics, AI, and Humanity

Science, Ethics, and Policy

 Springer

Editors
Joachim von Braun
Bonn University
Bonn, Germany

Margaret S. Archer
University of Warwick
Coventry, UK

Gregory M. Reichberg
Peace Research Institute
Oslo, Norway

Marcelo Sánchez Sorondo
Pontifical Academy of Sciences
Vatican City, Vatican

ISBN 978-3-030-54172-9   ISBN 978-3-030-54173-6   (eBook)
https://doi.org/10.1007/978-3-030-54173-6

This Springer imprint is published by the registered company Springer Nature Switzerland AG
The registered company address is: Gewerbestrasse 11, 6330 Cham, Switzerland

Photo credit: Gabriella C. Marino

Artificial intelligence is at the heart of the epochal change we are experiencing. Robotics can make a better world possible if it is joined to the common good. Indeed, if technological progress increases inequalities, it is not true progress. Future advances should be oriented towards respecting the dignity of the person and of Creation. Let us pray that the progress of robotics and artificial intelligence may always serve humankind . . . we could say, may it "be human".

Pope Francis, November Prayer Intention, 5 November 2020

Acknowledgements

This edited volume, including the suggestions for action, emerged from a Conference on "Robotics, AI and Humanity, Science, Ethics and Policy", organized jointly by the Pontifical Academy of Sciences (PAS) and the Pontifical Academy of Social Sciences (PASS), 16–17 May 2019, Casina Pio IV, Vatican City. Two related conferences had previously been held at Casina Pio IV, Vatican City: "Power and Limitations of Artificial Intelligence" (December 2016) and "Artificial Intelligence and Democracy" (March 2018). The presentations and discussions from these conferences are accessible on the website of the Pontifical Academy of Sciences www.pas.va/content/accademia/en.html. The contributions by all the participants in these conferences are gratefully acknowledged. This publication has been supported by the Center for Development Research (ZEF) at Bonn University and the Research Council of Norway.

Contents

AI, Robotics, and Humanity: Opportunities, Risks, and Implications for Ethics and Policy

Joachim von Braun, Margaret S. Archer, Gregory M. Reichberg, and Marcelo Sánchez Sorondo

Contents

Abstract

This introduction to the volume gives an overview of foundational issues in AI and robotics, looking into AI's computational basis, brain–AI comparisons, and conflicting positions on AI and consciousness. AI and robotics are changing the future of society in areas such as work, education, industry, farming, and mobility, as well as services like banking. Another important concern addressed in this volume are the impacts of AI and robotics on poor people and on inequality. These implications are being reviewed, including how to respond to challenges and how to build on the opportunities afforded by AI and robotics. An important area of new risks is robotics and AI implications for militarized conflicts. Throughout this introductory chapter and in the volume, AI/robot-human interactions, as well as the ethical and religious implications, are considered. Approaches for fruitfully managing the coexistence of humans and robots are evaluated. New forms of regulating AI and robotics are called

J. von Braun (✉)
Center for Development Research (ZEF) Bonn University, Bonn, Germany
e-mail: jvonbraun@uni-bonn.de

M. S. Archer
University of Warwick, Coventry, UK
e-mail: margaret.archer@warwick.ac.uk

G. M. Reichberg
Peace Research Institute Oslo (PRIO), Research School on Peace and Conflict | Political Science, University of Oslo, Grønland, Norway
e-mail: greg.reichberg@prio.org

M. Sánchez Sorondo
Pontifical Academy of Sciences, Vatican City, Vatican
e-mail: marcelosanchez@acdscience.va; pas@pas.va

J. von Braun et al. (eds.), *Robotics, AI, and Humanity*, https://doi.org/10.1007/978-3-030-54173-6_1

for which serve the public good but also ensure proper data protection and personal privacy.

Keywords

Artificial intelligence · Robotics · Consciousness · Labor markets · Services · Poverty · Agriculture · Militarized conflicts · Regulation

Introduction[1]

Advances in artificial intelligence (AI) and robotics are accelerating. They already significantly affect the functioning of societies and economies, and they have prompted widespread debate over the benefits and drawbacks for humanity. This fast-moving field of science and technology requires our careful attention. The emergent technologies have, for instance, implications for medicine and health care, employment, transport, manufacturing, agriculture, and armed conflict. Privacy rights and the intrusion of states into personal life is a major concern (Stanley 2019). While considerable attention has been devoted to AI/robotics applications in each of these domains, this volume aims to provide a fuller picture of their connections and the possible consequences for our shared humanity. In addition to examining the current research frontiers in AI/robotics, the contributors of this volume address the likely impacts on societal well-being, the risks for peace and sustainable development as well as the attendant ethical and religious dimensions of these technologies. Attention to ethics is called for, especially as there are also long-term scenarios in AI/robotics with consequences that may ultimately challenge the place of humans in society.

AI/robotics hold much potential to address some of our most intractable social, economic, and environmental problems, thereby helping to achieve the UN's Sustainable Development Goals (SDGs), including the reduction of climate change. However, the implications of AI/robotics for equity, for poor and marginalized people, are unclear. Of growing concern are risks of AI/robotics for peace due to their enabling new forms of warfare such as cyber-attacks or autonomous weapons, thus calling for new international

security regulations. Ethical and legal aspects of AI/robotics need clarification in order to inform regulatory policies on applications and the future development of these technologies.

The volume is structured in the following four sections:

- *Foundational issues in AI and robotics*, looking into AI's computational basis, brain–AI comparisons as well as AI and consciousness.
- *AI and robotics potentially changing the future of society* in areas such as employment, education, industry, farming, mobility, and services like banking. This section also addresses the impacts of AI and robotics on poor people and inequality.
- *Robotics and AI implications for militarized conflicts* and related risks.
- *AI/robot–human interactions* and ethical and religious implications: Here approaches for managing the coexistence of humans and robots are evaluated, legal issues are addressed, and policies that can assure the regulation of AI/robotics for the good of humanity are discussed.

Foundational Issues in AI and Robotics

Overview on Perspectives

The field of AI has developed a rich variety of theoretical approaches and frameworks on the one hand, and increasingly impressive practical applications on the other. AI has the potential to bring about advances in every area of science and society. It may help us overcome some of our cognitive limitations and solve complex problems.

In health, for instance, combinations of AI/robotics with brain–computer interfaces already bring unique support to patients with sensory or motor deficits and facilitate caretaking of patients with disabilities. By providing novel tools for knowledge acquisition, AI may bring about dramatic changes in education and facilitate access to knowledge. There may also be synergies arising from robot-to-robot interaction and possible synergies of humans and robots jointly working on tasks.

While vast amounts of data present a challenge to human cognitive abilities, Big Data presents unprecedented opportunities for science and the humanities. The translational potential of Big Data is considerable, for instance in medicine, public health, education, and the management of complex systems in general (biosphere, geosphere, economy). However, the science based on Big Data as such remains empiricist and challenges us to discover the underlying causal mechanisms for generating patterns. Moreover, questions remain whether the emphasis on AI's supra-human capacities for computation and compilation mask manifold limitations

[1]The conclusions in this section partly draw on the Concluding Statement from a Conference on "Robotics, AI and Humanity, Science, Ethics and Policy", organized jointly by the Pontifical Academy of Sciences (PAS) and the Pontifical Academy of Social Sciences (PASS), 16–17 May 2019, Casina Pio IV, Vatican City. The statement is available at http://www.casinapioiv.va/content/accademia/en/events/2019/robotics/statementrobotics.html including a list of participants provided via the same link. Their contributions to the statement are acknowledged.

of current artificial systems. Moreover, there are unresolved issues of data ownership to be tackled by transparent institutional arrangements.

In the first section of this volume (Chaps. 2–5), basic concepts of AI/robotics and of cognition are addressed from different and partly conflicting perspectives. Importantly, Singer (Chap. 2) explores the difference between natural and artificial cognitive systems. Computational foundations of AI are presented by Zimmermann and Cremers (Chap. 3). Thereafter the question "could robots be conscious?" is addressed from the perspective of cognitive neuro-science of consciousness by Dehaene et al., and from a philosophical perspective by Gabriel (Chaps. 4 and 5).

Among the foundational issues of AI/robotics is the question whether machines may hypothetically attain capabilities such as consciousness. This is currently debated from the contrasting perspectives of natural science, social theory, and philosophy; as such it remains an unresolved issue, in large measure because there are many diverse definitions of "consciousness." It should not come as a surprise that the contributors of this volume are neither presenting a unanimous position on this basic issue of robot consciousness nor on a robotic form of personhood (also see Russell 2019). The concept of this volume rather is to bring the different positions together. Most contributors maintain that robots cannot be considered persons, for which reason robots will not and should not be free agents or possess rights. Some, however, argue that "command and control" conceptions may not be appropriate to human–robotic relations, and others even ask if something like "electronic citizenship" should be considered.

Christian philosophy and theology maintain that the human soul is "Imago Dei" (Sánchez Sorondo, Chap. 14). This is the metaphysical foundation according to which human persons are free and capable of ethical awareness. Although rooted in matter, human beings are also spiritual subjects whose nature transcends corporeality. In this respect, they are imperishable ("incorruptible" or "immortal" in the language of theology) and are called to a completion in God that goes beyond what the material universe can offer. Understood in this manner, neither AI nor robots can be considered persons, so robots will not and should not possess human freedom; they are unable to possess a spiritual soul and cannot be considered "images of God." They may, however, be "images of human beings" as they are created by humans to be their instruments for the good of human society. These issues are elaborated in Sect. *AI/robot–Human interactions* of the volume from religious, social science, legal, and philosophical perspectives by Sánchez Sorondo (Chap. 14), Archer (Chap. 15), and Schröder (Chap. 16).

Intelligent Agents

Zimmermann and Cremers (Chap. 3) emphasize the tremendous progress of AI in recent years and explain the conceptual foundations. They focus on the problem of induction, i.e., extracting rules from examples, which leads to the question: What set of possible models of the data generating process should a learning agent consider? To answer this question, they argue, "it is necessary to explore the notion of all possible models from a mathematical and computational point of view." Moreover, Zimmermann and Cremers (Chap. 3) are convinced that effective universal induction can play an important role in causal learning by identifying generators of observed data.

Within machine-learning research, there is a line of development that aims to identify foundational justifications for the design of cognitive agents. Such justifications would enable the derivation of theorems characterizing the possibilities and limitations of intelligent agents, as Zimmermann and Cremers elaborate (Chap. 3). Cognitive agents act within an open, partially or completely unknown environment in order to achieve goals. Key concepts for a foundational framework for AI include agents, environments, rewards, local scores, global scores, the exact model of interaction between agents and environments, and a specification of the available computational resources of agents and environments. Zimmermann and Cremers (Chap. 3) define an intelligent agent as an agent that can achieve goals in a wide range of environments.[2]

A central aspect of learning from experience is the representation and processing of uncertain knowledge. In the absence of deterministic assumptions about the world, there is no nontrivial logical conclusion that can be drawn from the past for any future event. Accordingly, it is of interest to analyze the structure of uncertainty as a question in its own right.[3] Some recent results establish a tight connection between learnability and provability, thus reducing the question of what can be effectively learned to the foundational questions of mathematics with regard to set existence axioms. Zimmermann and Cremers (Chap. 3) also point to results of "reverse mathematics," a branch of mathematical logic analyzing theorems with reference to the set of existence axioms necessary to prove them, to illustrate the implications of machine learning frameworks. They stress that artificial intelligence has advanced to a state where ethical questions and the impact on society become pressing issues, and point to the need for algorithmic transparency, accountability, and

---

[2]For an overview of inductive processes that are currently employed by AI-systems, see Russell (2019, pp. 285–295). The philosophical foundations of induction as employed by AI were explored inter alia by Goodman (1954).

[3]Probability-based reasoning was extended to AI by Pearl (1988).

unbiasedness. Until recently, basic mathematical science had few (if any) ethical issues on its agenda. However, given that mathematicians and software designers are central to the development of AI, it is essential that they consider the ethical implications of their work.[4] In light of the questions that are increasingly raised about the trustworthiness of autonomous systems, AI developers have a responsibility—that ideally should become a legal obligation—to create trustworthy and controllable robot systems.

Consciousness

Singer (Chap. 2) benchmarks robots against brains and points out that organisms and robots both need to possess an internal model of the restricted environment in which they act and both need to adjust their actions to the conditions of the respective environment in order to accomplish their tasks. Thus, they may appear to have similar challenges but—Singer stresses—the computational strategies to cope with these challenges are different for natural and artificial systems. He finds it premature to enter discussions as to whether artificial systems can acquire functions that we consider intentional and conscious or whether artificial agents can be considered moral agents with responsibility for their actions (Singer, Chap. 2).

Dehaene et al. (Chap. 4) take a different position from Singer and argue that the controversial question whether machines may ever be conscious must be based on considerations of how consciousness arises in the human brain. They suggest that the word "consciousness" conflates two different types of information-processing computations in the brain: first, the selection of information for global broadcasting (consciousness in the first sense), and second, the self-monitoring of those computations, leading to a subjective sense of certainty or error (consciousness in the second sense). They argue that current AI/robotics mostly implements computations similar to unconscious processing in the human brain. They however contend that a machine endowed with consciousness in the first and second sense as defined above would behave as if it were conscious. They acknowledge that such a functional definition of consciousness may leave some unsatisfied and note in closing, "Although centuries of philosophical dualism have led us to consider consciousness as unreducible to physical interactions, the empirical evidence is compatible with the possibility that consciousness arises from nothing more than specific computations." (Dehaene et al., Chap. 4, pp. . . .).

It may actually be the diverse concepts and definitions of consciousness that make the position taken by Dehaene et al. appear different from the concepts outlined by Singer (Chap. 2) and controversial to others like Gabriel (Chap. 5), Sánchez Sorondo (Chap. 14), and Schröder (Chap. 16). At the same time, the long-run expectations regarding machines' causal learning abilities and cognition as considered by Zimmermann and Cremers (Chap. 3) and the differently based position of Archer (Chap. 15) both seem compatible with the functional consciousness definitions of Dehaene et al. (Chap. 4). This does not apply to Gabriel (Chap. 5) who is inclined to answer the question "could a robot be conscious?" with a clear "no," drawing his lessons selectively from philosophy. He argues that the human being is the indispensable locus of ethical discovery. "Questions concerning what we ought to do as morally equipped agents subject to normative guidance largely depend on our synchronically and diachronically varying answers to the question of "who we are." " He argues that robots are not conscious and could not be conscious ". . . if consciousness is what I take it to be: a systemic feature of the animal-environment relationship." (Gabriel, Chap. 5, pp. . . .).

AI and Robotics Changing the Future of Society

In the second section of this volume, AI applications (and related emergent technologies) in health, manufacturing, services, and agriculture are reviewed. Major opportunities for advances in productivity are noted for the applications of AI/robotics in each of these sectors. However, a sectorial perspective on AI and robotics has limitations. It seems necessary to obtain a more comprehensive picture of the connections between the applications and a focus on public policies that facilitates overall fairness, inclusivity, and equity enhancement through AI/robotics.

The growing role of robotics in industries and consequences for employment are addressed (De Backer and DeStefano, Chap. 6). Von Braun and Baumüller (Chap. 7) explore the implications of AI/robotics for poverty and marginalization, including links to public health. Opportunities of AI/robotics for sustainable crop production and food security are reported by Torero (Chap. 8). The hopes and threats of including robotics in education are considered by Léna (Chap. 9), and the risks and opportunities of AI in financial services, wherein humans are increasingly replaced and even judged by machines, are critically reviewed by Pasquale (Chap. 10). The five chapters in this section of the volume are closely connected as they all draw on current and fast emerging applications of AI/robotics, but the balance of opportunities and risks for society differ greatly among these domains of AI/robotics applications and penetrations.

---

[4]The ethical impact of mathematics on technology was groundbreakingly presented by Wiener (1960).

Work

Unless channeled for public benefit, AI may raise important concerns for the economy and the stability of society. Jobs may be lost to computerized devices in manufacturing, with a resulting increase in income disparity and knowledge gaps. Advances in automation and increased supplies of artificial labor particularly in the agricultural and industrial sectors can significantly reduce employment in emerging economies. Through linkages within global value chains, workers in low-income countries may be affected by growing reliance of industries and services in higher-income countries on robotics, which could reduce the need for outsourcing routine jobs to low-wage regions. However, robot use could also increase the demand for labor by reducing the cost of production, leading to industrial expansion. Reliable estimates of jobs lost or new jobs created in industries by robots are currently lacking. This uncertainty creates fears, and it is thus not surprising that the employment and work implications of robotics are a major public policy issue (Baldwin 2019). Policies should aim at providing the necessary social security measures for affected workers while investing in the development of the necessary skills to take advantage of the new jobs created.

The state might consider to redistribute the profits that are earned from the work carried out by robots. Such redistribution could, for instance, pay for the retraining of affected individuals so that they can remain within the work force. In this context, it is important to remember that many of these new technological innovations are being achieved with support from public funding. Robots, AI, and digital capital in general can be considered as a tax base. Currently this is not the case; human labor is directly taxed through income tax of workers, but robot labor is not. In this way, robotic systems are indirectly subsidized, if companies can offset them in their accounting systems, thus reducing corporate taxation. Such distortions should be carefully analyzed and, where there is disfavoring of human workers while favoring investment in robots, this should be reversed.

Returning to economy-wide AI/robotic effects including employment, De Backer and DeStefano (Chap. 6) note that the growing investment in robotics is an important aspect of the increasing digitalization of economy. They note that while economic research has recently begun to consider the role of robotics in modern economies, the empirical analysis remains overall too limited, except for the potential employment effects of robots. So far, the empirical evidence on effects of robotics on employment is mixed, as shown in the review by De Backer and DeStefano (Chap. 6). They also stress that the effects of robots on economies go further than employment effects, as they identify increasing impacts on the organization of production in global value chains. These change the division of labor between richer and poorer economies. An important finding of De Backer and DeStefano is the negative effect that robotics may have on the offshoring of activities from developed economies, which means that robotics seem to decrease the incentives for relocating production activities and jobs toward emerging economies. As a consequence, corporations and governments in emerging economies have also identified robotics as a determinant of their future economic success. Thereby, global spreading of automation with AI/robotics can lead to faster deindustrialization in the growth and development process. Low-cost jobs in manufacturing may increasingly be conducted by robots such that fewer jobs than expected may be on offer for humans even if industries were to grow in emerging economies.

AI/Robotics: Poverty and Welfare

Attention to robot rights seems overrated in comparison to attention to implications of robotics and AI for the poorer segments of societies, according to von Braun and Baumüller (Chap. 7). Opportunities and risks of AI/robotics for sustainable development and people suffering from poverty need more attention in research and in policy (Birhane and van Dijk 2020). Especially implications for low-income countries, marginalized population groups, and women need study and consideration in programs and policies. Outcomes of AI/robotics depend upon actual designs and applications. Some examples demonstrate this crosscutting issue:

- Big Data-based algorithms drawing patterns from past occurrences can perpetuate discrimination in business practices—or can detect such discrimination and provide a basis for corrective policy actions, depending on their application and the attention given to this issue. For instance, new financial systems (fintech) can be designed to include or to exclude (Chap. 10).
- AI/robotics-aided teaching resources offer opportunities in many low-income regions, but the potential of these resources greatly depends on both the teaching content and teachers' qualifications (Léna, Chap. 9).
- As a large proportion of the poor live on small farms, particularly in Africa and South and East Asia, it matters whether or not they get access to meaningful digital technologies and AI. Examples are land ownership certification through blockchain technology, precision technologies in land and crop management, and many more (Chaps. 7 and 8).
- Direct and indirect environmental impacts of AI/robotics should receive more attention. Monitoring through smart remote sensing in terrestrial and aquatic systems can be much enhanced to assess change in biodiversity and

impacts of interventions. However, there is also the issue of pollution through electronic waste dumped by industrialized countries in low-income countries. This issue needs attention as does the carbon footprint of AI/robotics.

Effects of robotics and AI for such structural changes in economies and for jobs will not be neutral for people suffering from poverty and marginalization. Extreme poverty is on the decline worldwide, and robotics and AI are potential game changers for accelerated or decelerated poverty reduction. Information on how AI/robotics may affect the poor is scarce. Von Braun and Baumüller (Chap. 7) address this gap. They establish a framework that depicts AI/robotics impact pathways on poverty and marginality conditions, health, education, public services, work, and farming as well as on the voice and empowerment of the poor. The framework identifies points of entry of AI/robotics and is complemented by a more detailed discussion of the pathways in which changes through AI/robotics in these areas may relate positively or negatively to the livelihoods of the poor. They conclude that the context of countries and societies play an important role in determining the consequences of AI/robotics for the diverse population groups at risk of falling into poverty. Without a clear focus on the characteristics and endowments of people, innovations in AI/robotics may not only bypass them but adversely impact them directly or indirectly through markets and services of relevance to their communities. Empirical scenario building and modelling is called for to better understand the components in AI/robotics innovations and to identify how they can best support livelihoods of households and communities suffering from poverty. Von Braun and Baumüller (Chap. 7) note that outcomes much depend on policies accompanying AI and robotics. Lee points to solutions with new government initiatives that finance care and creativity (Chap. 22).

Food and Agriculture

Closely related to poverty is the influence of AI/robotics on food security and agriculture. The global poor predominantly work in agriculture, and due to their low levels of income they spend a large shares of their income on food. Torero (Chap. 8) addresses AI/robotics in the food systems and points out that agricultural production—while under climate stress—still must increase while minimizing the negative impacts on ecosystems, such as the current decline in biodiversity. An interesting example is the case of autonomous robots for farm operations. Robotics are becoming increasingly scale neutral, which could benefit small farmers via wage and price effects (Fabregas et al. 2019). AI and robotics play a growing role in all elements of food value chains, where automation is driven

by labor costs as well as by demands for hygiene and food safety in processing.

Torero (Chap. 8) outlines the opportunities of new technologies for smallholder households. Small-size mechanization offers possibilities for remote areas, steep slopes or soft soil areas. Previously marginal areas could be productive again. Precision farming could be introduced to farmers that have little capital thus allowing them to adopt climate-smart practices. Farmers can be providers and consumers of data, as they link to cloud technologies using their smartphones, connecting to risk management instruments and track crop damage in real time.

Economic context may change with technologies. Buying new machinery may no longer mean getting oneself into debt thanks to better access to credit and leasing options. The reduced scale of efficient production would mean higher profitability for smallholders. Robots in the field also represent opportunities for income diversification for farmers and their family members as the need to use family labor for low productivity tasks is reduced and time can be allocated for more profit-generating activities. Additionally, robots can operate 24/7, allowing more precision on timing of harvest, especially for high-value commodities like grapes or strawberries.

Education

Besides health and caregiving, where innovations in AI/robotics have had a strong impact, in education and finance this impact is also likely to increase in the future. In education—be it in the classroom or in distance-learning systems, focused on children or on training and retraining of adults—robotics is already having an impact (Léna, Chap. 9). With the addition of AI, robotics offers to expand the reach of teaching in exciting new ways. At the same time, there are also concerns about new dependencies and unknown effects of these technologies on minds. Léna sees child education as a special case, due to it involving emotions as well as knowledge communicated between children and adults. He examines some of the modalities of teacher substitution by AI/robotic resources and discusses their ethical aspects. He emphasizes positive aspects of computer-aided education in contexts in which teachers are lacking. The technical possibilities combining artificial intelligence and teaching may be large, but the costs need consideration too. The ethical questions raised by these developments need attention, since children are extremely vulnerable human beings. As the need to develop education worldwide are so pressing, any reasonable solution which benefits from these technological advances can become helpful, especially in the area of computer-aided education.

Finance, Insurance, and Other Services

Turning to important service domains like finance and insurance, and real estate, some opportunities but also worrisome trends of applications of AI-based algorithms relying on Big Data are quickly emerging. In these domains, humans are increasingly assessed and judged by machines. Pasquale (Chap. 10) looks into the financial technology (Fintech) landscape, which ranges from automation of office procedures to new approaches of storing and transferring value, and granting credit. For instance, new services—e.g., insurance sold by the hour—are emerging, and investments on stock exchanges are conducted increasingly by AI systems, instead of by traders. These innovations in AI, other than industrial robotics, are probably already changing and reducing employment of (former) high-skill/high-income segments, but not routine tasks in manufacturing. A basis for some of the Fintech operations by established finance institutions and start-ups is the use of data sources from social media with algorithms to assess credit risk. Another area is financial institutions adopting distributed ledger technologies. Pasquale (Chap. 10) divides the Fintech landscape into two spheres, "incrementalist Fintech" and "futurist Fintech." Incrementalist Fintech uses new data, algorithms, and software to perform traditional tasks of existing financial institutions. Emerging AI/robotics do not change the underlying nature of underwriting, payment processing, or lending of the financial sector. Regulators still cover these institutions, and their adherence to rules accordingly assures that long-standing principles of financial regulation persist. Yet, futurist Fintech claims to disrupt financial markets in ways that supersede regulation or even render it obsolete. If blockchain memorializing of transactions is actually "immutable," the need for regulatory interventions to promote security or prevent modification of records may no longer be needed.

Pasquale (Chap. 10) sees large issues with futurist Fintech, which engages in detailed surveillance in order to get access to services. These can become predatory, creepy, and objectionable on diverse grounds, including that they subordinate inclusion, when they allow persons to compete for advantage in financial markets in ways that undermine their financial health, dignity, and political power (Pasquale, Chap. 10). Algorithmic accountability has become an important concern for reasons of discriminating against women for lower-paying jobs, discriminating against the aged, and stimulating consumers into buying things by sophisticated social psychology and individualized advertising based on "Phishing."[5] Pistor (2019) describes networks of obligation that even states find exceptionally difficult to break. Capital has imbricated into international legal orders that hide wealth and income from regulators and tax authorities. Cryptocurrency may become a tool for deflecting legal demands and serve the rich. Golumbia (2009) points at the potential destabilizing effects of cryptocurrencies for financial regulation and monetary policy. Pasquale (Chap. 10) stresses that both incrementalist and futurist Fintech expose the hidden costs of digital efforts to circumvent or co-opt state monetary authorities.

In some areas of innovations in AI/robotics, their future trajectories already seem quite clear. For example, robotics are fast expanding in space exploration and satellite systems observing earth,[6] in surgery and other forms of medical technology,[7] and in monitoring processes of change in the Anthropocene, for instance related to crop developments at small scales.[8] Paradigmatic for many application scenarios not just in industry but also in care and health are robotic hand-arm systems for which the challenges of precision, sensitivity, and robustness come along with safe grasping requirements. Promising applications are evolving in tele-manipulation systems in a variety of areas such as healthcare, factory production, and mobility. Depending on each of these areas, sound IP standards and/or open-source innovation systems should be explored systematically, in order to shape optimal innovation pathways. This is a promising area of economic, technological, legal, and political science research.

Robotics/AI and Militarized Conflict

Robotics and AI in militarized conflicts raise new challenges for building and strengthening peace among nations and for the prevention of war and militarized conflict in general. New political and legal principles and arrangements are needed but are evolving too slowly.

Within militarized conflict, AI-based systems (including robots) can serve a variety of purposes, inter alia, extracting wounded personnel, monitoring compliance with laws of war/rules of engagement, improving situational awareness/battlefield planning, and making targeting decisions. While it is the last category that raises the most challenging moral issues, in all cases the implications of lowered barriers of warfare, escalatory dangers, as well as systemic risks must be carefully examined before AI is implemented in battlefield settings.

---

[5] Relevant for insights in these issues are the analyses by Akerlof and Shiller (2015) in their book on "Phishing for Phools: The Economics of Manipulation and Deception."

[6] See for instance Martin Sweeting's (2020) review of opportunities of small satellites for earth observation.

[7] For a review on AI and robotics in health see for instance Erwin Loh (2018).

[8] On assessment of fossil fuel and anthrogpogenic emissions effects on public health and climate see Jos Lelieveld et al. (2019). On new ways of crop monitoring using AI see, for instance, Burke and Lobell (2017).

Worries about falling behind in the race to develop new AI military applications must not become an excuse for short-circuiting safety research, testing, and adequate training. Because weapon design is trending away from large-scale infrastructure toward autonomous, decentralized, and miniaturized systems, the destructive effects may be magnified compared to most systems operative today (Danzig 2018). AI-based technologies should be designed so they enhance (and do not detract from) the exercise of sound moral judgment by military personnel, which need not only more but also very different types of training under the changed circumstances. Whatever military advantages might accrue from the use of AI, human agents—political and military—must continue to assume responsibility for actions carried out in wartime.

International standards are urgently needed. Ideally, these would regulate the use of AI with respect to military planning (where AI risks to encourage pre-emptive strategies), cyberattack/defense as well as the kinetic battlefields of land, air, sea, undersea, and outer space. With respect to lethal autonomous weapon systems, given the present state of technical competence (and for the foreseeable future), no systems should be deployed that function in unsupervised mode. Whatever the battlefield—cyber or kinetic—human accountability must be maintained, so that adherence to internationally recognized laws of war can be assured and violations sanctioned.

Robots are increasingly utilized on the battlefield for a variety of tasks (Swett et al., Chap. 11). Human-piloted, remote-controlled fielded systems currently predominate. These include unmanned aerial vehicles (often called "drones"), unmanned ground, surface, and underwater vehicles as well as integrated air-defense and smart weapons. The authors recognize, however, that an arms race is currently underway to operate these robotic platforms as AI-enabled weapon systems. Some of these systems are being designed to act autonomously, i.e., without the direct intervention of a human operator for making targeting decisions. Motivating this drive toward AI-based autonomous targeting systems (Lethal Autonomous Weapons, or LAWS) brings about several factors, such as increasing the speed of decision-making, expanding the volume of information necessary for complex decisions, or carrying out operations in settings where the segments of the electromagnetic spectrum needed for secure communications are contested. Significant developments are also underway within the field of human–machine interaction, where the goal is to augment the abilities of military personnel in battlefield settings, providing, for instance, enhanced situational awareness or delegating to an AI-guided machine some aspect of a joint mission. This is the concept of human–AI "teaming" that is gaining ground in military planning. On this understanding, humans and AI function as tightly coordinated parts of a multi-agent team, requiring novel modes of communication and trust. The limitations of AI must be properly understood by system designers and military personnel if AI applications are to promote more, not less, adherence to norms of armed conflict.

It has long been recognized that the battlefield is an especially challenging domain for ethical assessment. It involves the infliction of the worst sorts of harm: killing, maiming, destruction of property, and devastation of the natural environment. Decision-making in war is carried out under conditions of urgency and disorder. This Clausewitz famously termed the "fog of war." Showing how ethics are realistically applicable in such a setting has long taxed philosophers, lawyers, and military ethicists. The advent of AI has added a new layer of complexity. Hopes have been kindled for smarter targeting on the battlefield, fewer combatants, and hence less bloodshed; simultaneously, warnings have been issued on the new arms race in "killer robots," as well as the risks associated with delegating lethal decisions to increasingly complex and autonomous machines. Because LAWS are designed to make targeting decisions without the direct intervention of human agents (who are "out of the killing loop"), considerable debate has arisen on whether this mode of autonomous targeting should be deemed morally permissible. Surveying the contours of this debate, Reichberg and Syse (Chap. 12) first present a prominent ethical argument that has been advanced in favor of LAWS, namely, that AI-directed robotic combatants would have an advantage over their human counterparts, insofar as the former would operate solely on the basis of rational assessment, while the latter are often swayed by emotions that conduce to poor judgment. Several counter arguments are then presented, inter alia, (i) that emotions have a positive influence on moral judgment and are indispensable to it; (ii) that it is a violation of human dignity to be killed by a machine, as opposed to being killed by a human being; and (iii) that the honor of the military profession hinges on maintaining an equality of risk between combatants, an equality that would be removed if one side delegates its fighting to robots. The chapter concludes with a reflection on the moral challenges posed by human–AI teaming in battlefield settings, and on how virtue ethics provide a valuable framework for addressing these challenges.

Nuclear deterrence is an integral aspect of the current security architecture and the question has arisen whether adoption of AI will enhance the stability of this architecture or weaken it. The stakes are very high. Akiyama (Chap. 13) examines the specific case of nuclear deterrence, namely, the possession of nuclear weapons, not specifically for battlefield use but to dissuade others from mounting a nuclear or conventional attack. Stable deterrence depends on a complex web of risk perceptions. All sorts of distortions and errors are possible, especially in moments of crisis. AI might contribute toward reinforcing the rationality of decision-making under these conditions (easily affected by the emotional disturbances and fallacious inferences to which human beings are prone), thereby preventing an accidental launch or unintended escalation. Conversely, judgments about what does or does not fit the "national interest" are not well suited to AI (at least in its current state of development). A purely logical reasoning process based on the wrong values could have disastrous consequences, which would clearly be the case if an AI-based machine were allowed to make the launch decision (which virtually all experts would emphatically exclude), but grave problems could similarly arise if a human actor relied too heavily on AI input.

Implications for Ethics and Policies

Major research is underway in areas that define us as humans, such as language, symbol processing, one-shot learning, self-evaluation, confidence judgment, program induction, conceiving goals, and integrating existing modules into an overarching, multi-purpose intelligent architecture (Zimmermann and Cremers, Chap. 3). Computational agents trained by reinforcement learning and deep learning frameworks demonstrate outstanding performance in tasks previously thought intractable. While a thorough foundation for a general theory of computational cognitive agents is still missing, the conceptual and practical advance of AI has reached a state in which ethical and safety questions and the impact on society overall become pressing issues. For example, AI-based inferences of persons' feelings derived from face recognition data are such an issue.

AI/Robotics: Human and Social Relations

The spread of robotics profoundly modifies human and social relations in many spheres of society, in the family as well as in the workplace and in the public sphere. These modifications can take on the character of hybridization processes between the human characteristics of relationships and the artificial ones, hence between analogical and virtual reality. Therefore, it is necessary to increase scientific research on issues concerning the social effects that derive from delegating relevant aspects of social organization to AI and robots. An aim of such research should be to understand how it is possible to govern the relevant processes of change and produce those relational goods that realize a virtuous human fulfillment within a sustainable and fair societal development.

We noted above that fast progress in robotics engineering is transforming whole industries (industry 4.0). The evolution of the internet of things (IoT) with communication among machines and inter-connected machine learning results in major changes for services such as banking and finance as reviewed above. Robot–robot and human–robot interactions are increasingly intensive; yet, AI systems are hard to test and validate. This raises issues of trust in AI and robots, and issues of regulation and ownership of data, assignment of responsibilities, and transparency of algorithms are arising and require legitimate institutional arrangements.

We can distinguish between mechanical robots, designed to accomplish routine tasks in production, and AI/robotics capacities to assist in social care, medical procedures, safe and energy efficient mobility systems, educational tasks, and scientific research. While intelligent assistants may benefit adults and children alike, they also carry risks because their impact on the developing brain is unknown, and because people may lose motivation in areas where AI appears superior.

Basically robots are instruments in the perspective of Sánchez Sorondo (Chap. 14) with the term "instrument" being used in various senses. "The primary sense is clearly that of not being a cause of itself or not existing by itself." Aristotle defines being free as the one that is a cause of himself or exists on its own and for himself, i.e., one who is cause of himself (*causa sui* or *causa sui ipsius*)." From the Christian perspective, "... for a being to be free and a cause of himself, it is necessary that he/she be a person endowed with a spiritual soul, on which his or her cognitive and volitional activity is based" (Sánchez Sorondo, Chap. 14, p. 173). An artificially intelligent robotic entity does not meet this standard. As an artifact and not a natural reality, the AI/robotic entity is invented by human beings to fulfill a purpose imposed by human beings. It can become a perfect entity that performs operations in quantity and quality more precisely than a human being, but it cannot choose for itself a different purpose from what was programmed in it for by a human being. As such, the artificially intelligent robot is a means at the service of humans.

The majority of social scientists have subscribed to a similar conclusion as the above. Philosophically, as distinct from theologically, this entails some version of "human essentialism" and "species-ism" that far from all would endorse in other contexts (e.g., social constructionists). The result is to reinforce Robophobia and the supposed need to protect humankind. Margaret S. Archer (Chap. 15) seeks to put the case for potential Robophilia based upon the positive properties and powers deriving from humans and AI co-working together in synergy. Hence, Archer asks "Can Human Beings and AI Robots be Friends?" She stresses the need to foreground social change (given this is increasingly morphogenetic rather than morphostatic) for structure, culture, and agency. Because of the central role the social sciences assign to agents and their "agency" this is crucial as we humans are continually "enhanced" and have since long increased their height and longevity. Human enhancement speeded up with medical advances from ear trumpets, to spectacles, to artificial insertions in the body, transplants, and genetic modification. In short, the constitution of most adult human bodies is no longer wholly organic. In consequence, the definition of "being human" is carried further away from naturalism and human essentialism. The old bifurcation into the "wet" and the "dry" is no longer a simple binary one. If the classical distinguishing feature of humankind was held to be possession of a "soul," this was never considered to be a biological organ. Today, she argues, with the growing capacities of AI robots, the tables are turned and implicitly pose the question, "so are they not persons too?" The paradox is that the public admires the AI who defeated Chess and Go world champions. They are content with AI roles in care of the elderly, with autistic children, and in surgical interventions, none of which are purely computational feats, but the fear of artificially intelligent robots "taking over" remains and repeats Asimov's (1950) protective laws. Perceiving this as a threat alone owes much to the influence of the Arts, especially sci-fi; Robophobia dominates Robophilia in popular imagination and academia. With AI capacities now including "error-detection," "self-elaboration of their pre-programming," and "adaptation to their environment," they have the potential for *active collaboration* with humankind, in research, therapy, and care. This would entail *synergy or co-working* between humans and AI beings.

Wolfgang Schröder (Chap. 16) also addresses robot–human interaction issues, but from positions in legal philosophy and ethics. He asks what normative conditions should apply to the use of robots in human society, and ranks the controversies about the moral and legal status of robots and of humanoid robots in particular among the top debates in recent practical philosophy and legal theory. As robots become increasingly sophisticated, and engineers make them combine properties of tools with seemingly psychological capacities that were thought to be reserved for humans, such considerations become pressing. While some are inclined to view humanoid robots as more than just tools, discussions are dominated by a clear divide: What some find appealing, others deem appalling, i.e., "robot rights" and "legal personhood" for AI systems. Obviously, we need to organize human–robot interactions according to ethical and juridical principles that optimize benefit and minimize mutual harm. Schröder concludes, based on a careful consideration of legal and philosophical positions, that, even the most human-like behaving robot will not lose its ontological machine character merely by being open to "humanizing" interpretations. However, even if they do not present an anthropological challenge, they certainly present an ethical one, because both AI and ethical frameworks are artifacts *of* our societies—and therefore subject to human choice and human control, Schröder argues. The latter holds for the moral status of robots and other AI systems, too. This status remains a choice, not a necessity. Schröder suggests that there should be no context of action where a complete absence of human respect for the integrity of other beings (natural or artificial) would be morally allowed or even encouraged. Avoiding disrespectful treatment of robots is ultimately for the sake of the humans, not for the sake of the robots. Maybe this insight can contribute to inspire an "overlapping consensus" as conceptualized by John Rawls (1987) in further discussions on responsibly coordinating human-robot interactions.

Human–robot interactions and affective computing's ethical implications are elaborated by Devillers (Chap. 17). The field of social robotics is fast developing and will have wide implications especially within health care, where much progress has been made toward the development of "companion robots." Such robots provide therapeutic or monitoring assistance to patients with a range of disabilities over a long timeframe. Preliminary results show that such robots may be particularly beneficial for use with individuals who suffer from neurodegenerative pathologies. Treatment can be accorded around the clock and with a level of patience rarely found among human healthcare workers. Several elements are requisite for the effective deployment of companion robots: They must be able to detect human emotions and in turn mimic human emotional reactions as well as having an outward appearance that corresponds to human expectations about their caregiving role. Devillers' chapter presents laboratory findings on AI-systems that enable robots to recognize specific emotions and adapt their behavior accordingly. Emotional perception by humans (how language and gestures are interpreted by us to grasp the emotional states of others) is being studied as a guide to programing robots so they can simulate emotions in their interactions with humans. Some of the relevant ethical issues are examined, particularly the use of "nudges," whereby detection of a human subject's cognitive biases enables the robot to initiate, through verbal or nonverbal cues, remedial measures to affect the subject's behavior in a beneficial direction. Whether this constitutes manipulation and is open to potential abuse merits closer study.

Taking the encyclical *Laudato si'* and its call for an "integral ecology" as its starting point, Donati (Chap. 18) examines how the processes of human enhancement that have been brought about by the digital revolution (including AI and robotics) have given rise to new social relationships. A central question consists in asking how the Digital Technological Mix, a hybridization of the human and nonhuman that issues from AI and related technologies, can promote human dignity. Hybridization is defined here as entanglements and interchanges between digital machines, their ways of operating, and human elements in social practices. The issue is not whether AI or robots can assume human-like characteristics, but how they interact with humans and affect their social relationships, thereby generating a new kind of society.

Advocating for the positive coexistence of humans and AI, Lee (Chap. 22) shares Donati's vision of a system that provides for all members of society, but one that also uses the wealth generated by AI to build a society that is more compassionate, loving, and ultimately human. Lee believes it is incumbent on us to use the economic abundance of the AI age to foster the values of volunteers who devote their time and energy toward making their communities more caring. As a practical measure, they propose to explore the creation not of a universal basic income to protect against AI/robotics' labor saving and job cutting effects, but a "social investment stipend." The stipend would be given to those who invest their time and energy in those activities that promote a kind, compassionate, and creative society, i.e., care work, community service, and education. It would put the economic bounty generated by AI to work in building a better society, rather than just numbing the pain of AI-induced job losses.

Joint action in the sphere of human–human interrelations may be a model for human–robot interactions. Human–human interrelations are only possible when several prerequisites are met (Clodic and Alami, Chap. 19), inter alia: (i) that each agent has a representation within itself of its distinction from the other so that their respective tasks can be coordinated; (ii) each agent attends to the same object, is aware of that fact, and the two sets of "attentions" are causally connected; and (iii) each agent understands the other's action as intentional, namely one where means are selected in view of a goal so that each is able to make an action-to-goal prediction about the other. The authors explain how human–robot interaction must follow the same threefold pattern. In this context, two key problems emerge. First, how can a robot be programed to recognize its distinction from a human subject in the same space, to detect when a human agent is attending to something, and make judgments about the goal-directedness of the other's actions such that the appropriate predictions can be made? Second, what must humans learn about robots so they are able to interact reliably with them in view of a shared goal? This dual process (robot perception of its human counterpart and human perception of the robot) is here examined by reference to the laboratory case of a human and a robot who team up in building a stack with four blocks.

Robots are increasingly prevalent in human life and their place is expected to grow exponentially in the coming years (van Wynsberghe, Chap. 20). Whether their impact is positive or negative will depend not only on how they are used, but also and especially on how they have been designed. If ethical use is to be made of robots, an ethical perspective must be made integral to their design and production. Today this approach goes by the name "responsible robotics," the parameters of which are laid out in the present chapter. Identifying lines of responsibility among the actors involved in a robot's development and implementation, as well as establishing procedures to track these responsibilities as they impact the robot's future use, constitutes the "responsibility attribution framework" for responsible robotics. Whereas Asimov's (1950) famous "three laws of robotics" focused on the behavior of the robot, current "responsible robotics" redirects our attention to the human actors, designers, and producers, who are involved in the development chain of robots. The robotics sector has become highly complex, with a wide network of actors engaged in various phases of development and production of a multitude of applications. Understanding the different sorts of responsibility—moral, legal, backward- and forward-looking, individual and collective—that are relevant within this space, enables the articulation of an adequate attribution framework of responsibility for the robotics industry.

Regulating for Good National and International Governance

An awareness that AI-based technologies have far outpaced the existing regulatory frameworks has raised challenging questions about how to set limits on the most dangerous developments (lethal autonomous weapons or surveillance bots, for instance). Under the assumption that the robotics industry cannot be relied on to regulate itself, calls for government intervention within the regulatory space—national and international—have multiplied (Kane, Chap. 21). The author recognizes how AI technologies offer a special difficulty to any regulatory authority, given their complexity (not easily understood by nonspecialists) and their rapid pace of development (a specific application will often be obsolete by the time needed untill regulations are finally established). The various approaches to regulating AI fall into two main categories. A sectoral approach looks to identify the societal risks posed by individual technologies, so that preventive or mitigating strategies can be implemented, on the assumption that the rules applicable to AI, in say the financial industry, would be very different from those relevant to heath care providers. A cross-sectoral approach, by contrast, involves the formulation of rules (whether norms adopted by industrial consensus or laws set down by governmental authority) that, as the name implies, would have application to AI-based technologies in their generality. After surveying some domestic and international initiatives that typify the two approaches, the chapter concludes with a list of 15 recommendations to guide reflection on the promotion of societally beneficial AI.

Toward Global AI Frameworks

Over the past two decades, the field of AI/robotics has spurred a multitude of applications for novel services. A particularly fast and enthusiastic development of AI/Robotics occurred in the first and second decades of the century around industrial applications and financial services. Whether or not the current decade will see continued fast innovation and expansion of AI-based commercial and public services is an open question. An important issue is and will become even more so, how the AI innovation fields are being dominated by national strategies especially in the USA and China, or if some global arrangement for standard setting and openness can be contemplated to serve the global common good along with justifiable protection of intellectual property (IP) and fair competition in the private sector. This will require numerous rounds of negotiation concerning AI/Robotics, comparable with the development of rules on trade and foreign direct investment. The United Nations could provide the framework. The European Union would have a strong interest in engaging in such a venture, too. Civil society may play key roles from the perspective of protection of privacy.

Whether AI may serve good governance or bad governance depends, inter alia, on the corresponding regulatory environment. Risks of manipulative applications of AI for shaping public opinion and electoral interference need attention, and national and international controls are called for. The identification and prevention of illegal transactions, for instance money received from criminal activities such as drug trafficking, human trafficking or illegal transplants, may serve positively, but when AI is in the hands of oppressive governments or unethically operating companies, AI/robotics may be used for political gain, exploitation, and undermining of political freedom. The new technologies must not become instruments to enslave people or further marginalize the people suffering already from poverty.

Efforts of publicly supported development of intelligent machines should be directed to the common good. The impact on public goods and services, as well as health, education, and sustainability, must be paramount. AI may have unexpected biases or inhuman consequences including segmentation of society and racial and gender bias. These need to be addressed within different regulatory instances—both governmental and nongovernmental—before they occur. These are national and global issues and the latter need further attention from the United Nations.

The war-related risks of AI/robotics need to be addressed. States should agree on concrete steps to reduce the risk of AI-facilitated and possibly escalated wars and aim for mechanisms that heighten rather than lower the barriers of development or use of autonomous weapons, and fostering the understanding that war is to be prevented in general. With respect to lethal autonomous weapon systems, no systems should be deployed that function in an unsupervised mode. Human accountability must be maintained so that adherence to internationally recognized laws of war can be assured and violations sanctioned.

Protecting People's and Individual Human Rights and Privacy

AI/robotics offer great opportunities and entail risks; therefore, regulations should be appropriately designed by legitimate public institutions, not hampering opportunities, but also not stimulating excessive risk-taking and bias. This requires a framework in which inclusive public societal discourse is informed by scientific inquiry within different disciplines. All segments of society should participate in the needed dialogue. New forms of regulating the digital economy are called for that ensure proper data protection and personal privacy. Moreover, deontic values such as "permitted," "obligatory," and "forbidden" need to be strengthened to navigate the web and interact with robots. Human rights need to be protected from intrusive AI.

Regarding privacy, access to new knowledge, and information rights, the poor are particularly threatened because of their current lack of power and voice. AI and robotics need to be accompanied by more empowerment of the poor through information, education, and investment in skills. Policies should aim for sharing the benefits of productivity growth through a combination of profit-sharing, not by subsidizing robots but through considering (digital) capital taxation, and a reduction of working time spent on routine tasks.

Developing Corporate Standards

The private sector generates many innovations in AI/robotics. It needs to establish sound rules and standards framed by public policy. Companies, including the large corporations developing and using AI, should create ethical and safety boards, and join with nonprofit organizations that aim to establish best practices and standards for the beneficial deployment of AI/ robotics. Appropriate protocols for AI/robotics' safety need to be developed, such as duplicated checking by independent design teams. The passing of ethical and safety tests, evaluating for instance the social impact or covert racial prejudice, should become a prerequisite for the release of new AI software. External civil boards performing recurrent and transparent evaluation of all technologies, including in the military, should be considered. Scientists and engineers, as the designers of AI and robot devices, have a responsibility to ensure that their inventions and innovations are safe and can be used for moral purposes (Gibney 2020). In this context, Pope Francis has called for the elaboration of ethical guidelines for the design of algorithms, namely an "algorethics." To this he adds that "it is not enough simply to trust in the moral sense of researchers and developers of devices and algorithms. There is a need to create intermediate social bodies that can incorporate and express the ethical sensibilities of users and educators." (Pope Francis 2020). Developing and setting such standards would help in mutual learning and innovation with international spillover effects. Standards for protecting people's rights for choices and privacy also apply and may be viewed differently around the world. The general standards, however, are defined for human dignity in the UN Human Rights codex.

References

Akerlof, G. A., & Shiller, R. J. (2015). *Phishing for phools: The economics of manipulation and deception*. Princeton, NJ: Princeton University Press.

Asimov, I. (1950). Runaround. In I. Asimov (Ed.), *I, Robot*. Garden City: Doubleday.

Baldwin, R. (2019). *The globotics upheaval: Globalization, robotics, and the future of work*. New York: Oxford Umiversity Press.

Birhane, A. & van Dijk, J. (2020). *Robot rights? Let's talk about human welfare instead*. Paper accepted to the AIES 2020 conference in New York, February 2020. Doi: https://doi.org/10.1145/3375627.3375855.

Burke, M., & Lobell, D. B. (2017). Satellite-based assessment of yield variation and its determinants in smallholder African systems. *PNAS, 114*(9), 2189–2194; first published February 15, 2017.. https://doi.org/10.1073/pnas.1616919114.

Danzig, R. (2018). *Technology roulette: Managing loss of control as many militaries pursue technological superiority*. Washington, D.C.: Center for a New American Security. Burke M.

Fabregas, R., Kremer, M., & Schilbach, F. (2019). Realizing the potential of digital development: The case of agricultural advice. *Science, 366*, 1328. https://doi.org/10.1126/science.aay3038.

Gibney, E. (2020). The Battle to embed ethics in AI research. *Nature, 577*, 609.

Golumbia, D. (2009). *The cultural logic of computation*. Cambridge, MA: Harvard University Press.

Goodman, N. (1954). *Fact, fiction, and forecast*. London: University of London Press.

Lelieveld, J., Klingmüller, K., Pozzer, A., Burnett, R. T., Haines, A., & Ramanathan, V. (2019). Effects of fossil fuel and total anthropogenic emission removal on public health and climate. *PNAS, 116*(15), 7192–7197. https://doi.org/10.1073/pnas.1819989116.

Loh, E. (2018). Medicine and the rise of the robots: A qualitative review of recent advances of artificial intelligence in health. *BMJ Leader, 2*, 59–63. https://doi.org/10.1136/leader-2018-000071.

Pearl, J. (1988). *Probabilistic reasoning in intelligent systems: Networks of plausible inference*. San Francisco: Morgan Kaufmann.

Pistor, K. (2019). *The code of capital: How the law creates wealth and inequality*. Princeton, NJ: Princeton University Press.

Pope Francis (2020). *Discourse to the general assembly of the Pontifical Academy for Life*. Retrieved February 28, from http://press.vatican.va/content/salastampa/it/bollettino/pubblico/2020/02/28/0134/00291.html#eng.

Rawls, J. (1987). The idea of an overlapping consensus. *Oxford Journal of Legal Studies, 7*(1), 1–25.

Russell, S. (2019). *Human compatible: AI and the problem of control*. New York: Viking.

Stanley, J. (2019). *The dawn of robot surveillance*. Available via American Civil Liberties Union. Retrieved March 11, 2019, from https://www.aclu.org/sites/default/files/field_document/061119-robot_surveillance.pdf.

Sweeting, M. (2020). Small satellites for earth observation—Bringing space within reach. In J. von Braun & M. Sánchez Sorondo (Eds.), *Transformative roles of science in society: From emerging basic science toward solutions for people's wellbeing Acta Varia 25*. Vatican City: The Pontifical Academy of Sciences.

Wiener, N. (1960). Some moral and technical consequences of automation. *Science, 131*, 1355–1358. https://doi.org/10.1126/science.131.3410.1355.

Part I

Foundational Issues in AI and Robotics

Differences Between Natural and Artificial Cognitive Systems

Wolf Singer

Contents

Abstract

This chapter identifies the differences between natural and artifical cognitive systems. Benchmarking robots against brains may suggest that organisms and robots both need to possess an internal model of the restricted environment in which they act and both need to adjust their actions to the conditions of the respective environment in order to accomplish their tasks. However, computational strategies to cope with these challenges are different for natural and artificial systems. Many of the specific human qualities cannot be deduced from the neuronal functions of individual brains alone but owe their existence to cultural evolution. Social interactions between agents endowed with the cognitive abilities of humans generate immaterial realities, addressed as social or cultural realities. Intentionality, morality, responsibility and certain aspects of consciousness such as the qualia of subjective experience belong to the immaterial dimension of social realities. It is premature to enter discussions as to whether artificial systems can acquire functions that we consider as intentional and conscious or whether artificial agents can be considered as moral agents with responsibility for their actions.

Keywords

Artificial intelligence · Cognitive systems · Brain · Robot · Cultural evolution · Neuronal functions · Consciousness

W. Singer (✉)
Max Planck Institute for Brain Research (MPI), Ernst Strüngmann Institute for Neuroscience (ESI) in Cooperation with Max Planck Society, Frankfurt, Germany

Frankfurt Institute for Advanced Studies (FIAS), Frankfurt, Germany
e-mail: wolf.singer@brain.mpg.de

Introduction

Organisms and robots have to cope with very similar challenges. Both need to possess an internal model of the restricted environment in which they act and both need to

J. von Braun et al. (eds.), *Robotics, AI, and Humanity*, https://doi.org/10.1007/978-3-030-54173-6_2

adjust their actions to the idiosyncratic conditions of the respective environment in order to accomplish particular tasks. However, the computational strategies to cope with these challenges exhibit marked differences between natural and artificial systems.

In natural systems the model of the world is to a large extent inherited, i.e. the relevant information has been acquired by selection and adaptation during evolution, is stored in the genes and expressed in the functional anatomy of the organism and the architecture of its nervous systems. This inborn model is subsequently complemented and refined during ontogeny by experience and practice. The same holds true for the specification of the tasks that the organism needs to accomplish and for the programs that control the execution of actions. Here, too, the necessary information is provided in part by evolution and in part by lifelong learning. In order to be able to evolve in an ever-changing environment, organisms have evolved cognitive systems that allow them to analyse the actual conditions of their embedding environment, to match them with the internal model, update the model, derive predictions and adapt future actions to the actual requirements.

In order to complement the inborn model of the world organisms rely on two different learning strategies: Unsupervised and supervised learning. The former serves to capture frequently occurring statistical contingencies in the environment and to adapt processing architectures to the efficient analysis of these contingencies. Babies apply this strategy for the acquisition of the basic building blocks of language. The unsupervised learning process is implemented by adaptive connections that change their gain (efficiency) as a function of the activity of the connected partners. If in a network two interconnected neurons are frequently coactivated, because the features to which they respond are often present simultaneously, the connections among these two neurons become more efficient. The neurons representing these correlated features become associated with one another. Thus, statistical contingencies between features get represented by the strength of neuronal interactions. "Neurons wire together if they fire together". Conversely, connections among neurons weaken, if these are rarely active together, i.e. if their activity is uncorrelated. By contrast, supervised learning strategies are applied when the outcome of a cognitive or executive process needs to be evaluated. An example is the generation of categories. If the system were to learn that dogs, sharks and eagles belong to the category of animals it needs to be told that such a category exists and during the learning process it needs to receive feedback on the correctness of the various classification attempts. In case of supervised learning the decision as to whether a particular activity pattern induces a change in coupling is made dependent not only on the

local activity of the coupled neurons but on additional gating signals that have a "now print" function. Only if these signals are available in addition can local activity lead to synaptic changes. These gating signals are generated by a few specialized centres in the depth of the brain and conveyed through widely branching nerve fibres to the whole forebrain. The activity of these value assigning systems is in turn controlled by widely distributed brain structures that evaluate the behavioural validity of ongoing or very recently accomplished cognitive or executive processes. In case the outcome is positive, the network connections whose activity contributed to this outcome get strengthened and if the outcome is negative they get weakened. This retrospective adjustment of synaptic modifications is possible, because activity patterns that potentially could change a connection leave a molecular trace at the respective synaptic contacts that outlasts the activity itself. If the "now print" signal of the gating systems arrives while this trace is still present, the tagged synapse will undergo a lasting change (Redondo and Morris 2011; Frey and Morris 1997). In this way, the network's specific activity pattern that led to the desired outcome will be reinforced. Therefore, this form of supervised learning is also addressed as reinforcement learning.

Comparing these basic features of natural systems with the organization of artificial "intelligent" systems already reveals a number of important differences.

Artificial systems have no evolutionary history but are the result of a purposeful design, just as any other tool humans have designed to fulfil special functions. Hence, their internal model is installed by engineers and adapted to the specific conditions in which the machine is expected to operate. The same applies for the programs that translate signals from the robot's sensors into action. Control theory is applied to assure effective coordination of the actuators. Although I am not a specialist in robotics I assume that the large majority of useful robots is hard wired in this way and lacks most of the generative, creative and self-organizing capacities of natural agents.

However, there is a new generation of robots with enhanced autonomy that capitalize on the recent progress in machine learning. Because of the astounding performance of these robots, autonomous cars are one example, and because of the demonstration that machines outperform humans in games such as Go and chess, it is necessary to examine in greater depth to which extent the computational principles realized in these machines resemble those of natural systems.

Over the last decades the field of artificial intelligence has been revolutionized by the implementation of computational strategies based on artificial neuronal networks. In the second half of the last century evidence accumulated that relatively simple neuronal networks, known as Perceptrons or Hopfield nets, can be trained to recognize and classify patterns and this fuelled intensive research in the domain of artificial intelligence. The growing availability of massive computing power and the design of ingenious training algorithms provided compelling evidence that this computational strategy is scalable. The early systems consisted of just three layers and a few dozens of neuron like nodes. The systems that have recently attracted considerable attention because they outperform professional Go players, recognize and classify correctly huge numbers of objects, transform verbal commands into actions and steer cars, are all designed according to the same principles as the initial three-layered networks. However, the systems now comprise more than hundred layers and millions of nodes which has earned them the designation "deep learning networks". Although the training of these networks requires millions of training trials with a very large number of samples, their amazing performance is often taken as evidence that they function according to the same principles as natural brains. However, as detailed in the following paragraph, a closer look at the organization of artificial and natural systems reveals that this is only true for a few aspects.

Strategies for the Encoding of Relations: A Comparison Between Artificial and Natural Systems

The world, animate and inanimate, is composed of a relatively small repertoire of elementary components that are combined at different scales and in ever different constellations to bring forth the virtually infinite diversity of objects. This is at least how the world appears to us. Whether we are caught in an epistemic circle and perceive the world as composite because our cognitive systems are tuned to divide wholes into parts or because the world is composite and our cognitive systems have adapted to this fact will not be discussed further. What matters is that the complexity of descriptions can be reduced by representing the components and their relations rather than the plethora of objects that result from different constellations of components. It is probably for this reason that evolution has optimized cognitive systems to exploit the power of combinatorial codes. A limited number of elementary features is extracted from the sensory environment and represented by the responses of feature selective neurons. Subsequently different but complementary strategies are applied to evaluate the relations between these features and to generate minimally overlapping representations of particular feature constellations for classification. In a sense this is the same strategy as utilized by human languages. In the Latin alphabet, 28 symbols suffice to compose the world literature.

Encoding of Relations in Feed-Forward Architectures

One strategy for the analysis and encoding of relations is based on convergent feed-forward circuits. This strategy is ubiquitous in natural systems. Nodes (neurons) of the input layer are tuned to respond to particular features of input patterns and their output connections are made to converge on nodes of the next higher layer. By adjusting the gain of these converging connections and the threshold of the target node it is assured that the latter responds preferentially to only a particular conjunction of features in the input pattern (Hubel and Wiesel 1968; Barlow 1972). In this way consistent relations among components become represented by the activity of conjunction-specific nodes (see Fig. 1). By iterating this strategy across multiple layers in hierarchically structured feed-forward architectures complex relational constructs (cognitive objects) can be represented by conjunction-specific nodes of higher order. This basic strategy for the encoding of relations has been realized independently several times during evolution in the nervous systems of different phyla (molluscs, insects, vertebrates) and reached the highest degree of sophistication in the hierarchical arrangement of processing levels in the cerebral cortex of mammals (Felleman and van Essen 1991; Glasser et al. 2016; Gross et al. 1972; Tsao et al. 2006; Hirabayashi et al. 2013; Quian Quiroga et al. 2005). This strategy is also the hallmark of the numerous versions of artificial neuronal networks designed for the recognition and classification of patterns (Rosenblatt 1958; Hopfield 1987; DiCarlo and Cox 2007; LeCun et al. 2015). As mentioned above, the highly successful recent developments in the field of artificial intelligence, addressed as "deep learning networks" (LeCun et al.

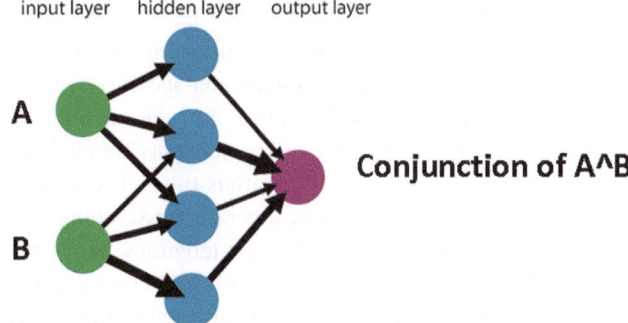

Fig. 1 The encoding of relations by conjunction-specific neurons (red) in a three-layered neuronal network. A and B refer to neurons at the input layer whose responses represent the presence of features A and B. Arrows indicate the flow of activity and their thickness the efficiency of the respective connections. The threshold of the conjunction-specific neuron is adjusted so that it responds only when A and B are active simultaneously

Fig. 2 Topology of a deep
learning network

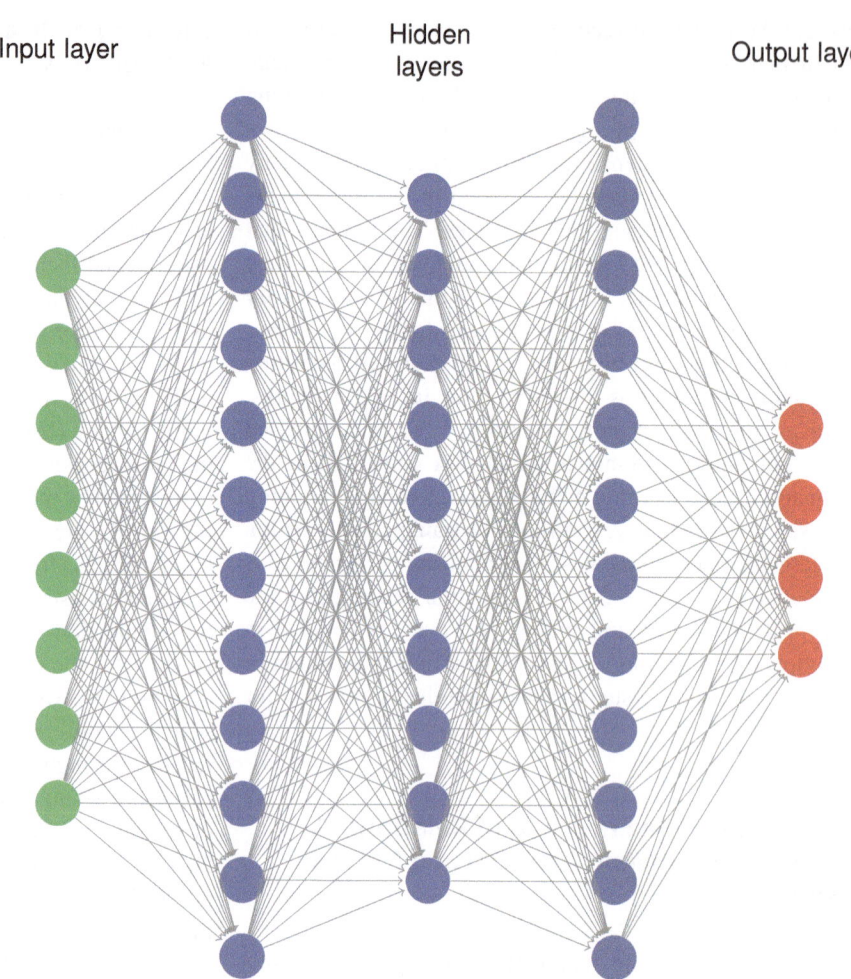

2015; Silver et al. 2017, 2018), capitalize on the scaling of this principle in large multilayer architectures (see Fig. 2).

Encoding of Relations by Assemblies

In natural systems, a second strategy for the encoding of relations is implemented that differs in important aspects from the formation of individual, conjunction-specific neurons (nodes) and requires a very different architecture of connections. In this case, relations among components are encoded by the temporary association of neurons (nodes) representing individual components into cooperating assemblies that respond collectively to particular constellations of related features. In contrast to the formation of conjunction-specific neurons by convergence of feed-forward connections, this second strategy requires recurrent (reciprocal) connections between the nodes of the same layer as well as feed-back connections from higher to lower levels of the processing hierarchy. In natural systems, these recurrent connections outnumber by far the feed-forward connections. As proposed by Donald Hebb as early as 1949, components (features) of

composite objects can not only be related to one another by the formation of conjunction-specific cells but also by the formation of functionally coherent assemblies of neurons. In this case, the neurons that encode the features that need to be bound together become associated into an assembly. Such assemblies, so the original assumption, are distinguished as a coherent whole that represents a particular constellation of components (features) because of the jointly enhanced activation of the neurons constituting the assembly. The joint enhancement of the neurons' activity is assumed to be caused by cooperative interactions that are mediated by the reciprocal connections between the nodes of the network. These connections are endowed with correlation-dependent synaptic plasticity mechanisms (Hebbian synapses, see below) and strengthen when the interconnected nodes are frequently co-activated. Thus, nodes that are often co-activated because the features to which they respond do often co-occur in the environment enhance their mutual interactions. As a result of these cooperative interactions, the vigour and/or coherence of the responses of the respective nodes is enhanced when they are activated by the respective feature constellation. In this way, consistent relations among the components of cognitive objects are translated into the weight distributions of the reciprocal connections between network nodes and become represented by the joint responses of a cooperating assembly of neurons. Accordingly, the information about the presence of a particular constellation of features is not represented by the activity of a single conjunction-specific neuron but by the amplified or more coherent or reverberating responses of a distributed assembly of neurons.

A Comparison Between the Two Strategies

Both relation-encoding strategies have advantages and disadvantages and evolution has apparently opted for a combination of the two. Feed-forward architectures are well suited to evaluate relations between simultaneously present features, raise no stability problems and allow for fast processing. However, encoding relations exclusively with conjunction-specific neurons is exceedingly expensive in terms of hardware requirements. Because specific constellations of (components) features have to be represented explicitly by conjunction-specific neurons via the convergence of the respective feed-forward connections and because the dynamic range of the nodes is limited, an astronomically large number of nodes and processing levels would be required to cope with the virtually infinite number of possible relations among the components (features) characterizing real-world objects, leave alone the representation of nested relations required to capture complex scenes. This problem is addressed as the "combinatorial explosion". Consequently, biological systems relying exclusively on feed-forward architectures are rare and can afford representation of only a limited number of behaviourally relevant relational constructs. Another serious disadvantage of networks consisting exclusively of feed-forward connections is that they have difficulties to encode relations among temporally segregated events (temporal relations) because they lack memory functions.

By contrast, assemblies of recurrently coupled, mutually interacting nodes (neurons) can cope very well with the encoding of temporal relations (sequences) because such networks exhibit fading memory due to reverberation and can integrate temporally segregated information. Assembly codes are also much less costly in terms of hardware requirements, because individual feature specific nodes can be recombined in flexible combinations into a virtually infinite number of different assemblies, each representing a different cognitive content, just as the letters of the alphabet can be combined into syllables, words, sentences and complex descriptions (combinatorial code). In addition, coding space is dramatically widened because information about the statistical contingencies of features can be encoded not only in the synaptic weights of feed forward connections but also in the weights of the recurrent and feed-back connections. Finally, the encoding of entirely new or the completion of incomplete relational constructs (associativity) is facilitated by the cooperativity inherent in recurrently coupled networks that allows for pattern completion and the generation of novel associations (generative creativity).

However, assembly coding and the required recurrent networks cannot easily be implemented in artificial systems for a number of reasons. First and above all it is extremely cumbersome to simulate the simultaneous reciprocal interactions between large numbers of interconnected nodes with conventional digital computers that can perform only sequential operations. Second, recurrent networks exhibit highly non-linear dynamics that are difficult to control. They can fall dead if global excitation drops below a critical level and they can engage in runaway dynamics and become epileptic if a critical level of excitation is reached. Theoretical analysis shows that such networks perform efficiently only if they operate in a dynamic regime close to criticality. Nature takes care of this problem with a number of self-regulating mechanisms involving normalization of synaptic strength (Turrigiano and Nelson 2004), inhibitory interactions (E/I balance) (Yizhar et al. 2011) and control of global excitability by modulatory systems, that keep the network within a narrow working range just below criticality (Plenz and Thiagarajan 2007; Hahn et al. 2010).

The third problem for the technical implementation of biological principles is the lack of hardware solutions for Hebbian synapses that adjust their gain as a function of the correlation between the activity of interconnected nodes. Most artificial systems rely on some sort of supervised learning in which temporal relations play only a minor role if at all. In these systems the gain of the feed-forward connections is iteratively adjusted until the activity patterns at the output layer represent particular input patterns with minimal overlap. To this end very large samples of input patterns are generated, deviations of the output patterns from the desired result are monitored as "errors" and backpropagated through the network in order to change the gain of those connections that contributed most to the error. In multilayer networks this is an extremely challenging procedure and the breakthroughs of recent developments in deep learning networks were due mainly to the design of efficient backpropagation algorithms. However, these are biologically implausible. In natural systems, the learning mechanisms exploit the fundamental role of consistent temporal relations for the definition of semantic relations. Simultaneously occurring events usually have a common cause or are interdependent because of interactions. If one event consistently precedes the other, the first is likely the cause of the latter, and if there are no temporal correlations between the events, they are most likely unrelated. Likewise, components (features) that often occur together are likely to be related, e.g. because their particular constellation is characteristic for a particular object or because they are part of a stereotyped sequence of events. Accordingly, the molecular mechanisms developed by evolution for the establishment of associations are exquisitely sensitive to temporal relations between the activity patterns of interconnected nodes. The crucial variable that determines the occurrence and polarity of gain changes of the connections is the *temporal relation* between discharges in converging presynaptic inputs and/or between the discharges of presynaptic afferents and the activity of the postsynaptic neuron. In natural systems most excitatory connections—feed forward, feed-back and recurrent—as well as the connections between excitatory and inhibitory neurons are adaptive and can change their gain as a function of the correlation between pre- and postsynaptic activity. The molecular mechanisms that translate electrical activity in lasting changes of synaptic gain evaluate correlation patterns with a precision in the range of tens of milliseconds and support both the experience-dependent generation of conjunction-specific neurons in feed-forward architectures and the formation of assemblies.

Assembly Coding and the Binding Problem

Although the backpropagation algorithm mimics in a rather efficient way the effects of reinforcement learning in deep learning networks it cannot be applied for the training of recurrent networks because it lacks sensitivity to temporal relations. However, there are efforts to design training algorithms applicable to recurrent networks and the results are promising (Bellec et al. 2019).

Another and particularly challenging problem associated with assembly coding is the *binding problem*. This problem arises whenever more than one object is present and when these objects and the relations among them need to be encoded within the same network layer. If assemblies were solely distinguished by enhanced activity of the constituting neurons, as proposed by Hebb (1949), it becomes difficult to distinguish which of the neurons with enhanced activity actually belong to which assembly, in particular, if objects share some common features and overlap in space. This condition is known as the *superposition catastrophe*. It has been proposed that this problem can be solved by multiplexing, i.e. by segregating the various assemblies in time (Milner 1992; von der Malsburg and Buhmann 1992; for reviews see Singer and Gray 1995; Singer 1999). Following the discovery that neurons in the cerebral cortex can synchronize their discharges with a precision in the millisecond range when they engage in high frequency oscillations (Gray et al. 1989), it has been proposed that the neurons temporarily bound into assemblies are distinguished not only by an increase of their discharge rate but also by the precise synchronization of their action potentials (Singer and Gray 1995; Singer 1999).

Synchronization is as effective in enhancing the efficiency of neuronal responses in down-stream targets as is enhancing discharge rate (Bruno and Sakmann 2006). Thus, activation of target cells at the subsequent processing stage can be assured by increasing either the rate or the synchronicity of discharges in the converging input connections. The advantage of increasing salience by synchronization is that integration intervals for synchronous inputs are very short, allowing for instantaneous detection of enhanced salience. Hence, information about the relatedness of responses can be read out very rapidly. In extremis, single discharges can be labelled as salient and identified as belonging to a particular assembly if synchronized with a precision in the millisecond range.

Again, however, it is not trivial to endow artificial recurrent networks with the dynamics necessary to solve the binding problem. It would require to implement oscillatory microcircuits and mechanisms ensuring selective synchronization of feature selective nodes. The latter, in turn, have to rely on Hebbian learning mechanisms for which there are yet no satisfactory hardware solutions. Hence, there are multiple reasons why the unique potential of recurrent networks is only marginally exploited by AI systems.

Computing in High-Dimensional State Space

Unlike contemporary AI systems that essentially rely on the deep learning algorithms discussed above, recurrent networks exhibit highly complex non-linear dynamics, especially if the nodes are configured as oscillators and if the coupling connections impose delays—as is the case for natural networks. These dynamics provide a very high-dimensional state space that can be exploited for the realization of functions that go far beyond those discussed above and are based on radically different computational strategies. In the following, some of these options will be discussed and substantiated with recently obtained experimental evidence.

The non-linear dynamics of recurrent networks are exploited for computation in certain AI systems, the respective strategies being addressed as "echo state, reservoir or liquid computing" (Lukoševičius and Jaeger 2009; Buonomano and Maass 2009; D'Huys et al. 2012; Soriano et al. 2013). In most cases, the properties of recurrent networks are simulated in digital computers, whereby only very few of the features of biological networks are captured. In the artificial systems the nodes act as simple integrators and the coupling connections lack most of the properties of their natural counterparts. They operate without delay, lack specific topologies and their gain is non-adaptive. Most artificial recurrent networks also lack inhibitory interneurons that constitute 20% of the neurons in natural systems and interact in highly selective ways with the excitatory neurons. Moreover, as the updating of network states has to be performed sequentially according to the clock cycle of the digital computer used to simulate the recurrent network, many of the analogue computations taking place in natural networks can only be approximated with iterations if at all. Therefore, attempts are made to emulate the dynamics of recurrent networks with analogue technology. An original and hardware efficient approach is based on optoelectronics. Laser diodes serve as oscillating nodes and these are reciprocally coupled through glass fibres whose variable length introduces variations of coupling delays (Soriano et al. 2013). All these implementations have in common to use the characteristic dynamics of recurrent networks as medium for the execution of specific computations.

Because the dynamics of recurrent networks resemble to some extent the dynamics of liquids—hence the term "liquid computing"—the basic principle can be illustrated by considering the consequences of perturbing a liquid. If objects impact at different intervals and locations in a pond of water, they generate interference patterns of propagating waves whose parameters reflect the size, speed, location and the time of impact of the objects. The wave patterns fade with a time constant determined by the viscosity of the liquid, interfere with one another and create a complex dynamic state. This state can be analysed by measuring at several locations in the pond the amplitude, frequency and phase of the respective oscillations and from these variables a trained classifier can subsequently reconstruct the exact sequence and nature of the impacting "stimuli". Similar effects occur in recurrent networks when subsets of nodes are perturbed by stimuli that have a particular spatial and temporal structure. The excitation of the stimulated nodes spreads across the network and creates a complex dynamic state, whose spatio-temporal structure is determined by the constellation of initially excited nodes and the functional architecture of the coupling connections. This stimulus-specific pattern continues to evolve beyond the duration of the stimulus due to reverberation and then eventually fades. If the activity has not induced changes in the gain of the recurrent connections the network returns to its initial state. This evolution of the network dynamics can be traced by assessing the activity changes of the nodes and is usually represented by time varying, high-dimensional vectors or trajectories. As these trajectories differ for different stimulus patterns, segments exhibiting maximal distance in the high-dimensional state space can be selected to train classifiers for the identification of the respective stimuli.

This computational strategy has several remarkable advantages: (1) low-dimensional stimulus events are projected into a high-dimensional state space where nonlinearly separable stimuli become linearly separable; (2) the high dimensionality of the state space can allow for the mapping of more complicated output functions (like the XOR) by simple classifiers, and (3) information about sequentially presented stimuli persists for some time in the medium (fading memory). Thus, information about multiple stimuli can be integrated over time, allowing for the representation of sequences; (4) information about the statistics of natural environments (the internal model) can be stored in the weight distributions and architecture of the recurrent connections for instantaneous comparison with incoming sensory evidence. These properties make recurrent networks extremely effective for the classification of input patterns that have both spatial and temporal structure and share overlapping features in low-dimensional space. Moreover, because these networks self-organize and produce spatio-temporally structured activity patterns, they have generative properties and can be used for pattern completion, the formation of novel associations and the generation of patterns for the control of movements. Consequently, an increasing number of AI systems now complement the feed-forward strategy implemented in deep learning networks with algorithms inspired by recurrent networks. One of these powerful and now widely used algorithms is the Long Short Term Memory (LSTM) algorithm, introduced decades ago by Hochreiter and Schmidhuber (1997) and used in systems such as AlphaGo, the network that outperforms professional GO players (Silver et al. 2017, 2018). The surprising efficiency of these systems that excels in certain domains human performance has nurtured the notion that brains operate in the same way. If one considers, however, how fast brains can solve certain tasks despite of their comparatively extremely slow components and how energy efficient they are, one is led to suspect implementation of additional and rather different strategies.

And indeed, natural recurrent networks differ from their artificial counterparts in several important features which is the likely reason for their amazing performance. In sensory cortices the nodes are feature selective, i.e. they can be activated only by specific spatio-temporal stimulus configurations. The reason is that they receive convergent input from selected nodes of the respective lower processing level and thus function as conjunction-specific units in very much the same way as the nodes in feed forward multilayer networks. In low areas of the visual system, for example, the nodes are selective for elementary features such as the location and orientation of contour borders, while in higher areas of the processing hierarchy the nodes respond to increasingly complex constellations of elementary features. In addition, the nodes of natural systems, the neurons, possess an immensely larger spectrum of integrative and adaptive functions than the nodes currently used in artificial recurrent networks. And finally the neurons and/or their embedding microcircuits are endowed with the propensity to oscillate.

The recurrent connections also differ in important respects from those implemented in most artificial networks. Because of the slow velocity of signals conveyed by neuronal axons interactions occur with variable delays. These delays cover a broad range and depend on the distance between interconnected nodes and the conduction velocity of the respective axons. This gives rise to exceedingly complex dynamics and permits exploitation of phase space for coding. Furthermore and most importantly, the connections are endowed with plastic synapses whose gain changes according to the correlation rules discussed above. Nodes tuned to features that often co-occur in natural environments tend to be more strongly coupled than nodes responding to features that rarely occur simultaneously. Thus, through both experience-dependent pruning of connections during early development and experience-dependent synaptic plasticity, statistical contingencies between features of the environment get internalized and stored not only in the synaptic weights of feed-forward connections to feature selective nodes but also in the weight distributions of the recurrent connections. Thus, in low levels of the processing hierarchy the weight distributions of the recurrent coupling connections reflect statistical contingencies of simple and at higher levels of more complex constellations of features. In other words, the hierarchy of reciprocally coupled recurrent networks contains a model of the world that reflects the frequency of co-occurrence of typical relations among the features/components of composite perceptual objects. Recent simulation studies have actually shown that performance of an artificial recurrent network is substantially improved if the recurrent connections are made adaptive and can "learn" about the feature contingencies of the processed patterns (Lazar et al. 2009; Hartmann et al. 2015).

Information Processing in Natural Recurrent Networks

Theories of perception formulated more than a hundred years ago (von Helmholtz 1867) and a plethora of experimental evidence indicate that perception is the result of a constructivist process. Sparse and noisy input signals are disambiguated and interpreted on the basis of an internal model of the world. This model is used to reduce redundancy, to detect characteristic relations between features, to bind signals evoked by features constituting a perceptual object, to facilitate segregation of figures from background and to eventually enable identification and classification. The store containing such an elaborate model must have an immense capacity, given that the interpretation of ever-changing sensory input patterns requires knowledge about the vast number of distinct feature conjunctions characterizing perceptual objects. Moreover, this massive amount of prior knowledge needs to be arranged in a configuration that permits ultrafast readout to meet the constraints of processing speed. Primates perform

on average four saccades per second. This implies that new visual information is sampled approximately every 250 ms (Maldonado et al. 2008; Ito et al. 2011) and psychophysical evidence indicates that attentional processes sample visual information at comparable rates (Landau 2018; Landau and Fries 2012). Thus, the priors required for the interpretation of a particular sensory input need to be made available within fractions of a second.

How the high-dimensional non-linear dynamics of delay-coupled recurrent networks could be exploited to accomplish these complex functions is discussed in the following paragraph.

A hallmark of natural recurrent networks such as the cerebral cortex is that they are spontaneously active. The dynamics of this resting activity reflects the weight distributions of the structured network and hence harbours the entirety of the stored "knowledge" about the statistics of feature contingencies, i.e. the latent priors used for the interpretation of sensory evidence. This predicts that resting activity is high dimensional and represents a vast but constrained manifold inside the universe of all theoretically possible dynamical states. Once input signals become available they are likely to trigger a cascade of effects: They drive in a graded way a subset of feature sensitive nodes *and* thereby perturb the network dynamics. If the evidence provided by the input patterns matches well the priors stored in the network architecture, the network dynamics will collapse to a specific substate that provides the best match with the corresponding sensory evidence. Such a substate is expected to have a lower dimensionality and to exhibit less variance than the resting activity, to possess a specific correlation structure and be metastable due to reverberation among nodes supporting the respective substate. Because these processes occur within a very-high-dimensional state space, substates induced by different input patterns are usually well segregated and therefore easy to classify. As the transition from the high-dimensional resting activity to substates follows stimulus-specific trajectories, classification of stimulus-specific patterns is possible once trajectories have sufficiently diverged and long before they reach a fix point and this could account for the extremely fast operations of natural systems.

Experimental studies testing such a scenario are still rare and have become possible only with the advent of massive parallel recordings from the network nodes. So far, however, the few predictions that have been subject to experimental testing appeared to be confirmed. For the sake of brevity, these experimental results are not discussed here. They have been reviewed recently in Singer (2019a). A simplified representation of the essential features of a delayed coupled oscillator network supposed to be realized in the superficial layers of the cerebral cortex is shown in Fig. 3 (adapted from Singer 2018).

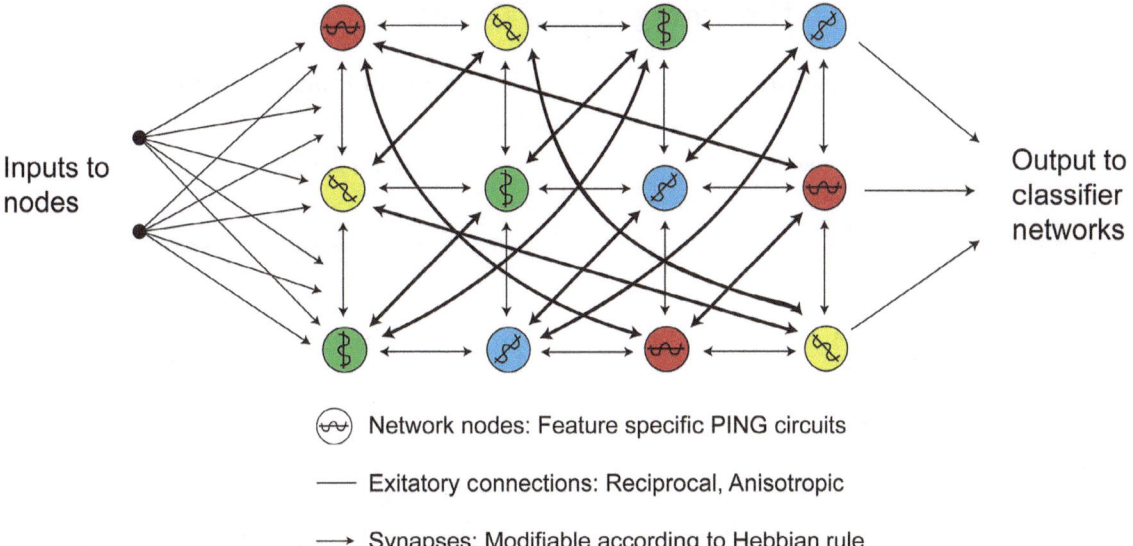

Inputs to nodes

Output to classifier networks

〜 Network nodes: Feature specific PING circuits

— Exitatory connections: Reciprocal, Anisotropic

→ Synapses: Modifiable according to Hebbian rule

Fig. 3 Schematic representation of wiring principles in supra-granular layers of the visual cortex. The coloured discs (nodes) stand for cortical columns that are tuned to specific features (here stimulus orientation) and have a high propensity to engage in oscillatory activity due to the intrinsic circuit motif of recurrent inhibition. These functional columns are reciprocally coupled by a dense network of excitatory connections that originate mainly from pyramidal cells and terminate both on pyramidal cells and inhibitory interneurons in the respective target columns. Because of the genetically determined span of these connections coupling decreases exponentially with the distance between columns. However, these connections undergo use-dependent selection during development and remain susceptible to Hebbian modifications of their gain in the adult. The effect is that the weight distributions of these connections and hence the coupling strength among functional columns (indicated by thickness of lines) reflect the statistical contingencies of the respective features in the visual environment (for further details see text). (From Singer W (2018) Neuronal oscillations: unavoidable and useful? Europ J Neurosci 48: 2389–2398)

Concluding Remarks

Despite considerable effort there is still no unifying theory of information processing in natural systems. As a consequence, numerous experimentally identified phenomena lack a cohesive theoretical framework. This is particularly true for the dynamic phenomena reviewed here because they cannot easily be accommodated in the prevailing concepts that emphasize serial feed-forward processing and the encoding of relations by conjunction-specific neurons. It is obvious, however, that natural systems exploit the computational power offered by the exceedingly complex, high-dimensional and non-linear dynamics that evolve in delay-coupled recurrent networks.

Here concepts have been reviewed that assign specific functions to oscillations, synchrony and the more complex dynamics emerging from a delay-coupled recurrent network and it is very likely that further computational principles are realized in natural systems that wait to be uncovered. In view of the already identified and quite remarkable differences between the computational principles implemented in artificial and natural systems it appears utterly premature to enter discussions as to whether artificial systems can acquire functions that we consider proper to natural systems such as intentionality and consciousness or whether artificial

agents can or should be considered as moral agents that are responsible for their actions. Even if we had a comprehensive understanding of the neuronal underpinnings of the cognitive and executive functions of human brains—which is by no means the case—we still would have to consider the likely possibility, that many of the specific human qualities cannot be deduced from the neuronal functions of individual brains alone but owe their existence to cultural evolution. As argued elsewhere (Singer 2019b), it is likely that most of the connotations that we associate with intentionality, responsibility, morality and consciousness are attributions to our self-model that result from social interactions of agents endowed with the cognitive abilities of human beings. In a nutshell the argument goes as follows: Perceptions—and this includes also the perception of oneself and other human beings—are the result of constructivist processes that depend on a match between sensory evidence and a-priory knowledge, so-called priors. Social interactions between agents endowed with the cognitive abilities of humans generate immaterial realities, addressed as social or cultural realities. This novel class of realities assume the role of implicit priors for the perception of the world and oneself. As a natural consequence perceptions shaped by these cultural priors impose a dualist classification of observables into material and immaterial phenomena, nurture the concept of ontological substance dualism and generate the epistemic conundrum of experiencing oneself

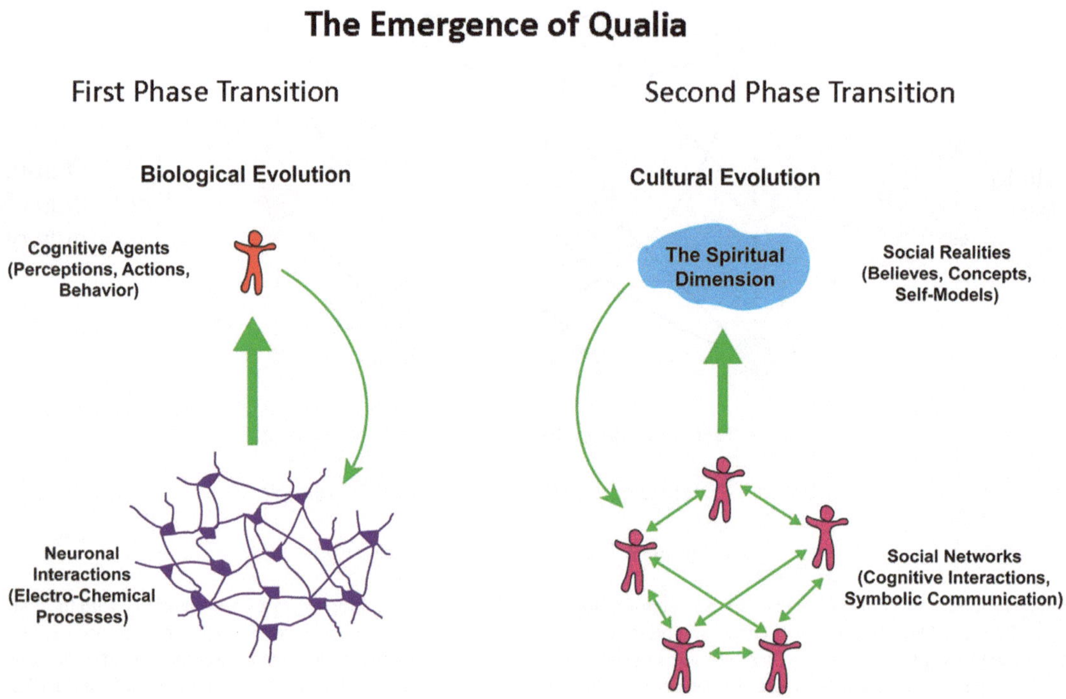

The Emergence of Qualia

Fig. 4 Phase transitions during biological (left) and cultural evolution (right) that lead to the emergence of new qualities. For details see text. (From Singer W (2019) A naturalistic approach to the hard problem of consciousness. Front Syst Neurosci 13: 58)

as existing in both a material and immaterial dimension. Intentionality, morality, responsibility and certain aspects of consciousness such as the qualia of subjective experience belong to this immaterial dimension of social realities.

This scenario is in agreement with the well-established phenomenon that phase transitions in complex systems can generate novel qualities that transcend the qualities of the systems' components. The proposal is that the specific human qualities (intentionality, consciousness, etc.) can only be accounted for by assuming at least two-phase transitions: One having occurred during biological evolution and the second during cultural evolution (Fig. 4). The first consists of the emergence of cognitive and executive functions from neuronal interactions during biological evolution and the second of the emergence of social realities from interactions between the cognitive agents that have been brought forth by the first phase transition. Accordingly, different terminologies (Sprachspiel) have to be used to capture the qualities of the respective substrates and the emergent phenomena, the neuronal interactions, the emerging cognitive and executive functions (behaviour), the social interactions among cognitive agents and the emerging social realities. If this evolutionary plausible scenario is valid, it predicts, that artificial agents, even if they should one day acquire functions resembling those of individual human brains,—and this is not going to happen tomorrow—will still lack the immaterial dimensions of our self-model. The only way to acquire this dimension—at least as far as I can see—would be for them

to be raised like children in human communities in order to internalize in their self-model our cultural achievements and attributions—and this would entail not only transmission of explicit knowledge but also emotional bonding. Or these man-made artefacts would have to develop the capacity and be given the opportunity to engage in their own social interactions and recapitulate their own cultural evolution.

References

Barlow, H. B. (1972). Single units and sensation: A neurone doctrine for perceptual psychology? *Perception, 1*, 371–394.

Bellec, G., Scherr, F., Hajek, E., Salaj, D., Legenstein, R., & Maass, W. (2019). Biologically inspired alternatives to backpropagation through time for learning in recurrent neural nets. *arXiv 2553450, 2019*, 1–34.

Bruno, R. M., & Sakmann, B. (2006). Cortex is driven by weak but synchronously active thalamocortical synapses. *Science, 312*, 1622–1627.

Buonomano, D. V., & Maass, W. (2009). State-dependent computations: Spatiotemporal processing in cortical networks. *Nature Reviews. Neuroscience, 10*, 113–125.

D'Huys, O., Fischer, I., Danckaert, J., & Vicente, R. (2012). Spectral and correlation properties of rings of delay-coupled elements: Comparing linear and nonlinear systems. *Physical Review E, 85*(056209), 1–5.

DiCarlo, J. J., & Cox, D. D. (2007). Untangling invariant object recognition. *Trends in Cognitive Sciences, 11*, 333–341.

Felleman, D. J., & van Essen, D. C. (1991). Distributed hierarchical processing in the primate cerebral cortex. *Cerebral Cortex, 1*, 1–47.

Frey, U., & Morris, R. G. M. (1997). Synaptic tagging and long-term potentiation. *Nature, 385*, 533–536.

Glasser, M. F., Coalson, T. S., Robinson, E. C., Hacker, C. D., Harwell, J., Yacoub, E., Ugurbil, K., Andersson, J., Beckmann, C. F., Jenkinson, M., Smith, S. M., & Van Essen, D. C. (2016). A multi-modal parcellation of human cerebral cortex. *Nature, 536*, 171–178.

Gray, C. M., König, P., Engel, A. K., & Singer, W. (1989). Oscillatory responses in cat visual cortex exhibit inter-columnar synchronization which reflects global stimulus properties. *Nature, 338*, 334–337.

Gross, C. G., Rocha-Miranda, C. E., & Bender, D. B. (1972). Visual properties of neurons in inferotemporal cortex of the macaque. *Journal of Neurophysiology, 35*, 96–111.

Hahn, G., Petermann, T., Havenith, M. N., Yu, Y., Singer, W., Plenz, D., & Nikolic, D. (2010). Neuronal avalanches in spontaneous activity in vivo. *Journal of Neurophysiology, 104*, 3312–3322.

Hartmann, C., Lazar, A., Nessler, B., & Triesch, J. (2015). Where's the noise? Key features of spontaneous activity and neural variability arise through learning in a deterministic network. *PLoS Computational Biology, 11*(12), e1004640, 1–35.

Hebb, D. O. (1949). *The organization of behavior*. New York: John Wiley & Sons.

Hirabayashi, T., Takeuchi, D., Tamura, K., & Miyashita, Y. (2013). Microcircuits for hierarchical elaboration of object coding across primate temporal areas. *Science, 341*, 191–195.

Hochreiter, S., & Schmidhuber, J. (1997). Long short-term memory. *Neural Computation, 9*, 1735–1780.

Hopfield, J. J. (1987). Learning algorithms and probability distributions in feed-forward and feed-back networks. *Proceedings of the National Academy of Sciences of the United States of America, 84*, 8429–8433.

Hubel, D. H., & Wiesel, T. N. (1968). Receptive fields and functional architecture of monkey striate cortex. *Journal of Physiology (London), 195*, 215–243.

Ito, J., Maldonado, P., Singer, W., & Grün, S. (2011). Saccade-related modulations of neuronal excitability support synchrony of visually elicited spikes. *Cerebral Cortex, 21*, 2482–2497.

Landau, A. N. (2018). Neuroscience: A mechanism for rhythmic sampling in vision. *Current Biology, 28*, R830–R832.

Landau, A. N., & Fries, P. (2012). Attention samples stimuli rhythmically. *Current Biology, 22*, 1000–1004.

Lazar, A., Pipa, G., & Triesch, J. (2009). SORN: A self-organizing recurrent neural network. *Frontiers in Computational Neuroscience, 3*(23), 1–9.

LeCun, Y., Bengio, Y., & Hinton, G. (2015). Deep learning. *Nature, 521*, 436–444.

Lukoševičius, M., & Jaeger, H. (2009). Reservoir computing approaches to recurrent neural network training. *Computer Science Review, 3*, 127–149.

Maldonado, P., Babul, C., Singer, W., Rodriguez, E., Berger, D., & Grün, S. (2008). Synchronization of neuronal responses in primary visual cortex of monkeys viewing natural images. *Journal of Neurophysiology, 100*, 1523–1532.

Milner, P. M. (1992). The functional nature of neuronal oscillations. *Trends in Neurosciences, 15*, 387.

Plenz, D., & Thiagarajan, T. C. (2007). The organizing principles of neuronal avalanches: Cell assemblies in the cortex? *Trends in Neurosciences, 30*, 99–110.

Quian Quiroga, R., Reddy, L., Kreiman, G., Koch, C., & Fried, I. (2005). Invariant visual representation by single neurons in the human brain. *Nature, 435*, 1102–1107.

Redondo, R. L., & Morris, R. G. M. (2011). Making memories last: The synaptic tagging and capture hypothesis. *Nature Reviews. Neuroscience, 12*, 17–30.

Rosenblatt, F. (1958). The perceptron. A probabilistic model for information storage and organization in the brain. *Psychological Review, 65*, 386–408.

Silver, D., Hubert, T., Schrittwieser, J., Antonoglou, I., Lai, M., Guez, A., Lanctot, M., Sifre, L., Kumaran, D., Graepel, T., Lillicrap, T., Simonyan, K., & Hassabis, D. (2018). A general reinforcement learning algorithm that masters chess, shogi, and go through self-play. *Science, 362*, 1140–1144.

Silver, D., Schrittwieser, J., Simonyan, K., Antonoglou, I., Huang, A., Guez, A., Hubert, T., Baker, L., Lai, M., Bolton, A., Chen, Y., Lillicrap, T., Hui, F., Sifre, L., van den Driessche, G., Graepel, T., & Hassabis, D. (2017). Mastering the game of go without human knowledge. *Nature, 550*, 354–359.

Singer, W. (1999). Neuronal synchrony: A versatile code for the definition of relations? *Neuron, 24*, 49–65.

Singer, W. (2018). Neuronal oscillations: Unavoidable and useful? *The European Journal of Neuroscience, 48*, 2389–2398.

Singer, W. (2019a). Cortical dynamics. In J. R. Lupp (Series Ed.) & W. Singer, T. J. Sejnowski, & P. Rakic (Vol. Eds.), Strüngmann Forum reports: Vol. 27. The neocortex. Cambridge, MA: MIT Press (in print). 167–194.

Singer, W. (2019b). A naturalistic approach to the hard problem of consciousness. *Frontiers in Systems Neuroscience, 13*, 58.

Singer, W., & Gray, C. M. (1995). Visual feature integration and the temporal correlation hypothesis. *Annual Review of Neuroscience, 18*, 555–586.

Soriano, M. C., Garcia-Ojalvo, J., Mirasso, C. R., & Fischer, I. (2013). Complex photonics: Dynamics and applications of delay-coupled semiconductors lasers. *Reviews of Modern Physics, 85*, 421–470.

Tsao, D. Y., Freiwald, W. A., Tootell, R. B. H., & Livingstone, M. S. (2006). A cortical region consisting entirely of face-selective cells. *Science, 311*, 670–674.

Turrigiano, G. G., & Nelson, S. B. (2004). Homeostatic plasticity in the developing nervous system. *Nature Reviews. Neuroscience, 5*, 97–107.

von der Malsburg, C., & Buhmann, J. (1992). Sensory segmentation with coupled neural oscillators. *Biological Cybernetics, 67*, 233–242.

von Helmholtz, H. (1867). *Handbuch der Physiologischen Optik*. Hamburg, Leipzig: Leopold Voss Verlag.

Yizhar, O., Fenno, L. E., Prigge, M., Schneider, F., Davidson, T. J., O'Shea, D. J., Sohal, V. S., Goshen, I., Finkelstein, J., Paz, J. T., Stehfest, K., Fudim, R., Ramakrishnan, C., Huguenard, J. R., Hegemann, P., & Deisseroth, K. (2011). Neocortical excitation/inhibition balance in information processing and social dysfunction. *Nature, 477*, 171–178.

Foundations of Artificial Intelligence and Effective Universal Induction

Jörg Zimmermann and Armin B. Cremers

Contents

Abstract

The term Artificial Intelligence was coined in 1956. Since then, this new research area has gone through several cycles of fast progress and periods of apparent stagnation. Today, the field has broadened and deepened significantly, and developed a rich variety of theoretical approaches and frameworks on the one side, and increasingly impressive practical applications on the other side. While a thorough foundation for a general theory of cognitive agents is still missing, there is a line of development within AI research which aims at foundational justifications for the design of cognitive agents, enabling the derivation of theorems characterizing the possibilities and limitations of computational cognitive agents.

J. Zimmermann (✉) · A. B. Cremers
Institute of Computer Science, University of Bonn, Bonn, Germany
e-mail: jz@cs.uni-bonn.de

Keywords

Artificial intelligence · Machine learning · Computational cognitive agents · Universal induction · Algorithmic transparency

J. von Braun et al. (eds.), *Robotics, AI, and Humanity*, https://doi.org/10.1007/978-3-030-54173-6_3

Introduction

In its most general form, artificial intelligence is an area of computer science which is concerned with the design and analysis of agents acting within an open, partially, or completely unknown environment. The agent and the environment are coupled by observations and actions, i.e., the agent observes the environment and executes actions which can affect the environment. Additionally, the agent has an internal state, which can serve as memory and as a resource for internal reflection. The environment, too, has a state, which in general is not directly accessible by the agent. Only by observations the agent gets indirect and partial information about the state of the environment.

In total, the agent–environment system is a coupled dynamical system, which can be described by the following two functions:

$$E : In_E \times State_E \rightarrow State_E \times Out_E,$$

$$A : In_A \times State_A \rightarrow State_A \times Out_A,$$

where E is the function defining the dynamics of the environment and A is the function defining the agent. These two functions are coupled by setting $Out_E = In_A$ and $Out_A = In_E$. Typically, the elements of the input set of the agent are called percepts, and the elements of the output set of the agent actions. The agent function is often referred to as *agent policy* (Fig. 1).

In order to define good or even optimal agent policies, it is necessary to introduce the concept of goal or reward.

An agent policy is optimal if it reaches a goal with minimal resources or maximizes reward. Ans *intelligent agent* is now defined as an agent which achieves goals in a wide range of environments. This definition was extracted by Legg and Hutter from more than 70 informal definitions occurring in cognitive science and AI research (Legg & Hutter, 2007a). In Legg and Hutter (2007b) they introduce the first general, formal definition of the intelligence of a computational agent. With the Υ-functional and its successors, e.g. for the incorporation of spatio-temporal aspects, see Orseau and Ring (2012), there are finally formal definitions of the core concept of artificial intelligence. The formal definition of intelligence by Legg and Hutter is briefly discussed in section "Defining Intelligence".

Learning from Data: The Problem of Induction

The problem of induction, which can be informally described as extracting rules from examples, leads to the following question:

- What set of possible models of the data generating process should a learning agent consider?

To answer this question in its full generality, it is necessary to explore the notion of "all possible models" from a mathematical and computational point of view, and discuss the question of effective learnability in the context of

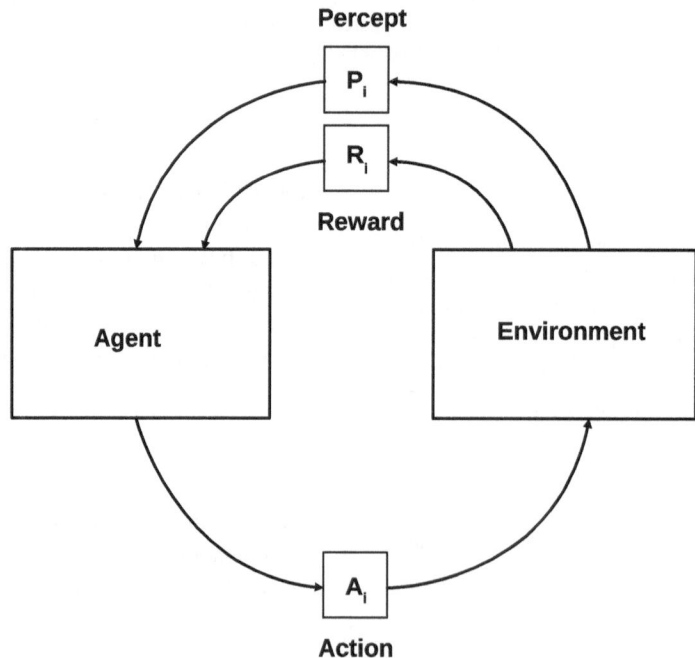

Fig. 1 Reinforcement learning agent

such generic model spaces. In Zimmermann and Cremers (2012) we showed that within the learning framework introduced by Solomonoff (1964a,b), Li and Vitányi (2008) the notion of "all possible models" cannot be defined in an absolute sense, but only with regard to a reference proof system. This dependence is used to establish a relationship between the *time* complexity of the data generating process and the *logical* complexity—defined as the proof-theoretic strength of a background axiom system—of the algorithmic learning system, thus shedding new light on the undecidability of the induction scheme introduced by Solomonoff.

The incomputability of Solomonoff induction can be traced back to the fact that the learning system does not know how much time has passed between two observations, i.e., how much time the data generating process has "invested" in order to produce the next observation. Such learning frameworks, where the generator and the learner are suspended while the other one is busy, will be called *asynchronous learning frameworks*. If one introduces a synchrony condition, which couples the time scales of the generator and the learner, one gets a *synchronous learning framework* and we will show that within such a learning framework effective and universal induction is possible, i.e., every effectively generated data sequence can be effectively learned.

Learning Frameworks

Every formal analysis of learning has to define a framework which specifies the exact type of learning problems considered and what successful learning means within this framework. The details of such a learning framework can have major implications for the question which learning tasks are solvable and which are not. In the following we will introduce two learning frameworks and we will show that these frameworks answer the same question— are universality and effectivity compatible properties?— differently.

The Asynchronous Learning Framework

A widely used model for analyzing sequential learning or decision tasks is, for example, defined in Hutter (2005), p. 126:

Definition 1 An agent is a system that interacts with an environment in cycles $k = 1, 2, 3, \ldots$. In cycle k the action (output) $y_k \in \mathcal{Y}$ of the agent is determined by a policy p that depends on the I/O history $y_1 x_1 \cdots y_{k-1} x_{k-1}$. The environment reacts to this action, and leads to a new perception (input) $x_k \in \mathcal{X}$ determined by a deterministic function q or probability distribution μ, which depends on the history $y_1 x_1 \cdots y_{k-1} x_{k-1} y_k$. Then the next cycle $k + 1$ starts.

Here $\mathcal{X}$ is a set containing all possible perceptions and $\mathcal{Y}$ is a set containing all possible actions of the agent. If the actions affect the future observations, then we call the above model an *asynchronous agent framework*, and if the actions are predictions which do not affect future observations, we call it an *asynchronous learning framework*.

In these asynchronous frameworks the resources, especially time, needed for generating the perceptions or the actions and predictions by the environment (the data generating process) or the agent are not modeled. This, for example, does imply that an agent does not know whether a new observation has arrived after 1 s or after one billion years, or, more importantly, that it has to wait longer and longer for each new observation. This last implication means that the time scales of the environment and the agent are not coupled, that, in a way, they belong to different universes. This decoupling of time scales is the reason why we call the framework asynchronous, and we will see that this property has deep implications.

Figure 2 illustrates the coupling of a learning system and an environment in the asynchronous learning framework.

The following notions are based on definitions in Zimmermann and Cremers (2012). Real-valued probabilistic learning systems are a specific type of learning system within the asynchronous learning framework:

Definition 2 A *real-valued probabilistic learning system* is a function

$$\Lambda : \{0, 1\}^* \times \{0, 1\} \to [0, 1]_{\mathbf{R}}, \quad \text{with } \Lambda(x, 0) + \Lambda(x, 1)$$
$$= 1 \text{ forall } x \in \{0, 1\}^*.$$

A real-valued probabilistic learning system has bits as perceptions and the predictions are probabilities for the next bit. One can extend the prediction horizon of Λ by feeding it with its own predictions. This leads to a learning system $\Lambda^{(k)}$ which makes probabilistic predictions for the next k bits (xy is the concatenation of strings x and y):

$$\Lambda^{(1)} = \Lambda,$$
$$\Lambda^{(k+1)}(x, y1) = \Lambda^{(k)}(x, y) \cdot \Lambda(xy, 1), \quad x \in \{0, 1\}^*, y \in \{0, 1\}^k,$$
$$\Lambda^{(k+1)}(x, y0) = \Lambda^{(k)}(x, y) \cdot \Lambda(xy, 0).$$

Finally, the learnability of an infinite bit sequence s ($s_{i:j}$ is the subsequence of s starting with bit i and ending with bit j) is defined as follows:

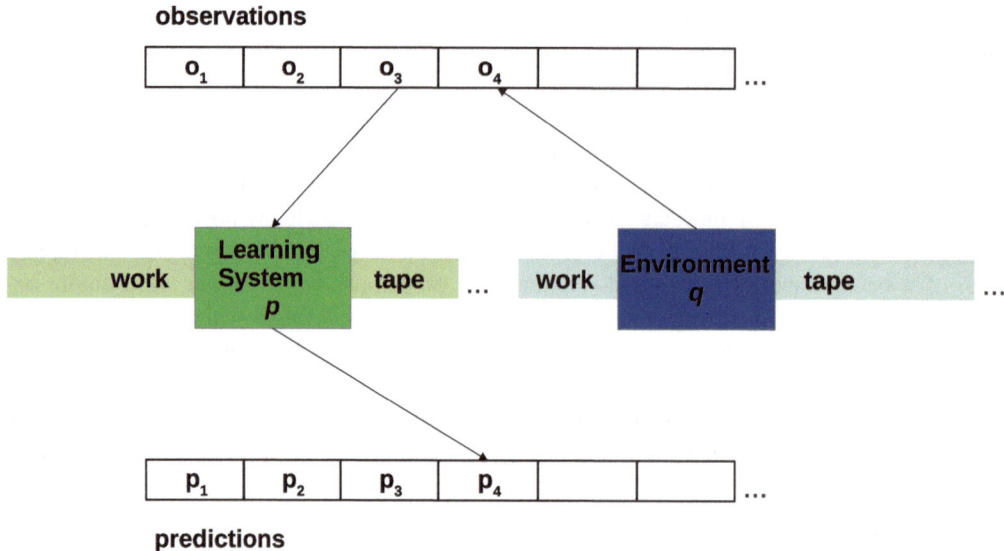

Fig. 2 Asynchronous learning framework

Definition 3 An infinite bit sequence s is learnable in the limit by the probabilistic learning system Λ, if for all $\epsilon > 0$ there is an n_0 so that for all $n \geq n_0$ and all $k \geq 1$:

$$\Lambda^{(k)}(s_{1:n}, s_{n+1:n+k}) > 1 - \epsilon.$$

This type of learnability criterion (learning in the limit) only requires that the learning system eventually will be nearly correct, but says nothing about the prediction accuracy on initial segments of the bit sequence.

Solomonoff Induction

The induction scheme introduced by Solomonoff (1964a,b) can be seen as a real-valued probabilistic learning system within an asynchronous learning framework. Solomonoff induction can learn (in the sense of Definition 3) all bit sequences generated by Turing machines. In this sense it is universal. In the following we will analyze the incomputability of Solomonoff induction and discuss why this incomputability cannot be resolved within the asynchronous learning framework.

The possible environments for Solomonoff induction can be described as programs p (represented as finite binary strings) executed by a fixed universal Turing machine U. Specifically, the universal Turing machine U has a one-way read-only input tape, some work tapes, and a one-way write-only output tape (such Turing machines are called monotone). The choice of the specific universal Turing machine affects space complexity only by a constant factor and time complexity at most by a logarithmic factor (Arora & Barak, 2009). Since the resources for generating the percepts are

not modeled in an asynchronous learning framework, these effects are irrelevant and we can use any universal Turing machine as our reference machine. The program strings are chosen to be prefix-free, i.e. no program string is the prefix of another program string. This is advantageous from a coding point of view, and does not restrict universality (Li & Vitányi, 2008).

A program p is a generator of a possible world, if it outputs an infinite stream of bits when executed by U. Unfortunately, it is not decidable whether a given program p has this well-definedness property. This is the reason why Solomonoff induction is incomputable: the inference process uses the whole set of programs (program space) as possible generators, even the programs which are not well-defined in the above sense. It follows that either one restricts the model space to a decidable set of well-defined programs, which leads to an effective inference process but ignores possibly meaningful programs, or one keeps all well-defined programs, but at the price of necessarily keeping ill-defined programs as well.

The Synchronous Learning Framework

We will now introduce a learning framework where the learning system gets information about the time the data generating process has used in order to produce the next observation. This concept is inspired by an analysis of real-world sequential learning situations, where both the environment and the learning system are not suspended while the other one is busy. But first we need the notion of the *generator time function*, generator function for short, of a program p (see Zimmermann & Cremers 2012):

Definition 4 The *generator time function* $G_p^{(U)} : \mathbf{N} \to \mathbf{N} \cup \{\infty\}$ of a program p wrt. the universal reference machine U assigns every $n \in \mathbf{N}$ the number of transitions needed to generate the first n bits by the reference machine U executing p. If n_0 is the smallest number for which p does not generate a new bit, then $G_p^{(U)}(n) = \infty$ for all $n \geq n_0$.

Further we call two programs p and q *observation equivalent* if they generate the same bit sequence s. The equivalence class of all programs corresponding to an infinite bit sequence s will be denoted by $[s]$. According to the Oxford Dictionaries Online (2013), *synchrony* can be defined as:

> The state of operating or developing according to the same time scale as something else.

This is a good description of what we have in mind, so we call bit sequences having the following property *synchronous*:

Definition 5 s is *synchronous* (wrt. U) if $\limsup_{n \to \infty} \frac{G_p^{(U)}(n)}{n} < \infty$ for at least one $p \in [s]$.

As stated in section "Solomonoff Induction", the time complexity between different universal Turing machines can vary by a logarithmic factor, so we have to define the notion of synchrony relative to a fixed universal Turing machine U. A bit sequence s is called *synchronous*, if there is a universal Turing machine U so that s is synchronous wrt. U.

Synchrony entails that the time scales of the learning system and the environment are coupled, that they cannot ultimately drift apart. As long as one not assumes a malicious environment, i.e., an environment that decelerates the computing speed of the learning system more and more, synchrony seems to be a natural property. A setting where observable bit sequences can be assumed to be synchronous will be called a *synchronous learning framework*.

Effective Universal Induction

We will now show that the problem of universal induction in the synchronous learning framework is effective and discuss implications of this result. The first step is formulated by the following theorem:

Theorem 1 *All synchronous bit sequences are learnable in the limit by an effective learning system.*

Proof This can be shown straightforward by using the generator-predictor theorem proved in Zimmermann and Cremers (2012), which states that a bit sequence s is learnable in the limit by a learning system $\Lambda(\Sigma)$, if Σ (a background axiom system) proves the totality of a recursive functions

which dominates the generator function of at least one program in $[s]$.

Now combining the synchrony condition wrt. a specific universal Turing machine and the fact that the time complexities of different universal Turing machines vary at most by a logarithmic factor, it suffices to find a background axiom system which proves the totality of a function which dominates $c \cdot n \cdot \log(n)$ for all positive constants c. Because the function n^2 will eventually be greater than $c \cdot n \cdot \log(n)$ for all fixed c, and the axiom system RCA_0 (Recursive Comprehension Axiom, see Zimmermann & Cremers 2012) proves the totality of n^2, the effective learning system $\Lambda(RCA_0)$ will learn all synchronous bit sequences in the limit. $\square$

The next idea is that via a process called *clockification* an arbitrary computable bit sequence can be transformed into a synchronous one (see Fig. 3). Clockification is a process by which a learning system extends in regular time intervals (measured by its internal transitions) an observed bit sequence s by inserting "clock signals" (coding a clock signal by "00" and the original observed bits by "10" and "11") marking the passing of time. The resulting bit sequence is a synchronous one.

Theorem 2 *Within a synchronous learning framework, all effectively generated bit sequences can be effectively learned in the limit.*

Proof By combining clockification and Theorem 1 we will get the desired result. $\square$

Caveats

The previous section has established an important result: all effective generators can eventually be effectively learned within the synchronous learning framework. This implies, for example, that if a universe can be described by a Turing machine, and we assume the assumptions of the synchronous learning framework as valid, then there is an effective learning system Λ which would converge to the "theory of everything" (TOE). This is an interesting result, but here is a list of caveats which help to put this theorem into perspective:

1. Λ converges to the TOE, but we will never know when this has happened or how close the current predictions are to the truth.
2. The true model probably is not useful, learnability and predictability fall apart, i.e., the true model could be extremely complex, its evaluation would take so long that its predictions would only arrive after the fact.
3. Even having a TOE does not mean that one can answer all questions: there are cellular automata like "Game of Life"

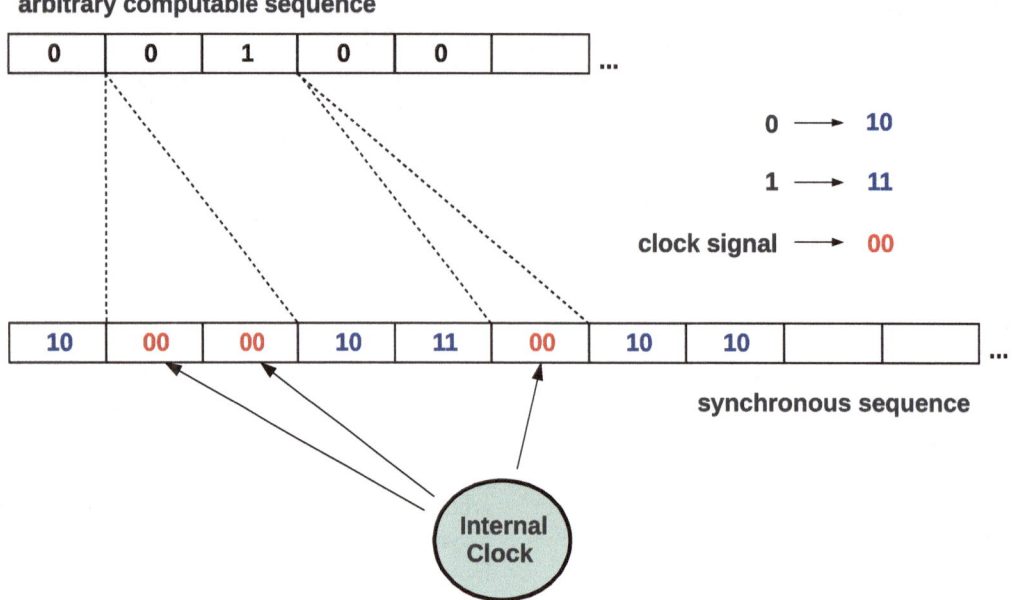

Fig. 3 Clockification: using an internal clock transforms all computable bit sequences into synchronous bit sequences

(Berlekamp, Conway, & Guy, 2001) or "Langton's Ants" (Langton, 1986) which can be seen as possible universes, and the transition rules define the TOE of these universes. But questions like "Are there self-reproducing patterns?" or "Does this ant build a highway (i.e., a certain repetitive pattern)?" cannot be answered in general, despite the fact that we know the TOE of the "Game of Life" and the ant world.

4. Finally, the information content of the universe could be infinite: imagine a Turing machine which has a work tape initialized to an infinite random bit sequence. Then the transition process is effective, but the output stream could still be incomputable by using ever more bits of the random bit sequence.

The second caveat can be termed the "postdiction problem": one can in principle predict the future exactly, but the resources needed to compute the predictions are prohibitive: they would arrive long after the predicted event has happened. This situation, where the notions of determinism and predictability fall apart, is discussed, for example, in Rummens and Cuypers (2010).

In summary, the compatibility of universality and effectiveness of inductive inference within the synchronous learning framework is an interesting theoretical finding, but has no immediate practical implications. However, it can shed some light on the path towards learning systems which are both efficient and extremely general at the same time.

The Structure of Uncertainty

One central aspect of learning from experience is the representation and processing of uncertain knowledge. In the absence of assumptions about the world, there is no nontrivial logical conclusion which can be drawn from the past on any future event. Accordingly, it is of foundational interest to analyze the structure of uncertainty as a question in its own right, and it has spawned a subfield of research within artificial intelligence and philosophy. A plethora of approaches has emerged over the last century to address this question, for example, Dempster–Shafer theory (Dempster, 1967; Shafer, 1976), Possibility theory (Dubois & Prade, 1988; Dubois, 2006), Revision theory (Gärdenfors, 1992), Ranking theory (Spohn, 1999, 2009), and non-monotonic logic (Ginsberg, 1987). A survey and discussion of many of the existing approaches is given in Huber and Schmidt-Petri (2009).

In the following we discuss an approach to reasoning under uncertainty by introducing a small axiom system describing necessary conditions for uncertainty measures. Furthermore, this axiom system does not define the structure of uncertainty explicitly, e.g. that uncertainty can be measured by one real number, but entails the algebraic structure of uncertainty values. This approach, which can be called *algebraic uncertainty theory*, enables a unifying perspective on reasoning under uncertainty. A good overview and a discussion with examples of this algebraic approach can be found in Arnborg (2016).

Formalizing Uncertainty

First we have to discuss a subtle issue of terminology. Above we have used the notion "uncertainty values" to denote generalized truth values. Unfortunately, there is the following problem when using this term in a formalized context: no uncertainty about a proposition can be identified with sure knowledge, but maximal uncertainty about a proposition is *not* certainty with regard to the negation of the proposition. The domains of truth values we want to axiomatize contain a greatest and a least element, where the greatest element should represent certainty and the least element impossibility, i.e. certainty of the negated proposition. For this reason, we adopt the notion "confidence measure" instead of uncertainty measure in the following definitions and axioms.

The Algebra of Truth Bearers

Before delving into the structure of uncertainty, we have to define the objects and their relations which are capable to take on truth values, the *truth bearers*. In a context of crisp events, i.e., after the fact it is unambiguously decidable if the event has occurred or not, the algebra of truth bearers is normally considered to be a Boolean algebra, but when truth bearers are not crisp, then another proposition algebra has to be considered, i.e., a fuzzy logic where the law of complementation is not valid: $x \vee \neg x \neq 1$, or quantum logic. The propositional algebra in quantum logic is "formally indistinguishable from the calculus of linear subspaces of a Hilbert space with respect to set products, linear sums, and orthogonal complements" corresponding to the roles of *and*, *or*, and *not* in a Boolean algebra. These linear subspaces form orthomodular lattices which in general do not satisfy the distributivity laws, see Padmanabhan and Rudeanu (2008), page 128ff. The investigation of uncertainty measures for non-Boolean proposition algebras is open to future research.

Uncertainty: The Boolean Case

A *conditional confidence measure* for a Boolean Algebra **U** and a domain of confidence values $\mathcal{C}$ is a mapping Γ : $\mathbf{U} \times \mathbf{U} \setminus \{\perp\} \rightarrow \mathcal{C}$. Let $A, B \in \mathbf{U}$, then the expression $\Gamma(A|B)$ reads: "the confidence value of A given B (wrt. Γ)". The domain of confidence values is partially ordered and has a greatest ($\top$) and a least ($\perp$) element. A *confidence space* is a triple $(\mathbf{U}, \Gamma, \mathcal{C})$. One of the following axioms (Extensibility) for confidence measures deals with relations between confidence spaces defined over different Boolean algebras. Thus it is necessary to introduce a *set of confidence spaces* all sharing the same domain of confidence values. Such a set of confidence spaces we will call a *confidence universe*, and the following axiom system is concerned with such confidence universes, and not single confidence spaces. This seemingly technical shift in perspective is essential for the formalization

of natural properties like extensibility, which plays a crucial role as an intuitive axiom complementing Cox's assumptions.

In Zimmermann (2012) seven axioms are introduced, which can be grouped in three connective axioms, two order axioms, and two "infrastructure axioms," where the connective axioms concern properties of the logical connectives, the order axioms relate the order structures of a proposition algebra and the confidence domain, and the infrastructure axioms deal with the combinability of confidence spaces and a closure property. Here we only state two of the seven axioms as examples; for a complete list of axioms and a discussion, see Zimmermann (2012).

Axioms for Uncertainty

In the following, we use $\Gamma(A)$ as an abbreviation for $\Gamma(A|\top)$.

(Not) For all $(\mathbf{U_1}, \Gamma_1, \mathcal{C})$ and $(\mathbf{U_2}, \Gamma_2, \mathcal{C})$: If $\Gamma_1(A_1) = \Gamma_2(A_2)$, then $\Gamma_1(\bar{A}_1) = \Gamma_2(\bar{A}_2)$.

The axiom **Not** expresses that the information in the confidence value of a statement A is sufficient to determine the confidence value of $\bar{A}$. This is justified by the requirement that every piece of information which is relevant for the confidence value of A is relevant for the confidence value of $\bar{A}$ and vice versa.

The other two connective axioms concern similar properties for the conjunction of two propositions. The next axiom states that if a proposition A implies a proposition B (the implication relation defines an order relation on a proposition algebra), denoted by $A \leq B$, then the confidence in B is at least as high as the confidence in A.

(Order₁) For all $(\mathbf{U}, \Gamma, \mathcal{C})$ and all $A, B \in \mathbf{U}$: If $A \leq B$, then $\Gamma(A) \leq \Gamma(B)$.

The order axioms connect the implication ordering of the proposition algebra with the ordering on the confidence domain, where **Order₁** specifies the forward direction and a second order axiom specifies the backward direction (Fig. 4).

The infrastructure axioms require the extensibility of domains of discourse, i.e., two independently defined confidence spaces shall be embeddable into one frame of reference, and a closure property of conditioning which assures that for every confidence measure conditioned on some background knowledge there is an equivalent unconditional confidence measure.

For the justification of the axioms it is important to interpret the expression $\Gamma(A|B)$ as: "*all* that can be said about the confidence of A given B (wrt. Γ)." Given this interpretation, the common justification of the connective axioms is that a violation of these axioms will necessarily lead to a loss of relevant information. Note that the axioms use only equations

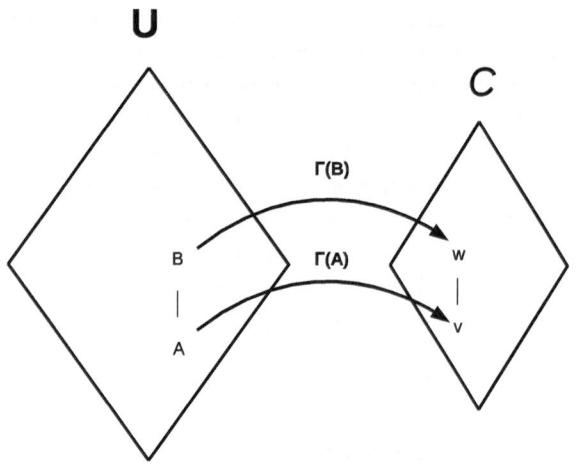

Fig. 4 Ordered confidence values v and w with corresponding propositions in a suitably chosen confidence space $(\mathbf{U}, \Gamma, \mathcal{C})$

and inequalities between confidence values, because there are no algebraic operations defined on the domain of confidence values yet.

It is now possible to characterize the algebraic structure of a confidence domain as the [0, 1]-interval of a partially ordered ring. Rings are algebraic structures which generalize fields. For example, the real numbers with addition and multiplication form a field. In a field all elements except zero have a multiplicative inverse, in a ring this is not required, i.e., a ring can contain elements other than 0 which are not invertible. Confidence measures satisfy the analogs of the axioms of probability, but with regard to the ring operations. This is stated by the following theorem:

Ring Theorem The domain of confidence values $\mathcal{C}$ of a confidence universe satisfying the connectivity, order, and infrastructure axioms can be embedded into a partially ordered commutative ring. All confidence measures Γ of the confidence universe satisfy:

$$\hat{\Gamma}(\top) = 1\,, \tag{1}$$

$$\hat{\Gamma}(A \vee B) = \hat{\Gamma}(A) \oplus \hat{\Gamma}(B)\,, \qquad \text{if } A \wedge B = \bot\,, \tag{2}$$

$$\hat{\Gamma}(A \wedge B) = \hat{\Gamma}(A|B) \odot \hat{\Gamma}(B)\,. \tag{3}$$

In the next chapter we discuss a model for a general computational agent called AIXI, which was introduced by Hutter (2005). This agent satisfies certain optimality conditions with regard to its long-term behavior within the class of computational agents. AIXI combines Solomonoff induction and reinforcement learning, which captures also interactions of an agent with the environment generating its perceptions. AIXI, like Solomonoff induction, uses the Bayesian frame-

work for representing and processing uncertainty, which does not utilize the full generality of uncertainty calculi discussed in this chapter, like infinitesimal or incomparable uncertainty values, but Bayesian inference is a possible model of the axioms introduced in Zimmermann (2012). How uncertainty calculi using the full expressiveness of confidence domains can be combined and integrated with the AIXI agent model is open to future research.

A General Agent Architecture: AIXI

The framework of universal induction introduced by Solomonoff only treats the modeling and predicting aspect of learning, but the agent does not act based on its predictions, so in the Solomonoff framework the environment affects the learning agent, but not vice versa. In this sense, the loop between agent and environment is not closed (no senso-motoric loop). Enhancing the Solomonoff framework in order to incorporate the possibility of actions leads to a framework introduced by Hutter (2005), which can be seen as an integration of the reinforcement learning framework (Sutton, 1984) and the framework of Solomonoff. Now the agent acts based on its predictions, and these actions can affect the environment and change its future course, thus also changing future observations of the agent. In order to define the quality of an agent policy, we need generalization of the loss function used to evaluate the predictions of learning agents. Success is now defined by the environment and is the second feedback channel, besides the percepts, from the environment to the agent.

The search for optimal policies in this framework leads to a generalization of Solomonoff induction, and agents following such an optimal policy are called AIXI agents. AIXI is a reinforcement learning agent which maximizes the expected total rewards received from an environment. It simultaneously considers every computable environment as a possible generator of its perceptions. In each time step, it looks at every computable environment and evaluates how many rewards that environment generates depending on the next action taken. The expected rewards are then weighted by the subjective belief that this program constitutes the true environment. This belief is computed from the length of the program describing the environment: longer programs are considered less likely, in line with Occam's razor. AIXI then selects the action that has the highest expected total reward in the weighted sum of all these programs.

However, in Leike and Hutter (2015) it is shown, that a bad prior for inductive inference can affect the agent behavior indefinitely, because it does not sufficiently incite the agent to explorative behavior. Accordingly, no theorem comparable to the invariance theorem for Solomonoff induction is available, and the choice of the reference machine becomes crucial. Unfortunately, investigations into suitable reference machines are still in an early stage and have not yet resulted in a clear candidate for a reference machine on which to base a general cognitive agent.

Defining Intelligence

Legg and Hutter (2007b) used the combination of a general reinforcement learning agent and Solomonoff induction to define an intelligence functional Υ by assigning every agent policy π an intelligence score describing its expected reward averaged over all possible computable environments. It is an attempt to translate their informal definition of intelligence, "the ability to achieve goals in a wide range of environments," in a quantitative intelligence measure.

Let $\mathcal{X}$ be a set of perceptions, $\mathcal{R}$ be a set of rewards, and $\mathcal{Y}$ be a set of actions of an agent. A deterministic agent policy assigns to all possible sequences of percepts from $\mathcal{X}$ and rewards from $\mathcal{R}$ an action from $\mathcal{Y}$. A probabilistic policy assigns to all percept/reward sequences a probability distribution on the action set $\mathcal{Y}$. The total reward $V_\mu(\pi)$ of a policy π for an environment μ is the accumulated reward an agent executing policy π in environment μ collects during its lifetime.

Now the computable environment μ can be seen as a binary program running on a suitable universal Turing machine used as a reference machine. Solomonoff induction assumes that the prior probability of an environment μ is proportional to $2^{-|\mu|}$, where $|\mu|$ is the length of the binary program describing μ (Li & Vitányi, 2008). Thus simpler environments, meaning that there is a shorter program to describe them, get a higher prior probability. These prior probabilities are used to define the expected reward of policy π over all computable environments:

$$\Upsilon(\pi) := \sum_{\mu \in E} 2^{-|\mu|} \cdot V_\mu(\pi),$$

where E is the set of all computable environments. Legg and Hutter call $\Upsilon(\pi)$ the *universal intelligence* of an agent using policy π. The first aspect of their informal definition of intelligence, "achieving goals," is encoded in the value $V_\mu(\pi)$ of policy π with regard to each environment, the second aspect, "in a wide range of environments," is represented by averaging over all computable environments. This measure was the first formal definition of the intelligence of a general computational agent, and thus represents an important milestone in the foundations of artificial intelligence.

The Quest for a Standard Reference Machine

The results of Leike and Hutter (2015) made it abundantly clear that in order to make progress in understanding the simplicity or complexity of finite objects it is necessary to reach a consensus on a meaningful reference machine, i.e., which operations are available and executable in unit time. Such a consensus on a reference machine could serve as a standard for measuring descriptive and computational complexity. Like today's physical units, such a standard reference machine would contain contingent elements, but if it is chosen in a "natural" way it could nevertheless be tremendously useful.

Reference Machines and Initial Complexity

In order to analyze the computational complexity (or simplicity) of a computational object (algorithm, model, agent), it is necessary to define a reference machine which executes the computations. The first precisely defined mathematical model of computation, an abstract machine, was introduced by Alan Turing in 1936. There were many different attempts to define a model of computation, for example, the λ-calculus or Markov algorithms, but they were all found to equivalent to or weaker than Turing machines. This led to the formulation of the Church–Turing thesis, that all conceivable mathematical models of computation are equivalent to the Turing machine. The thesis found widespread acceptance, and today Turing machines are seen as defining an absolute notion of computability. Turing also showed that there are incomputable problems, of which the halting problem is the most famous. Another important discovery was the existence of universal Turning machines, i.e., Turning machines which are capable to simulate all other Turing machines. For a discussion of the Church–Turing thesis, universal Turing machines, and related topics, see Herken (1994).

If one is only interested whether a problem can be solved by computation or not, one can use any universal Turing machine U as a reference machine and if there is a program for U which solves the problem, then the problem is computable, otherwise not. So for questions of computability any universal Turing machine can be used and will lead to the same answers. But things become much more complicated when one is not only interested in computability, but also in complexity, i.e. the resources needed to actually execute the computations. Typically, one is interested in time and space complexity, and a central theorem relates the time and space complexity of a universal Turing machine to any Turing machine (Arora & Barak, 2009):

Theorem *There exists a TM U such that for every x, p ∈ {0, 1}*, with U(x, p) = M_p(x), where M_p denotes the TM represented by p.*

Furthermore, if M_p halts on input x within T steps, then $U(x, p)$ halts within $C \cdot T \cdot log(T)$ steps, where C is a number independent of $|x|$ and depending only on M_p's alphabet size, number of tapes, and number of states.

This means if one is interested only in the general growth of the time complexity with the input length, i.e., with the asymptotic behavior, a suitably chosen UTM can serve as a reference machine for analyzing the time complexity of computational problems. Current computational complexity theory tries to classify problems with regard to the asymptotic complexity, and for this goal the above specification of a reference machine is sufficient. For example, one of the most important problem classes, P, i.e., the problems solvable in polynomial time, does not change when one changes from one UTM U_1 to another UTM U_2, provided they can simulate all other TM's within polynomial time. This has led to a very successful theory of computational complexity, which can help to classify the hardness of a computational problem. The famous $P = NP$ problem is one of the major open questions of this field, and problems which can be shown to be $NP - hard$ are generally believed to have no efficient algorithms to solve them (Arora & Barak, 2009).

For questions aiming at the asymptotic growth of needed resources depending on the size of the input, this is a suitable resolution of computational complexity. But for questions regarding the computational complexity of a finite problem, like the computational complexity of a good strategy for a game like Go, or for deciding which of two binary strings has a shorter description, we need to look closer at the reference machine.

Iterated Boolean Circuits

We now introduce a proposal for a reference machine inspired by the basic functionality of current computing devices, but also by striving for mathematical simplicity. Current computing devices can be seen as the iterative application of a hardwired Boolean circuit to a vector of bits. Accordingly, an *iterated Boolean circuit* is defined as a Boolean function on B^n, the set of n-bit vectors, which then is applied iteratively, generating a sequence of bit vectors. Additionally, the Boolean circuit is build entirely of NAND-gates, i.e., the Boolean function which is false if both inputs are true and otherwise true. The NAND-gate is a Boolean base, so all Boolean functions can be expressed entirely with NAND-gates. Interestingly, a similar machine model was already defined by Alan Turing in a National Physical Laboratory Report "Intelligent Machinery" published in 1948. He

Fig. 5 A Boolean circuit consisting of 4 NAND-gates

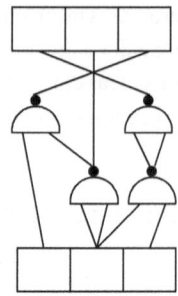

called networks of binary nodes connected by NAND-gates "Unorganized Machines," and introduced them as a possible model for information processing in the human brain. This report is reproduced in Cooper and Leeuwen (2013), pp. 501–516.

These iterated Boolean circuits are now used to generate sequences of output bits, and for an observed bit sequence the learning problem is to find a small (measured by the number of NAND-gates) Boolean circuit which, when iterated, generates the observed bit sequence. As an example, consider the following bit sequence: 00010001000. There is a Boolean circuit with 4 NAND-gates which generates this sequence, see Fig. 5. The leftmost bit is considered the output bit. In Fig. 6 the sequence of output bits generated by the Boolean circuit after 1 and after 11 iterations is depicted. Finally, when the output sequence matches the observed sequence, we can just continue with the iterated applications of the Boolean circuit to generate predictions, see Fig. 7. In this case, the prediction for the 12th bit is "1."

The problem of finding a generating Boolean circuit matching an observed sequence can be seen as an inversion problem. Inversion problems often lead to a combinatorial search space, where no exhaustive strategy is applicable. We now discuss an approach to deal with such combinatorial search problems based on recent advances in machine learning.

Outlook: Search in Circuit Space

The number of possible circuits grows like $2^{O(n^2)}$, i.e., super-exponentially fast with the number n of gates. Even for small numbers (like 10) an exhaustive search is not possible anymore. The current advancements in combining deep learning, a variant of artificial neural networks using many hidden layers, with reinforcement learning can lead the way how to explore huge combinatorial search spaces with limited resources (Silver et al., 2018). In March 2016 a Go program based on deep learning and a self-play loop won against one of the best professional Go-players. This progress of Computer Go was not expected within the next decade, which

Fig. 6 Left: output sequence after one iteration. Right: output sequence after 11 iterations is matching the observed sequence

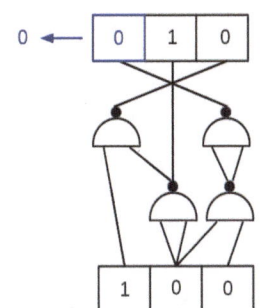

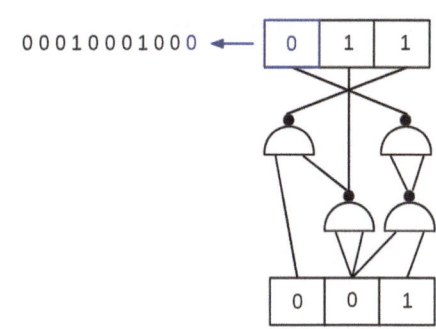

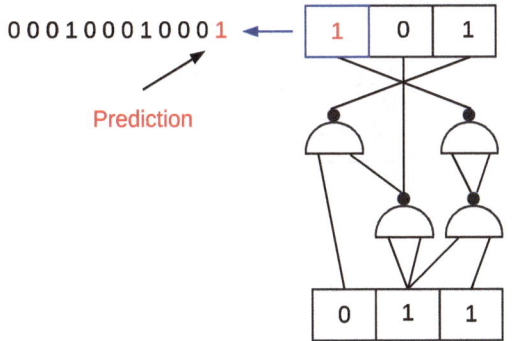

Prediction

Fig. 7 The twelfth bit is the prediction generated by the iterated Boolean circuit

is a reason to hope that the basic principles of AlphaGo, and its subsequent generalization AlphaZero, can be applied to other combinatorial search problems as well. The core idea is to use deep reinforcement learning to focus the exploration of combinatorial search spaces on the most promising parts (Fig. 8).

By introducing operators on circuit space (like adding a gate, removing a gate, rewire a connection,…) the inversion problem can be transformed into a reachability problem for graphs and will thus be accessible to AlphaZero-like search strategies (Fig. 9).

Conclusions and Outlook

Despite foundational results on learnability within the synchronous and asynchronous learning frameworks, an axiomatization of uncertain reasoning, a formal definition of intelligence, and many results on general reinforcement learning agents, there is still no unifying axiomatization of general cognitive agents comparable, for example, to the axiomatic foundations of probability theory or set theory. Especially the topics of a standard reference machine and cognition with bounded resources have to be explored much further in order to reach a meaningful and integrated foundational framework for artificial intelligence.

Nevertheless, the theoretical and practical advance of artificial intelligence has reached a state where ethical questions and the impact on society become pressing issues. In the following outlook we will discuss the emerging landscape of ethical and social questions arising from the expansion of AI systems to increasingly critical applications.

Algorithmic Accountability, Transparency, and Fairness

The increase of computational resources and available data on the one side, and the latest advancements in machine learning, notably deep learning, on the other side have now reached a critical level where AI systems start to leave highly specialized and controlled environments and become part—now or in the foreseeable future—of our daily lives, on an individual and a societal level. Examples are autonomous driving, natural language processing, and applications in the judicial system. The prospect of general AI systems which are not limited to narrow applications has led to growing concerns about safety and trustworthiness. See Everitt, Lea, and Hutter (2018) for a comprehensive review of current literature.

The potential impact of AI applications on individuals and society as a whole leads to an increased need for transparency and accountability of AI systems which keeps pace with the technical development. For example, autonomous driving can lead to moral dilemmas when during an accident the loss of human life becomes unavoidable, but the autonomous driving system still can influence whose life will be endangered (Awad et al., 2018). Natural language processing can be used to facilitate fraud or to wield political influence, e.g. via bots in social networks (Simonite, 2019). One especially controversial decision support system already used by the US judicial system is COMPAS, a system which assesses the likelihood of a defendant becoming a recidivist. These risk assessments can inform decisions about who will be set free and who is not. Even if the race of the defendant is not part of the variables considered by COMPAS, reports

Fig. 8 The search strategy of
AlphaGo

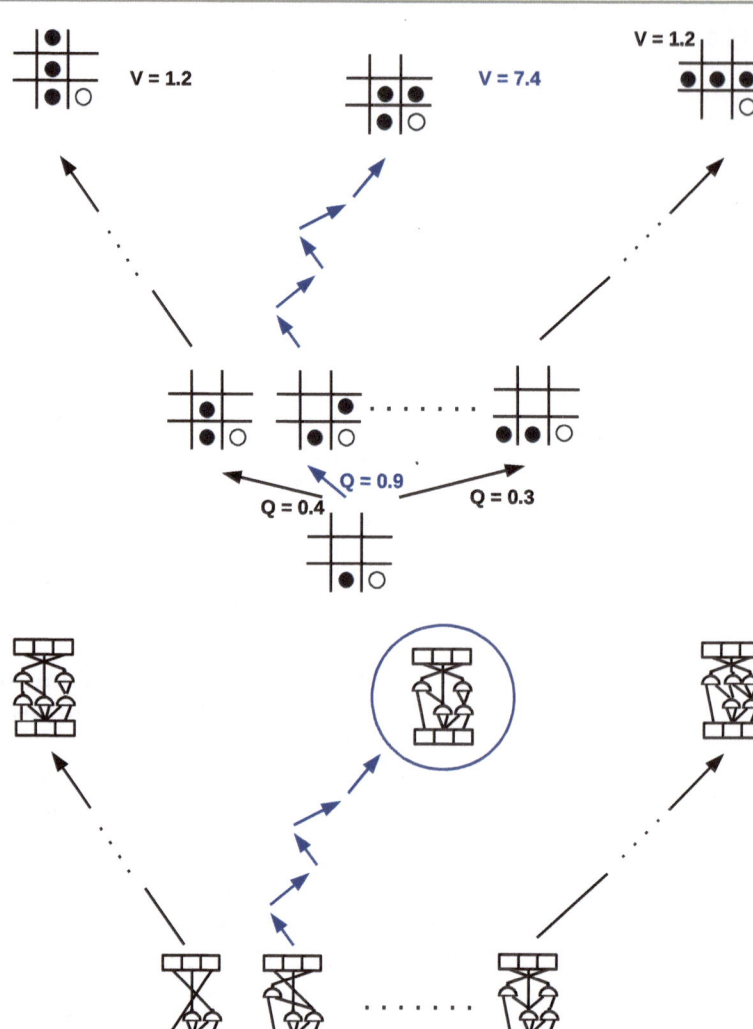

Fig. 9 The exploration of
Boolean circuits using an
AlphaZero-like search strategy

A = Add Gate
D = Delete Gate
R = Rewire

have emerged that COMPAS risk levels are racially biased (Angwin, Larson, Mattu, & Kirchner, 2016). A closer look shows that the exact definition of unbiasedness or fairness is instrumental, and different definitions can lead to different outcomes (Corbett-Davies, Pierson, Feller, & Goel, 2016). In this case, no decision system can be simultaneously unbiased or fair with regard to all desirable definitions of unbiasedness or fairness, and only an emerging consensus on which definition is the "right" or the "least problematic" one can mitigate this dilemma.

From Association Learning to Causal Learning

The need for algorithmic transparency, accountability, and unbiasedness adds new urgency to a topic which has affected machine learning and statistics from the beginning:

the learned relationships are in general only association relations and not causal relations, i.e., the observed covariation between two variables A and B is caused by an unknown third variable C. When actions based on predictions significantly feed back into the observed system, association learning cannot answer important questions arising with regard to the consequences of the executed actions. In order to develop and apply standards of transparency, accountability, and unbiasedness, the result of learning has to identify the *causal* factors that determine the predictions. The notion of causality and the detection of causal relationships is a longstanding problem in machine learning and statistics, but recently there has been some progress, most notably the theory of causal inference by Pearl, Glymour, and Jewell (2016), but also attribution science (Otto, 2017) and causal deconvolution (Zenil, Kiani, Zea, & Tegnér, 2019) are interesting developments.

Attribution science, or probabilistic event attribution (PEA), is an emerging field that assigns probabilities to possible causes for observed effects, especially in the context of climate change, but is still in an early stage and the validation of its claims is subject to further research.

We are convinced that effective universal induction can play an important role in causal learning by identifying generators of observed data and not only associations within the observed data. The importance of universal induction was emphasized by one of the founding figures of artificial intelligence, Marvin Minsky, during a discussion panel in 2010:

"It seems to me that the most important discovery since Gödel was the discovery by Chaitin, Solomonoff, and Kolmogorov of the concept called Algorithmic Probability which is a fundamental new theory of how to make predictions given a collection of experiences and this is a beautiful theory, everybody should learn it, but it has got one problem, that is, that you cannot actually calculate what this theory predicts because it is too hard, it requires an infinite amount of work. However, it should be possible to make practical approximations to the Chaitin, Kolmogorov, Solomonoff theory that would make better predictions than anything we have today. Everybody should learn all about that and spend the rest of their lives working on it."

References

Angwin, J., Larson, J., Mattu, S., & Kirchner, L. (2016). Machine bias. *ProPublica*.

Arnborg, S. (2016). Robust Bayesian analysis in partially ordered plausibility calculi. *International Journal of Approximate Reasoning, 78*, 1–14.

Arora, S., & Barak, B. (2009). *Complexity theory: A modern approach*. Cambridge: Cambridge University Press.

Awad, E., Dsouza, S., Kim, R., Schulz, J., Henrich, J., Shariff, A., et al. (2018). The moral machine experiment. *Nature, 536*, 59–64.

Berlekamp, E. R., Conway, J. H., & Guy, R. K. (2001). *Winning ways for your mathematical plays* (2nd ed.). Natick: A K Peters.

Cooper, S. B., & van Leeuwen, J. (2013). *Alan Turing: His work and impact*. Amsterdam: Elsevier Science.

Corbett-Davies, S., Pierson, E., Feller, A., & Goel, S. (2016). A computer program used for bail and sentencing decisions was labeled biased against blacks. It's actually not that clear. *The Washington Post*.

Dempster, A. P. (1967). Upper and lower probabilities induced by a multivalued mapping. *The Annals of Mathematical Statistics, 38*(2), 325–339.

Dubois, D. (2006). Possibility theory and statistical reasoning. *Computational Statistics & Data Analysis, 51*(1), 47–69.

Dubois, D., & Prade, H. (1988). *Possibility theory*. New York: Plenum Press.

Herken, R. (Ed.). (1994). *The universal Turing machine: A half-century survey*. Berlin: Springer.

Everitt, T., Lea, G., & Hutter, M. (2018). AGI safety literature review. In *Proceedings of the 27th International Joint Conference on Artificial Intelligence (IJCAI'18), Stockholm, Sweden* (pp. 5441–5449).

Gärdenfors, P. (Ed.). (1992). *Belief revision*. Cambridge: Cambridge University Press.

Ginsberg, M. (Ed.). (1987). *Readings in nonmonotonic reasoning*. Los Altos, CA: Morgan Kauffman.

Huber, F., & Schmidt-Petri, C. (Ed.) (2009). *Degrees of belief*. Berlin: Springer.

Hutter, M. (2005). *Universal artificial intelligence: Sequential decisions based on algorithmic probability*. Berlin: Springer.

Langton, C. G. (1986). Studying artificial life with cellular automata. *Physica D: Nonlinear Phenomena, 22*(1–3), 120–149.

Legg, S., & Hutter, M. (2007a). A collection of definitions of intelligence. In B. Goertzel & P. Wang (Eds.), *Advances in artificial general intelligence: Concepts, architectures and algorithms*. Frontiers in Artificial Intelligence and Applications (Vol. 157, pp. 17–24). Amsterdam, NL: IOS Press.

Legg, S., & Hutter, M. (2007b). Universal intelligence: A definition of machine intelligence. *Minds & Machines, 17*(4), 391–444.

Leike, J., Hutter, M. (2015). Bad universal priors and notions of optimality. *Journal of Machine Learning Research, 40*, 1244–1259.

Li, M., & Vitányi, P. M. B. (2008). *An introduction to Kolmogorov complexity and its applications*. Graduate texts in computer science (3rd ed.). Berlin: Springer.

Orseau, L., & Ring, M. (2012). Space-time embedded intelligence. In *Proceedings of the 5th International Conference on Artificial General Intelligence, AGI'12* (pp. 209–218). Berlin: Springer.

Otto, F. E. L. (2017). Attribution of weather and climate events. *Annual Review of Environment and Resources, 42*, 627–646.

Oxford Dictionaries Online. (2013). *Synchrony, noun*. Oxford: Oxford University Press. Retrieved May 11, 2013 from http://oxforddictionaries.com/definition/english/synchrony

Padmanabhan, R., & Rudeanu, S. (2008). *Axioms for Lattices and Boolean algebras*. Singapore: World Scientific.

Pearl, J., Glymour, M., & Jewell, N. P. (2016). *Causal inference in statistics: A primer*. Hoboken: Wiley.

Rummens, S., & Cuypers, S. E. (2010). Determinism and the paradox of predictability. *Erkenntnis, 72*(2), 233–249.

Shafer, G. (1976). *Mathematical theory of evidence*. Princeton: Princeton University Press.

Silver, D., Hubert, T., Schrittwieser, J., Antonoglou, I., Lai, M., Guez, A., et al. (2018). A general reinforcement learning algorithm that masters chess, shogi, and go through self-play. *Science, 362*(6419), 1140–1144.

Simonite, T. (2019). The AI text generator that's too dangerous to make public. *Wired*.

Solomonoff, R. (1964a). A formal theory of inductive inference, part I. *Information and Control, 7*(1), 1–22.

Solomonoff, R. (1964b). A formal theory of inductive inference, part II. *Information and Control, 7*(2), 224–254.

Spohn, W. (1999). Ranking functions, AGM style. In B. Hansson, S. Halldén, N.-E. Sahlin, & W. Rabinowicz (Eds.), *Internet festschrift for Peter Gärdenfors*, Lund. http://www.lucs.lu.se/spinning

Spohn, W. (2009). A survey of ranking theory. In F. Huber & C. Schmidt-Petri (Eds.), *Degrees of belief*. Berlin: Springer.

Sutton, R. S. (1984). *Temporal Credit Assignment in Reinforcement Learning*. PhD Thesis, University of Massachusetts, Amherst, MA.

Zenil, H., Kiani, N. A., Zea, A. A., & Tegnér, J. (2019). Causal deconvolution by algorithmic generative models. *Nature Machine Intelligence, 1*, 58–66.

Zimmermann, J. (2012). *Algebraic Uncertainty Theory*. PhD Thesis, Rheinische Friedrich-Wilhelms-Universität Bonn.

Zimmermann, J., & Cremers, A. B. (2012). Making Solomonoff induction effective or you can learn what you can bound. In S. B. Cooper, A. Dawar, and B. Löwe (Eds.), *How the world computes*. Lecture Notes in Computer Science (Vol. 7318). Berlin: Springer.

What Is Consciousness, and Could Machines Have It?

Stanislas Dehaene, Hakwan Lau, and Sid Kouider

Contents

S. Dehaene (✉)
Experimental Cognitive Psychology, Collège de France, Paris, France

Cognitive Neuroimaging Unit, CEA, INSERM, Université Paris-Sud, Université Paris-Saclay, NeuroSpin Center, Gif/Yvette, France
e-mail: stanislas.dehaene@cea.fr

H. Lau
Department of Psychology, Brain Research Institute, UCLA, Los Angeles, CA, USA

Department of Psychology, The University of Hong Kong, Kowloon, Hong Kong

S. Kouider
Brain and Consciousness Group (ENS, EHESS, CNRS), Département d'Études Cognitives,École Normale Supérieure—PSL Research University, Paris, France

© The Author(s) 2021
J. von Braun et al. (eds.), *Robotics, AI, and Humanity*, https://doi.org/10.1007/978-3-030-54173-6_4

Abstract

The controversial question of whether machines may ever be conscious must be based on a careful consideration of how consciousness arises in the only physical system that undoubtedly possesses it: the human brain. We suggest that the word "consciousness" conflates two different types of information-processing computations in the brain: the selection of information for global broadcasting, thus making it flexibly available for computation and report (C1, consciousness in the first sense), and the self-monitoring of those computations, leading to a subjective sense of certainty or error (C2, consciousness in the second sense). We argue that despite their recent successes, current machines are still mostly implementing computations that reflect unconscious processing (C0) in the human brain. We review the psychological and neural science of unconscious (C0) and conscious computations (C1 and C2) and outline how they may inspire novel machine architectures.

Keywords

Consciousness · Mind · Brain · Perception · Metacognition

Imagine that you are driving when you suddenly realize that the fuel-tank light is on. What makes you, a complex assembly of neurons, aware of the light? And what makes the car, a sophisticated piece of electronics and engineering, unaware of it? What would it take for the car to be endowed with a consciousness similar to our own? Are those questions scientifically tractable?

Alan Turing and John von Neumann, the founders of the modern science of computation, entertained the possibility that machines would ultimately mimic all of the brain's abilities, including consciousness. Recent advances in artificial intelligence (AI) have revived this goal. Refinements in machine learning, inspired by neurobiology, have led to artificial neural networks that approach or, occasionally, surpass humans (Silver et al. 2016; Lake et al. 2017). Although those networks do not mimic the biophysical properties of actual brains, their design benefitted from several neurobiological insights, including non-linear input-output functions, layers with converging projections, and modifiable synaptic weights. Advances in computer hardware and training algorithms now allow such networks to operate on complex problems (e.g., machine translation) with success rates previously thought to be the privilege of real brains. Are they on the verge of consciousness?

We argue that the answer is negative: the computations implemented by current deep-learning networks correspond mostly to nonconscious operations in the human brain. However, much like artificial neural networks took their inspiration from neurobiology, artificial consciousness may progress by investigating the architectures that allow the human brain to generate consciousness, then transferring those insights into computer algorithms. Our aim is to foster such progress by reviewing aspects of the cognitive neuroscience of consciousness that may be pertinent for machines.

Multiple Meanings of Consciousness

The word "consciousness," like many pre-scientific terms, is used in widely different senses. In a medical context, it is often used in an intransitive sense (as in "the patient was no longer conscious"), in the context of assessing vigilance and wakefulness. Elucidating the brain mechanisms of vigilance is an essential scientific goal with major consequences for our understanding of sleep, anesthesia, coma, or vegetative state. For lack of space, we do not deal with this aspect here, however, because its computational impact seems minimal: obviously, a machine must be properly turned on for its computations to unfold normally.

We suggest that it is useful to distinguish two other essential dimensions of conscious computation. We label them using the terms global availability (C1) and self-monitoring (C2).

- *C1: Global availability.* This corresponds to the transitive meaning of consciousness (as in "The driver is conscious *of* the light"). It refers to the relationship between a cognitive system and a specific object of thought, such as a mental representation of "the light." This object appears to be selected for further processing, including verbal and nonverbal report. Information which is conscious in this sense becomes globally available to the organism: we can recall it, act upon it, speak about it, etc. This sense is synonymous with "having the information in mind": among the vast repertoire of thoughts that can become conscious at a given time, only that which is globally available constitutes the content of C1-consciousness.
- *C2: Self-monitoring.* Another meaning of consciousness is reflexive. It refers to a self-referential relationship in which the cognitive system is able to monitor its own processing and obtain information about itself. Human beings know a lot about themselves, including such diverse information as the layout and position of their body, whether they know or perceive something, or whether they just made an error. This sense ofconsciousness corresponds

to what is commonly called introspection, or what psychologists call "meta-cognition"—the ability to conceive and make use of internal representations of one's own knowledge and abilities.

We propose that C1 and C2 constitute orthogonal dimensions of conscious computations. This is not to say that C1 and C2 do not involve overlapping physical substrates; in fact, as we review below, in the human brain, both depend on prefrontal cortex. But we argue that, empirically and conceptually, the two may come apart, as there can be C1 without C2, for instance when reportable processing is not accompanied by accurate metacognition, or C2 without C1, for instance when a self-monitoring operation unfolds without being consciously reportable. As such, it is advantageous to consider these computations separately before we consider their synergy. Furthermore, many computations involve neither C1 nor C2 and therefore properly called "unconscious" (or C0 for short). It was Turing's original insight that even sophisticated information processing can be realized by a mindless automaton. Cognitive neuroscience confirms that complex computations such as face or speech recognition, chess-game evaluation, sentence parsing, and meaning extraction occur unconsciously in the human brain, i.e., under conditions that yield neither global reportability nor self-monitoring (Table 1). The brain appears to operate, in part, as a juxtaposition of specialized processors or "modules" that operate nonconsciously and, we argue, correspond tightly to the operation of current feedforward deep-learning networks.

We now review the experimental evidence for how human and animal brains handle C0-, C1-, and C2-level computations—before returning to machines and how they could benefit from this understanding of brain architecture.

Unconscious Processing (C0): Where Most of Our Intelligence Lies

Probing Unconscious Computations

"We cannot be conscious of what we are not conscious of" (Jaynes 1976). This truism has deep consequences. Because we are blind to our unconscious processes, we tend to underestimate their role in our mental life. However, cognitive neuroscientists developed various means of presenting images or sounds without inducing any conscious experience (Fig. 1), and then used behavioral and brain-imaging to probe their processing depth.

The phenomenon of priming illustrates the remarkable depth of unconscious processing. A highly visible target stimulus, such as the written word "four," is processedmore

efficiently when preceded by a related prime stimulus, such as the Arabic digit "4," even when subjects do not notice the presence of the prime and cannot reliably report its identity. Subliminal digits, words, faces, or objects can be invariantly recognized and influence motor, semantic, and decision levels of processing (Table 1). Neuroimaging methods reveal that the vast majority of brain areas can be activated nonconsciously.

Unconscious View-Invariance and Meaning Extraction in the Human Brain

Many of the difficult perceptual computations, such as invariant face recognition or speaker-invariant speech recognition, that were recently addressed by AI, correspond to nonconscious computations in the human brain (Dupoux et al. 2008; Kouider and Dehaene 2007; Qiao et al. 2010). For instance, processing someone's face is facilitated when it is preceded by the subliminal presentation of a totally different view of the same person, indicating unconscious invariant recognition (Fig. 1). Subliminal priming generalizes across visual-auditory modalities (Faivre et al. 2014; Kouider and Dehaene 2009), revealing that cross-modal computations that remain challenging for AI software (e.g., extraction of semantic vectors, speech-to-text) also involve unconscious mechanisms. Even the semantic meaning of sensory input can be processed without awareness by the human brain. Compared to related words (e.g., animal-dog), semantic violations (e.g., furniture-dog) generate a brain response as late as 400 ms after stimulus onset in temporal-lobe language networks, even if one of the two words cannot be consciously detected (Luck et al. 1996; van Gaal et al. 2014).

Unconscious Control and Decision-Making

Unconscious processes can reach even deeper levels of the cortical hierarchy. For instance, subliminal primes can influence prefrontal mechanisms of cognitive control involved in the selection of a task (Lau and Passingham 2007) or the inhibition of a motor response (van Gaal et al. 2010). Neural mechanisms of decision-making involve accumulating sensory evidence that affects the probability of the various choices, until a threshold is attained. This accumulation of probabilistic knowledge continues to happen even with subliminal stimuli (de Lange et al. 2011; Vorberg et al. 2003; Dehaene et al. 1998a; Vlassova et al. 2014). Bayesian inference and evidence accumulation, which are cornerstone computations for AI (Lake et al. 2017), are basic unconscious mechanisms for humans.

Table 1 Examples of computations pertaining to information-processing levels C0, C1, and C2 in the human brain

Computation	Examples of experimental findings	References
C0: Unconscious processing		
Invariant visual recognition	Subliminal priming by unseen words and faces, invariant for font, size or viewpoint.	Kouider and Dehaene (2007)
	fMRI and single-neuron response to unseen words and faces	Sergent et al. (2005), Kreiman et al. (2002), Dehaene et al. (2001), Vuilleumier et al. (2001)
	Unconscious judgment of chess-game configurations	Kiesel et al. (2009)
Access to meaning	N400 response to unseen out-of-context words	Luck et al. (1996), van Gaal et al. (2014)
Cognitive control	Unconscious inhibition or task set preparation by an unseen cue	Lau and Passingham (2007), van Gaal et al. (2010)
Reinforcement learning	Subliminal instrumental conditioning by unseen shapes	Pessiglione et al. (2008)
Consciousness in the first sense (C1): global availability of information		
All-or-none selection and broadcasting of a relevant content	Conscious perception of a single picture during visual rivalry	Moreno-Bote et al. (2011)
	Conscious perception of a single detail in a picture or stream	Vul et al. (2009), Aly and Yonelinas (2012)
	All-or-none memory retrieval	Harlow and Yonelinas (2016)
	Attentional blink: conscious perception of item A prevents the simultaneous perception of item B	Asplund et al. (2014), Vul et al. (2008), Pincham et al. (2016), Sergent and Dehaene (2004)
	All-or-none "ignition" of event-related potentials and fMRI signals, only on trials with conscious perception	Sergent et al. (2005), Marti et al. (2012), Marti et al. (2015), Del Cul et al. (2007), Marois et al. (2004), Moutard et al. (2015)
	All-or-none firing of neurons coding for the perceived object in prefrontal cortex and other higher areas	Panagiotaropoulos et al. (2012), Logothetis (1998), Kreiman et al. (2002), Quiroga et al. (2008), Rey et al. (2014)
Stabilization of short-lived information for off-line processing	Brain states are more stable when information is consciously perceived; unconscious information quickly decays (1 s)	King et al. (2016), Schurger et al. (2015)
	Conscious access may occur long after the stimulus is gone	Sergent et al. (2013)
Flexible routing of information	Only conscious information can be routed through a series of successive operations (e.g., successive calculations $3 \times 4 + 2$)	Sackur and Dehaene (2009)
Sequential performance of several tasks	Psychological refractory period: conscious processing of item A delays conscious processing of item B	Marti et al. (2012), Marois and Ivanoff (2005)
	Serial calculations or strategies require conscious perception	de Lange et al. (2011), Sackur and Dehaene (2009)
	Serial organization of spontaneous brain activity during conscious thought in the "resting state"	Barttfeld et al. (2015)
Consciousness in the second sense (C2): self-monitoring		
Self-confidence	Humans accurately report subjective confidence, i.e., a probabilistic estimate in the accuracy of a decision or computation	Meyniel et al. (2015), Fleming et al. (2010)
Evaluation of one's knowledge	Humans and animals can ask for help or "opt out" when unsure	Smith (2009), Goupil and Kouider (2016), Goupil et al. (2016)
	Humans and animals know when they don't know or remember	Dunlosky and Metcalfe (2008), Smith (2009)
Error detection	Anterior cingulate response to self-detected errors	Charles et al. (2013), Goupil and Kouider (2016), Gehring et al. (1993)
Listing one's skills	Children know the arithmetic procedures at their disposal, their speed, and error rate.	Siegler (1988)
Sharing one's confidence with others	Decision-making improves when two persons share knowledge	Bahrami et al. (2010)

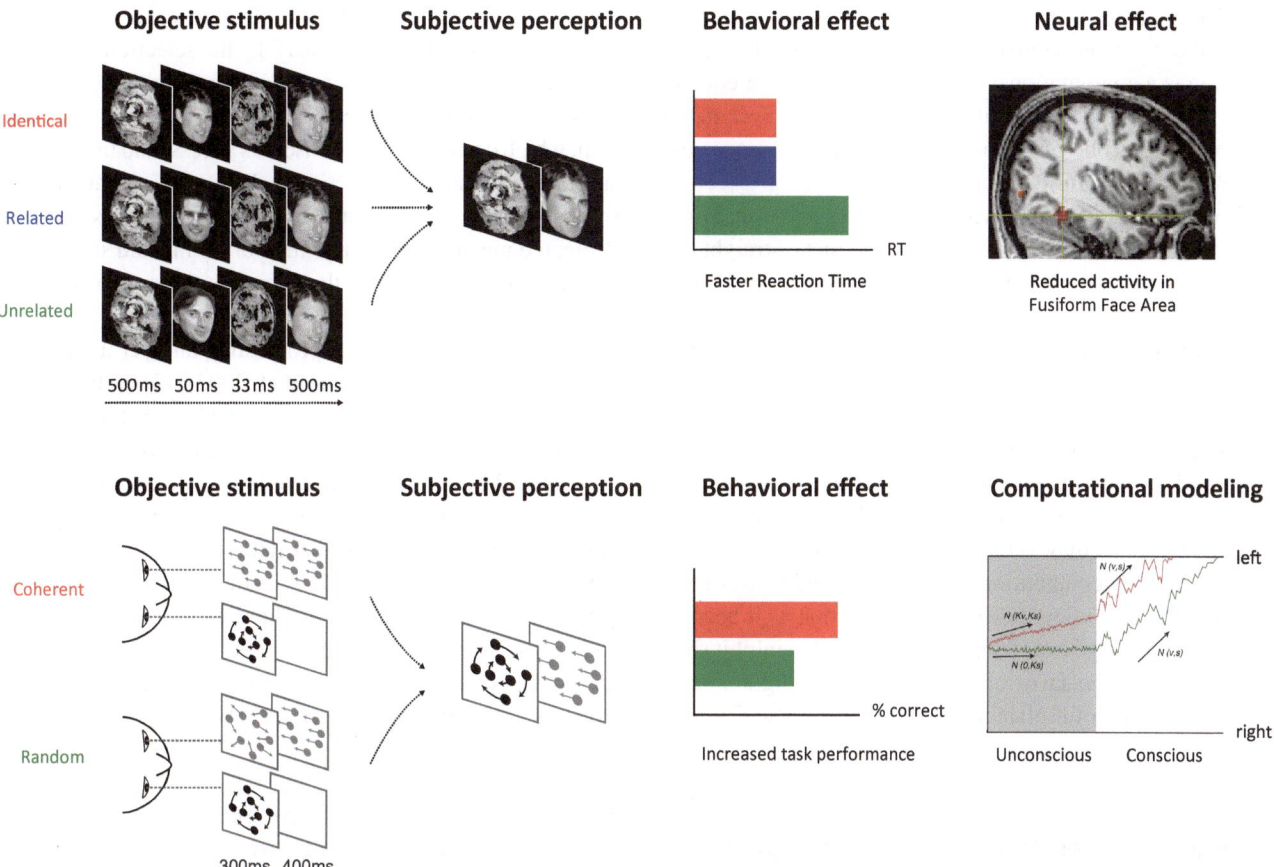

Fig. 1 *Examples of paradigms probing unconscious processing (C0).* (Top) Subliminal view-invariant face recognition (Kouider et al. 2009). On each trial, a prime face is briefly presented (50 ms), surrounded by masks that make it invisible, followed by a visible target face (500 ms). Although subjective perception is identical across conditions, processing is facilitated whenever the two faces represent the same person, in same or different view. At the behavioral level, this view-invariant unconscious priming is reflected by reduced reaction time in recognizing the target face. At the neural level, it is reflected by reduced cortical response to the target face (i.e., repetition suppression) in the Fusiform Face Area of human inferotemporal cortex. (Bottom) Subliminal accumulation of evidence during interocular suppression (Vlassova et al. 2014). Presentation of salient moving dots in one eye prevents the conscious perception of paler moving dots in the opposite eye. Despite their invisibility, the gray dots facilitate performance when they moved in the same direction as a subsequent dot-display, an effect proportional to their amount of motion coherence. This facilitation only affects a first-order task (judging the direction of motion), not a second-order metacognitive judgment (rating the confidence in the first response). A computational model of evidence accumulation proposes that subliminal motion information gets added to conscious information, thus biasing and shortening the decision

Unconscious Learning

Reinforcement learning algorithms, which capture how humans and animals shape their future actions based on the history of past rewards, have excelled in attaining suprahuman AI performance in several applications, such as playing Go (Silver et al. 2016). Remarkably, in humans, such learning appears to proceed even when the cues, reward, or motivation signals are presented below the consciousness threshold (Pessiglione et al. 2008, 2007).

In summary, complex unconscious computations and inferences routinely occur in parallel within various brain areas. Many of these C0 computations have now been captured by AI, particularly using feedforward convolutional neural networks (CNNs). We now consider what additional computations are required for conscious processing.

Consciousness in the First Sense (C1): Global Availability of Relevant Information

The Need for Integration and Coordination

The organization of the brain into computationally specialized subsystems is efficient, but this architecture also raises a specific computational problem: the organism as a whole cannot stick to a diversity of probabilistic interpretations—it must *act*, and therefore cut through the multiple possibilities and decide in favor of a single course of action. Integrating all of the available evidence to converge towards a single decision is a computational requirement which, we contend, must be faced by any animal or autonomous AI system, and corresponds to our first functional definition of consciousness: global availability (C1).

For instance, elephants, when thirsty, manage to determine the location of the nearest water hole and move straight to it, from a distance of 5 to 50 km (Polansky et al. 2015). Such decision-making requires a sophisticated architecture for (1) efficiently pooling over all available sources of information, including multisensory and memory cues; (2) considering the available options and selecting the best one based on this large information pool; (3) sticking to this choice over time; and (4) coordinating all internal and external processes towards the achievement of that goal. Primitive organisms, such as bacteria, may achieve such decision solely through an unconscious competition of uncoordinated sensorimotor systems. This solution, however, fails as soon as it becomes necessary to bridge over temporal delays and to inhibit short-term tendencies in favor of longer-term winning strategies. Coherent, thoughtful planning required a specific C1 architecture.

Consciousness as Access to an Internal Global Workspace

We hypothesize that consciousness in the first sense (C1) evolved as an information-processing architecture that addresses this information-pooling problem (Baars 1988; Dehaene et al. 1998b; Dennett 2001; Dehaene and Naccache 2001). In this view, the architecture of C1 evolved to break the modularity and parallelism of unconscious computations. On top of a deep hierarchy of specialized modules, a "global neuronal workspace," with limited capacity, evolved to select a piece of information, hold it over time, and share it across modules. We call "conscious" whichever representation, at a given time, wins the competition for access to this mental arena and gets selected for global sharing and decision-making. Consciousness is therefore manifested by the temporary dominance of a thought or train of thoughts over mental processes, such that it can guide a broad variety of behaviors. These behaviors include not only physical actions, but also mental ones such as committing information to episodic memory or routing it to other processors.

Relation Between Consciousness and Attention

William James described attention as "the taking possession by the mind, in clear and vivid form, of one out of what seem several simultaneously possible objects or trains of thought" (James 1890). This definition is close to what we mean by consciousness in the first sense (C1): the selection of a single piece of information for entry into the global workspace. There is, however, a clear-cut distinction between this final step, which corresponds to conscious access, and the previous stages of attentional selection, which can operate unconsciously. Many experiments have established the existence of dedicated mechanisms of attention orienting and shown that, like any other processors, they can operate nonconsciously: (1) in the top-down direction, attention can be oriented towards an object, amplify its processing, and yet fail to bring it to consciousness (Naccache et al. 2002); (2) in the bottom-up direction, attention can be attracted by a flash even if this stimulus ultimately remains unconscious (Kentridge et al. 1999). What we call attention is a hierarchical system of sieves that operate unconsciously. Such unconscious systems compute with probability distributions, but only a single sample, drawn from this probabilistic distribution, becomes conscious at a given time (Asplund et al. 2014; Vul et al. 2009). We may become aware of several alternative interpretations, but only by sampling their unconscious distributions over time (Moreno-Bote et al. 2011; Vul et al. 2008).

Evidence for All-Or-None Selection in a Capacity-Limited System

The primate brain comprises a conscious bottleneck and can only consciously access a single item at a time (see Table 1). For instance, rivalling pictures or ambiguous words are perceived in an all-or-none manner: at any given time, we subjectively perceive only a single interpretation out of many possible ones (even though the others continue to be processed unconsciously (Panagiotaropoulos et al. 2012; Logothetis 1998)). The serial operation of consciousness is attested by phenomena such as the attentional blink and the psychological refractory period, whereby conscious access to a first item A prevents or delays the perception of a second competing item B (Luck et al. 1996; Asplund et al. 2014; Vul et al. 2008; Sergent et al. 2005; Marti et al. 2012, 2015). Such interference with the perception of B is triggered by the mere conscious perception of A, even if no task is performed (Nieuwenstein et al. 2009). Thus, C1-consciousness is causally responsible for a serial information-processing bottleneck.

Evidence for Integration and Broadcasting

Brain-imaging in humans and neuronal recordings in monkeys indicate that the conscious bottleneck is implemented by a network of neurons which is distributed through the cortex, but with a stronger emphasis on high-level associative areas.

Table 1 lists some of the publications that have evidenced an all-or-none "ignition" of this network during conscious perception, using a variety of brain-imaging techniques. Single-cell recordings indicate that each specific conscious percept, such as a person's face, is encoded by the all-or-none firing of a subset of neurons in high-level temporal and prefrontal cortices, while others remain silent (Fig. 2) (Panagiotaropoulos et al. 2012; Logothetis 1998; Kreiman et al. 2002; Quiroga et al. 2008).

Stability as a Feature of Consciousness

Direct contrasts between seen and unseen pictures or words confirm that such ignition occurs only for the conscious percept. As explained earlier, nonconscious stimuli may reach into deep cortical networks and influence higher levels of processing and even central executive functions, but these effects tend to be small, variable, and short-lived (although nonconscious information decays at a slower rate than initially expected (King et al. 2016; Trübutschek et al. 2017)). By contrast, the stable, reproducible representation of high-quality information by a distributed activity pattern in higher cortical areas is a feature of conscious processing (Table 1). Such transient "meta-stability" seems to be necessary for the nervous system to integrate information from a variety of modules and then broadcast it back to them, thereby achieving flexible cross-module routing.

C1 Consciousness in Human and Nonhuman Animals

C1 consciousness is an elementary property which is present in human infants (Kouider et al. 2013) as well as in animals. Nonhuman primates exhibit similar visual illusions (Panagiotaropoulos et al. 2012; Logothetis 1998), attentional blink (Maloney et al. 2013), and central capacity limits (Watanabe and Funahashi 2014) as human subjects. Prefrontal cortex appears to act as a central information sharing device and serial bottleneck in both human and nonhuman primates (Watanabe and Funahashi 2014). The considerable expansion of prefrontal cortex in the human lineage may have resulted in a greater capacity for multimodal convergence and integration (Elston 2003; Neubert et al. 2014; Wang et al. 2015). Furthermore, humans possess additional circuits in inferior prefrontal cortex for verbally formulating and reporting information to others. The capacity to report information through language is universally considered as one of the clearest signs of conscious perception, because once information has reached this level of representation in humans, it is necessarily available for sharing across mental modules, and therefore conscious in the C1 sense. Thus,

while language is not required for conscious perception and processing, the emergence of language circuits in humans may have resulted in a considerable increase in the speed, ease, and flexibility of C1-level information sharing.

Consciousness in the Second Sense (C2): Self-Monitoring

While C1-consciousness reflects the capacity to access external, objective information, consciousness in the second sense (C2) is characterized by the ability to reflexively represent oneself (Cleeremans et al. 2007; Cleeremans 2014; Dunlosky and Metcalfe 2008; Clark and Karmiloff-Smith 1993). A substantial amount of research in cognitive neuroscience and psychology has addressed self-monitoring under the term of "metacognition," roughly defined as cognition about cognition or knowing about knowing. Below, we review the mechanisms by which the primate brain monitors itself, while stressing their implications for building self-reflective machines.

A Probabilistic Sense of Confidence

When taking a decision, humans feel more or less confident about their choice. Confidence can be defined as a sense of the probability that a decision or computation is correct (Meyniel et al. 2015). Almost anytime the brain perceives or decides, it also estimates its degree of confidence. Learning is also accompanied by a quantitative sense of confidence: humans evaluate how much trust they have in what they have learned, and use it to weigh past knowledge versus present evidence (Meyniel and Dehaene 2017). Confidence can be assessed nonverbally, either retrospectively, by measuring whether humans persist in their initial choice, or prospectively, by allowing them to opt out from a task without even attempting it. Both measures have been used in nonhuman animals to show that they too possess metacognitive abilities (Smith 2009). By contrast, most current neural networks lack them: although they can learn, they generally lack meta-knowledge of the reliability and limits of what has been learned. A noticeable exception is biologically constrained models that rely on Bayesian mechanisms to simulate the integration of multiple probabilistic cues in neural circuits (Ma et al. 2006). These models have been fruitful in describing how neural populations may automatically compute the probability that a given process is performed successfully. Although these implementations remain rare and have not addressed the same range of computational problems as traditional AI, they offer a promising venue for incorporating uncertainty monitoring in deep-learning networks.

Objective stimulus

Subjective perception / single-cell activity

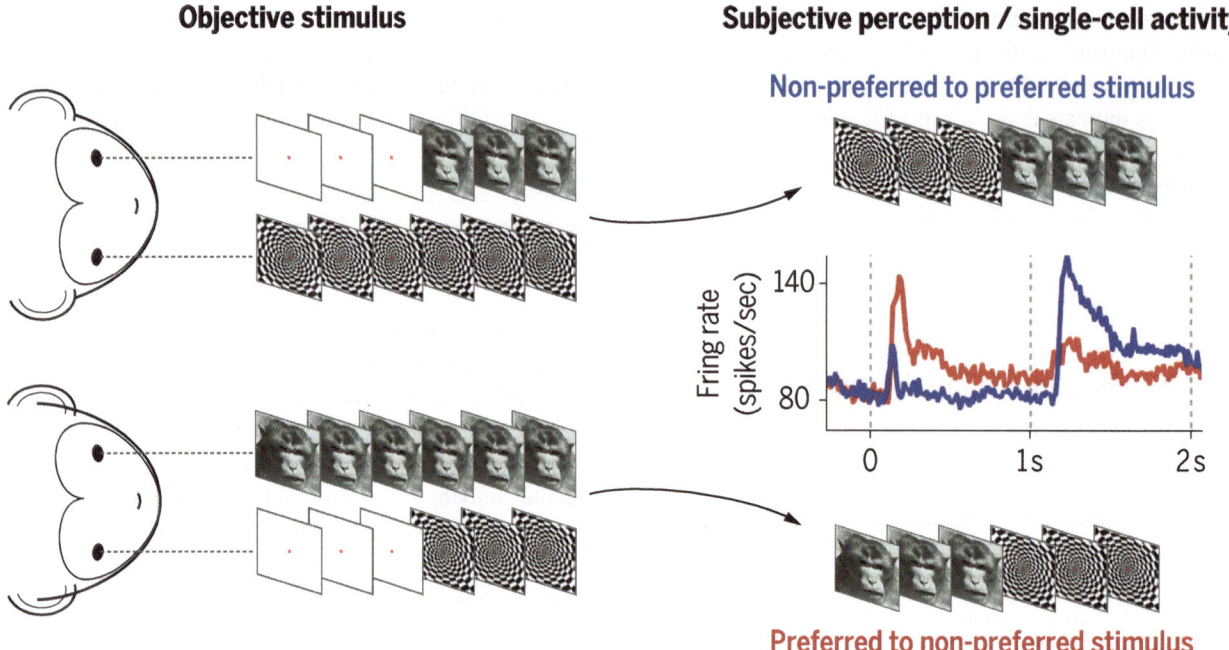

Non-preferred to preferred stimulus

Preferred to non-preferred stimulus

Response of neuron selective to World Trade Center

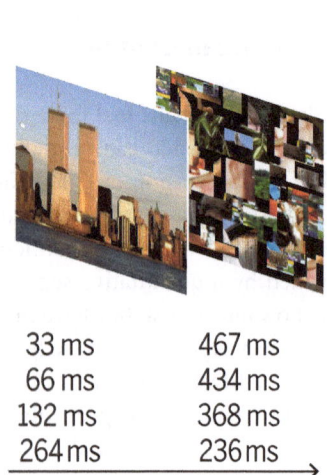

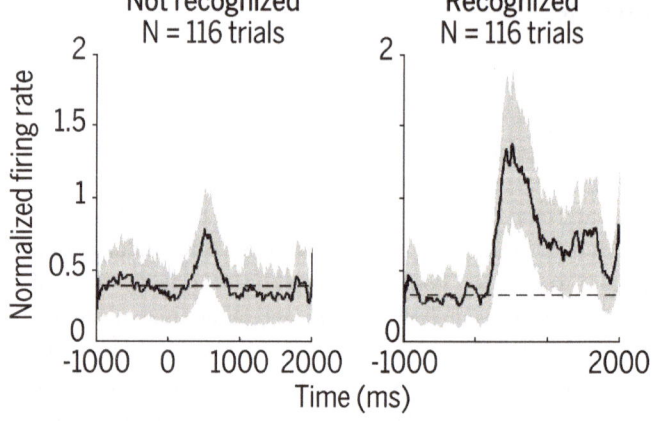

Fig. 2 Global availability: consciousness in the first sense (C1): Conscious subjective percepts are encoded by the sudden firing of stimulus-specific neural populations distributed in interconnected, high-level cortical areas, including lateral prefrontal cortex, anterior temporal cortex, and hippocampus. (Top) During binocular flash suppression, the flashing of a picture to one eye suppresses the conscious perception of a second picture presented to the other eye. As a result, the same physical stimulus can lead to distinct subjective percepts. This example illustrates a prefrontal neuron sensitive to faces and unresponsive to checkers, whose firing shoots up in tight association with the sudden onset of subjective face perception (Panagiotaropoulos et al. 2012). (Bottom) During masking, a flashed image, if brief enough and followed by a longer "mask," can remain subjectively invisible. Shown is a neuron in the entorhinal cortex firing selectively to the concept of "World Trade Center." Rasters in red indicate trials where the subject reported recognizing the picture (blue = no recognition). Under masking, when the picture is presented for only 33 ms there is little or no neural activity—but once presentation time is longer than the perceptual threshold (66 ms or larger), the neuron fires substantially only on recognized trials. Overall, even for identical objective input (same duration), spiking activity is higher and more stable for recognized trials (Quiroga et al. 2008)

Explicit Confidence in Prefrontal Cortex

According to Bayesian accounts, each local cortical circuit may represent and combine probability distributions in order to estimate processing uncertainty (Ma et al. 2006). However, additional neural circuits may be required in order to explicitly extract and manipulate confidence signals. MRI studies in humans and physiological recordings in primates and even in rats have specifically linked such confidence processing to the prefrontal cortex (Fleming et al. 2010; Miyamoto et al. 2017; Kepecs et al. 2008). Inactivation of prefrontal cortex can induce a specific deficit in second-order (i.e., metacognitive) judgments while sparing performance on the first-order task (Miyamoto et al. 2017; Rounis et al. 2010). Thus, circuits in prefrontal cortex may have evolved to monitor the performance of other brain processes.

Error Detection: Reflecting on One's Own Mistakes

Error detection provides a particularly clear example of self-monitoring: just after responding, we sometimes realize that we made an error and change our mind. Error detection is reflected by two components of EEG activity, the error-relativity negativity (ERN) and the positivity upon error (Pe), which emerge in cingulate and medial prefrontal cortex just after a wrong response, but before any feedback is received. How can the brain make a mistake and detect it? One possibility is that the accumulation of sensory evidence continues after a decision is made, and an error is inferred whenever this further evidence points in the opposite direction (Resulaj et al. 2009). A second possibility, more compatible with the remarkable speed of error detection, is that two parallel circuits, a low-level sensory-motor circuit and a higher-level intention circuit, operate on the same sensory data and signal an error whenever their conclusions diverge (Charles et al. 2014, 2013).

Meta-Memory

Humans don't just know things about the world—they actually know that they know, or that they don't know. A familiar example is having a word "on the tip of the tongue." The term "meta-memory" was coined to capture the fact that humans report feelings of knowing, confidence, and doubts on their memories. Meta-memory is thought to involve a second-order system that monitors internal signals (e.g., the strength and quality of a memory trace) to regulate behavior. Meta-memory is associated with prefrontal structures whose phar-

macological inactivation leads to a metacognitive impairment while sparing memory performance itself (Miyamoto et al. 2017). Meta-memory is crucial to human learning and education, by allowing learners to develop strategies such as increasing the amount of study or adapting the time allocated to memory encoding and rehearsal (Dunlosky and Metcalfe 2008).

Reality Monitoring

In addition to monitoring the quality of sensory and memory representations, the human brain must also distinguish self-generated versus externally driven representations. Indeed, we can perceive things, but also conjure them from imagination or memory. Hallucinations in schizophrenia have been linked to a failure to distinguish whether sensory activity is generated by oneself or by the external world (Frith 1992). Neuroimaging studies have linked this kind of reality monitoring to the anterior prefrontal cortex (Simons et al. 2017). In nonhuman primates, neurons in the prefrontal cortex distinguish between normal visual perception and active maintenance of the same visual content in memory (Mendoza-Halliday and Martinez-Trujillo 2017).

Foundations of C2 Consciousness in Infants

Self-monitoring is such a basic ability that it is already present during infancy (Fig. 3). The ERN, indicating error monitoring, was observed when one-year-old infants made a wrong choice in a perceptual decision task (Goupil and Kouider 2016). Similarly, after $1\frac{1}{2}$-year-old infants pointed to one of two boxes in order to obtain a hidden toy, they waited longer for an upcoming reward (e.g., a toy) when their initial choice was correct than when it was wrong, suggesting that they monitored the likelihood that their decision was right (Kepecs et al. 2008; Goupil and Kouider 2016). Moreover, when given the opportunity to ask (nonverbally) their parents for help instead of pointing, they chose this opt-out option specifically on trials where they were likely to be wrong, revealing a prospective estimate of their own uncertainty (Goupil et al. 2016). The fact that infants can communicate their own uncertainty to other agents further suggests that they consciously experience metacognitive information. Thus, infants are already equipped with the ability to monitor their own mental states. Facing a world where everything remains to be learned, C2 mechanisms allow them to actively orient towards domains that they know they don't know—a mechanism that we call "curiosity."

First-order decision
Memory recall

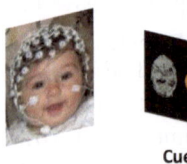 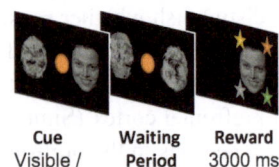

Evidence	Delay	Pointing
Toy location	Task difficulty	Decision

Second-order measure
Manual Search persistence

Longer searching time
when correct

Second-order measure
Opt-out

Opt-out by asking for help
to avoid errors

First-order decision
Perceptual choice

Second-order measure
Eye Fixation persistence

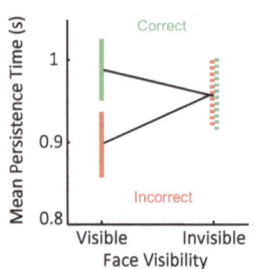

Second-order measure
Error-specific neural signal

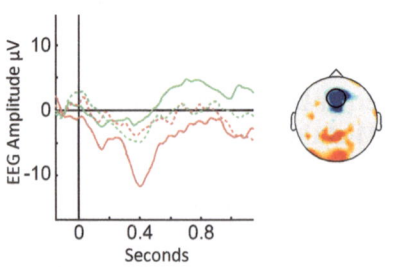

Cue	Waiting	Reward
Visible /	Period	3000 ms
Invisible	2500 ms	

Fig. 3 Self-monitoring: consciousness in the second sense (C2): Self-monitoring (also called "meta-cognition"), the capacity to reflect on one's own mental state, is available early during infancy. (Top) One-and-half-year-old infants, after deciding to point to the location of a hidden toy, exhibit two types of evidence for self-monitoring of their decision: (1) they persist longer in searching for the hidden object within the selected box when their initial choice was correct than when it was incorrect. (2) When given the opportunity to ask for help, they use this option selectively to reduce the probability of making an error. (Bottom) One-year-old infants were presented with either a meaningless pattern or a face that was either visible or invisible (depending on its duration) and then decided to gaze left or right in anticipation of face reappearance. As for manual search, post-decision persistence in waiting at the same gaze location increased for correct compared to incorrect initial decisions. Moreover, EEG signals revealed the presence of the error-related negativity over fronto-central electrodes when infants make an incorrect choice. These markers of metacognition were elicited by visible but not by invisible stimuli, as also shown in adults (Charles et al. 2013)

Dissociations Between C1 and C2

According to our analysis, C1 and C2 are largely orthogonal and complementary dimensions of what we call consciousness. On one side of this double dissociation, self-monitoring can exist for unreportable stimuli (C2 without C1). Automatic typing provides a good example: subjects slow down after a typing mistake, even when they fail to consciously notice the error (Logan and Crump 2010). Similarly, at the neural level, an ERN can occur for subjectively undetected errors (Nieuwenhuis et al. 2001). On the other side of this dissociation, consciously reportable contents sometimes fail to be accompanied by an adequate sense of confidence (C1 without C2). For instance, when we retrieve a memory, it pops into consciousness (C1) but sometimes without any accurate evaluation of its confidence (C2), leading to false memories. As noted by Marvin Minsky, "what we call consciousness [in the C1 sense] is a very imperfect summary in one part of the brain of what the rest is doing." The

imperfection arises in part from the fact that the conscious global workspace reduces complex parallel sensory streams of probabilistic computation to a single conscious sample (Asplund et al. 2014; Vul et al. 2009; Moreno-Bote et al. 2011). Thus, probabilistic information is often lost on the way, and subjects feel over-confident in the accuracy of their perception.

Synergies Between C1 and C2 Consciousness

Because C1 and C2 are orthogonal, their joint possession may have synergistic benefits to organisms. In one direction, bringing probabilistic metacognitive information (C2) into the global workspace (C1) allows it to be held over time, integrated into explicit long-term reflection, and shared with others. Social information sharing improves decisions: by sharing their confidence signals, two persons achieve a better performance in collective decision-making than either person alone (Bahrami et al. 2010). In the converse direction, the possession of an explicit repertoire of one's own abilities (C2) improves the efficiency with which C1 information is processed. During mental arithmetic, children can perform a C2-level evaluation of their available competences (e.g., counting, adding, multiplying, memory retrieval . . .) and use this information to evaluate how to best face a given arithmetic problem (Siegler 1988). This functionality requires a single "common currency" for confidence across difference modules, which humans appear to possess (de Gardelle and Mamassian 2014).

Endowing Machines with C1 and C2

How could machines be endowed with C1 and C2 computations? Let us return to the car light example. In current machines, the "low gas" light is a prototypical example of an unconscious modular signal (C0). When the light flashes, all other processors in the machine remain uninformed and unchanged: fuel continues to be injected in the carburetor, the car passes gas stations without stopping (although they might be present on the GPS map), etc. Current cars or cell phones are mere collections of specialized modules that are largely "unaware" of each other. Endowing this machine with global information availability (C1) would allow these modules to share information and collaborate to address the impending problem (much like humans do when they become aware of the light, or elephants of thirst).

While AI has met considerable success in solving specific problems, implementing multiple processes in a single system and flexibly coordinating them remain difficult problems. In the 1960s, computational architectures called "blackboard systems" were specifically designed to post information and make it available to other modules in a flexible and interpretable manner, similar in flavor to a global workspace (Baars 1988). A recent architecture called Pathnet uses a genetic algorithm to learn which path through its many specialized neural networks is most suited to a given task (Fernando et al. 2017). This architecture exhibits robust, flexible performance and generalization across tasks, and may constitute a first step towards primate-like conscious flexibility.

To make optimal use of the information provided by the fuel-gauge light, it would also be useful for the car to possess a database of its own capacities and limits. Such self-monitoring (C2) would include an integrated image of itself, including its current location, fuel consumption, etc., as well as its internal databases (e.g., "knowing" that it possesses a GPS map that can locate gas stations). A self-monitoring machine would keep a list of its subprograms, compute estimates of their probabilities of succeeding at various tasks, and constantly update them (e.g., noticing if a part fails).

Most present-day machine-learning systems are devoid of any self-monitoring: they compute (C0) without representing the extent and limits of their knowledge or the fact that others may have a different viewpoint than their own. There are a few exceptions: Bayesian networks (Ma et al. 2006) or programs (Tenenbaum et al. 2011) compute with probability distributions and therefore keep track of how likely they are to be correct. Even when the primary computation is performed by a classical CNN, and is therefore opaque to introspection, it is possible to train a second, hierarchically higher neural network to predict the first one's performance (Cleeremans et al. 2007). This approach, whereby a system re-describes itself, has been claimed to lead to "the emergence of internal models that are metacognitive in nature and (. . .) make it possible for an agent to develop a (limited, implicit, practical) understanding of itself" (Cleeremans 2014). Pathnet (Fernando et al. 2017) uses a related architecture to track which internal configurations are most successful at a given task and use this knowledge to guide subsequent processing. Robots have also been programed to monitor their learning progress, and use it to orient resources towards the problems that maximize information gain, thus implementing a form of curiosity (Gottlieb et al. 2013).

An important element of C2 which has received relatively little attention is reality monitoring. Bayesian approaches to AI (Lake et al. 2017; Tenenbaum et al. 2011) have recognized the usefulness of learning generative models that can be jointly used for actual perception (present), prospective planning (future), and retrospective analysis (past). In humans, the same sensory areas are involved in both perception and imagination. As such, some mechanisms are needed to tell apart self-generated versus externally triggered activity. A powerful method for training generative models, called adversarial learning (Goodfellow et al. 2014) involves having a secondary network "compete" against a generative network, to critically evaluate the authenticity of self-generated representations. When such reality monitoring (C2) is coupled with C1 mechanisms, the resulting machine may more closely mimic human consciousness in terms of affording global access to perceptual representations while having an immediate sense that their content is a genuine reflection of the current state of the world.

Concluding Remarks

Our stance is based on a simple hypothesis: what we call "consciousness" results from specific types of information-processing computations, physically realized by the hardware of the brain. It differs from other theories in being resolutely computational—we surmise that mere information-theoretic quantities (Tononi et al. 2016) do not suffice to define consciousness unless one also considers the nature and depth of the information being processed.

We contend that a machine endowed with C1 and C2 would behave as if it were conscious—for instance, it would know that it is seeing something, would express confidence in it, would report it to others, could suffer hallucinations when its monitoring mechanisms break down, and may even experience the same perceptual illusions as humans. Still, such a purely functional definition of consciousness may leave some readers unsatisfied. Are we "over-intellectualizing" consciousness, by assuming that some high-level cognitive functions are necessary tied to consciousness? Are we leaving aside the experiential component ("what it is like" to be conscious)? Does subjective experience escape a computational definition?

While those philosophical questions lie beyond the scope of the present paper, we close by noting that, empirically, in humans, the loss of C1 and C2 computations co-varies with a loss of subjective experience. For example, in humans, damage to the primary visual cortex may lead to a neurological condition called "blindsight," in which the patients report being blind in the affected visual field. Remarkably, those patients can localize visual stimuli in their blind field, but they cannot report them (C1) nor can they effectively assess their likelihood of success (C2)—they believe that they are merely "guessing." In this example at least, subjective experience appears to cohere with possession of C1 and C2. Although centuries of philosophical dualism have led us to consider consciousness as unreducible to physical interactions, the empirical evidence is compatible with the possibility that consciousness arises from nothing more than specific computations.

Acknowledgments This work was supported by funding from INSERM, CEA, CIFAR, Collège de France, the European Research Council (ERC project NeuroConsc to S.D. and ERC project METAWARE to S.K), the French Agence National de la Recherche (ANR-10-LABX-0087 and ANR-10-IDEX-0001-02), the US National Institute of Health (National Institute of Neurological Disorders and Stroke R01NS088628 to H.L.), and the US Air Force Office of Scientific Research (FA9550-15-1-0110 to H.L.)

References

Aly, M., & Yonelinas, A. P. (2012). Bridging consciousness and cognition in memory and perception: Evidence for both state and strength processes. *PLoS One, 7,* e30231.

Asplund, C. L., Fougnie, D., Zughni, S., Martin, J. W., & Marois, R. (2014). The attentional blink reveals the probabilistic nature of discrete conscious perception. *Psychological Science, 25,* 824–831.

Baars, B. (1988). *A cognitive theory of consciousness.* Cambridge, MA: Cambridge University Press.

Bahrami, B., et al. (2010). Optimally interacting minds. *Science, 329,* 1081–1085.

Barttfeld, P., et al. (2015). Signature of consciousness in the dynamics of resting-state brain activity. *Proceedings of the National Academy of Sciences of the United States of America, 112,* 887–892.

Charles, L., King, J.-R., & Dehaene, S. (2014). Decoding the dynamics of action, intention, and error detection for conscious and subliminal stimuli. *Journal of Neuroscience: The Official Journal of the Society for Neuroscience, 34,* 1158–1170.

Charles, L., Van Opstal, F., Marti, S., & Dehaene, S. (2013). Distinct brain mechanisms for conscious versus subliminal error detection. *NeuroImage, 73,* 80–94.

Clark, A., & Karmiloff-Smith, A. (1993). The Cognizer's innards: A psychological and philosophical perspective on the development of thought. *Mind & Language, 8,* 487–519.

Cleeremans, A. (2014). Connecting conscious and unconscious processing. *Cognitive Science, 38,* 1286–1315.

Cleeremans, A., Timmermans, B., & Pasquali, A. (2007). Consciousness and metarepresentation: A computational sketch. *Neural Networks, 20,* 1032–1039.

de Gardelle, V., & Mamassian, P. (2014). Does confidence use a common currency across two visual tasks? *Psychological Science, 25,* 1286–1288.

de Lange, F. P., van Gaal, S., Lamme, V. A., & Dehaene, S. (2011). How awareness changes the relative weights of evidence during human decision-making. *PLoS Biology, 9,* e1001203.

Dehaene, S., Kerszberg, M., & Changeux, J. P. (1998b). A neuronal model of a global workspace in effortful cognitive tasks. *Proceedings of the National Academy of Sciences of the United States of America, 95,* 14,529–14,534.

Dehaene, S., & Naccache, L. (2001). Towards a cognitive neuroscience of consciousness: Basic evidence and a workspace framework. *Cognition, 79,* 1–37.

Dehaene, S., et al. (1998a). Imaging unconscious semantic priming. *Nature, 395,* 597–600.

Dehaene, S., et al. (2001). Cerebral mechanisms of word masking and unconscious repetition priming. *Nature Neuroscience, 4,* 752–758.

Del Cul, A., Baillet, S., & Dehaene, S. (2007). Brain dynamics underlying the nonlinear threshold for access to consciousness. *PLoS Biology, 5,* e260.

Dennett, D. (2001). Are we explaining consciousness yet? *Cognition, 79,* 221–237.

Dunlosky, J., & Metcalfe, J. (2008). *Metacognition.* Thousand Oaks, CA: Sage Publications.

Dupoux, E., de Gardelle, V., & Kouider, S. (2008). Subliminal speech perception and auditory streaming. *Cognition, 109,* 267–273.

Elston, G. N. (2003). Cortex, cognition and the cell: New insights into the pyramidal neuron and prefrontal function. *Cerebral Cortex, 13,* 1124–1138.

Faivre, N., Mudrik, L., Schwartz, N., & Koch, C. (2014). Multisensory integration in complete unawareness: Evidence from audiovisual congruency priming. *Psychological Science, 25,* 2006–2016.

Fernando, C., Banarse, D., Blundell, C., Zwols, Y., Ha, D. R., Rusu, A. A., Pritzel, A., & Wierstra D. (2017). PathNet: Evolution channels gradient descent in super neural networks. *ArXiv170108734 Cs.* Available at http://arxiv.org/abs/1701.08734.

Fleming, S. M., Weil, R. S., Nagy, Z., Dolan, R. J., & Rees, G. (2010). Relating introspective accuracy to individual differences in brain structure. *Science, 329,* 1541–1543.

Frith, C. D. (1992). *The cognitive neuropsychology of schizophrenia.* New York: Psychology Press.

Gehring, W. J., Goss, B., Coles, M. G. H., Meyer, D. E., & Donchin, E. (1993). A neural system for error detection and compensation. *Psychological Science, 4,* 385–390.

Goodfellow, I. J., Pouget-Abadie, J., Mirza, M., Xu, B., Warde-Farley, D., Ozair, S., Courville, A., & Bengio, Y. (2014). Generative adversarial networks. *ArXiv14062661 Cs Stat.* Retrieved from http://arxiv.org/abs/1406.2661.

Gottlieb, J., Oudeyer, P.-Y., Lopes, M., & Baranes, A. (2013). Information-seeking, curiosity, and attention: Computational and neural mechanisms. *Trends in Cognitive Sciences, 17*, 585–593.

Goupil, L., & Kouider, S. (2016). Behavioral and neural indices of metacognitive sensitivity in preverbal infants. *Current Biology, 26*, 3038–3045.

Goupil, L., Romand-Monnier, M., & Kouider, S. (2016). Infants ask for help when they know they don't know. *Proceedings of the National Academy of Sciences of the United States of America, 113*, 3492–3496.

Harlow, I. M., & Yonelinas, A. P. (2016). Distinguishing between the success and precision of recollection. *Memory (Hove, England), 24*, 114–127.

James, W. (1890). *The principles of psychology*. New York: Holt.

Jaynes, J. (1976). *The origin of consciousness in the breakdown of the bicameral mind*. New York: Houghton Mifflin Company.

Kentridge, R. W., Heywood, C. A., & Weiskrantz, L. (1999). Attention without awareness in blindsight. *Proceedings of the Royal Society of London—Series B: Biological Sciences, 266*, 1805–1811.

Kepecs, A., Uchida, N., Zariwala, H. A., & Mainen, Z. F. (2008). Neural correlates, computation and behavioural impact of decision confidence. *Nature, 455*, 227–231.

Kiesel, A., Kunde, W., Pohl, C., Berner, M. P., & Hoffmann, J. (2009). Playing chess unconsciously. *Journal of Experimental Psychology. Learning, Memory, and Cognition, 35*, 292–298.

King, J.-R., Pescetelli, N., & Dehaene, S. (2016). Brain mechanisms underlying the brief maintenance of seen and unseen sensory information. *Neuron, 92*, 1122–1134.

Kouider, S., & Dehaene, S. (2007). Levels of processing during non-conscious perception: A critical review of visual masking. *Philosophical Transactions of the Royal Society of London. Series B, Biological Sciences, 362*, 857–875.

Kouider, S., & Dehaene, S. (2009). Subliminal number priming within and across the visual and auditory modalities. *Experimental Psychology, 56*, 418–433.

Kouider, S., Eger, E., Dolan, R., & Henson, R. N. (2009). Activity in face-responsive brain regions is modulated by invisible, attended faces: Evidence from masked priming. *Cerebral Cortex, 19*, 13–23.

Kouider, S., et al. (2013). A neural marker of perceptual consciousness in infants. *Science, 340*, 376–380.

Kreiman, G., Fried, I., & Koch, C. (2002). Single-neuron correlates of subjective vision in the human medial temporal lobe. *Proceedings of the National Academy of Sciences of the United States of America, 99*, 8378–8383.

Lake, B. M., Ullman, T. D., Tenenbaum, J. B., & Gershman, S. J. (2017). Building machines that learn and think like people. *Behavioral and Brain Science, 40*, 1–101.

Lau, H. C., & Passingham, R. E. (2007). Unconscious activation of the cognitive control system in the human prefrontal cortex. *The Journal of Neuroscience, 27*, 5805–5811.

Logan, G. D., & Crump, M. J. (2010). Cognitive illusions of authorship reveal hierarchical error detection in skilled typists. *Science, 330*, 683–686.

Logothetis, N. K. (1998). Single units and conscious vision. *Philosophical Transactions of the Royal Society of London. Series B, Biological Sciences, 353*, 1801–1818.

Luck, S. J., Vogel, E. K., & Shapiro, K. L. (1996). Word meanings can be accessed but not reported during the attentional blink. *Nature, 383*, 616–618.

Ma, W. J., Beck, J. M., Latham, P. E., & Pouget, A. (2006). Bayesian inference with probabilistic population codes. *Nature Neuroscience, 9*, 1432–1438.

Maloney, R. T., Jayakumar, J., Levichkina, E. V., Pigarev, I. N., & Vidyasagar, T. R. (2013). Information processing bottlenecks in macaque posterior parietal cortex: An attentional blink? *Experimental Brain Research, 228*, 365–376.

Marois, R., & Ivanoff, J. (2005). Capacity limits of information processing in the brain. *Trends in Cognitive Sciences, 9*, 296–305.

Marois, R., Yi, D. J., & Chun, M. M. (2004). The neural fate of consciously perceived and missed events in the attentional blink. *Neuron, 41*, 465–472.

Marti, S., King, J.-R., & Dehaene, S. (2015). Time-resolved decoding of two processing chains during dual-task interference. *Neuron, 88*, 1297–1307.

Marti, S., Sigman, M., & Dehaene, S. (2012). A shared cortical bottleneck underlying attentional blink and psychological refractory period. *NeuroImage, 59*, 2883–2898.

Mendoza-Halliday, D., & Martinez-Trujillo, J. C. (2017). Neuronal population coding of perceived and memorized visual features in the lateral prefrontal cortex. *Nature Communications, 8*, 15471. https://doi.org/10.1038/ncomms15471.

Meyniel, F., & Dehaene, S. (2017). Brain networks for confidence weighting and hierarchical inference during probabilistic learning. *Proceedings of the National Academy of Sciences of the United States of America, 114*, E3859–E3868.

Meyniel, F., Schlunegger, D., & Dehaene, S. (2015). The sense of confidence during probabilistic learning: A normative account. *PLoS Computational Biology, 11*, e1004305.

Miyamoto, K., et al. (2017). Causal neural network of metamemory for retrospection in primates. *Science, 355*, 188–193.

Moreno-Bote, R., Knill, D. C., & Pouget, A. (2011). Bayesian sampling in visual perception. *Proceedings of the National Academy of Sciences of the United States of America, 108*, 12491–12496.

Moutard, C., Dehaene, S., & Malach, R. (2015). Spontaneous fluctuations and non-linear ignitions: Two dynamic faces of cortical recurrent loops. *Neuron, 88*, 194–206.

Naccache, L., Blandin, E., & Dehaene, S. (2002). Unconscious masked priming depends on temporal attention. *Psychological Science, 13*, 416–424.

Neubert, F.-X., Mars, R. B., Thomas, A. G., Sallet, J., & Rushworth, M. F. S. (2014). Comparison of human ventral frontal cortex areas for cognitive control and language with areas in monkey frontal cortex. *Neuron, 81*, 700–713.

Nieuwenhuis, S., Ridderinkhof, K. R., Blom, J., Band, G. P., & Kok, A. (2001). Error-related brain potentials are differentially related to awareness of response errors: Evidence from an antisaccade task. *Psychophysiology, 38*, 752–760.

Nieuwenstein, M., Van der Burg, E., Theeuwes, J., Wyble, B., & Potter, M. (2009). Temporal constraints on conscious vision: on the ubiquitous nature of the attentional blink. *Journal of Vision, 9*, 18.1–18.14.

Panagiotaropoulos, T. I., Deco, G., Kapoor, V., & Logothetis, N. K. (2012). Neuronal discharges and gamma oscillations explicitly reflect visual consciousness in the lateral prefrontal cortex. *Neuron, 74*, 924–935.

Pessiglione, M., et al. (2007). How the brain translates money into force: A neuroimaging study of subliminal motivation. *Science, 316*, 904–906.

Pessiglione, M., et al. (2008). Subliminal instrumental conditioning demonstrated in the human brain. *Neuron, 59*, 561–567.

Pincham, H. L., Bowman, H., & Szucs, D. (2016). The experiential blink: Mapping the cost of working memory encoding onto conscious perception in the attentional blink. *Cortex, 81*, 35–49.

Polansky, L., Kilian, W., & Wittemyer, G. (2015). Elucidating the significance of spatial memory on movement decisions by African savannah elephants using state-space models. *Proceedings of the Royal Society B Biological Sciences, 282*, 20143042. https://doi.org/10.1098/rspb.2014.3042.

Qiao, E., et al. (2010). Unconsciously deciphering handwriting: Subliminal invariance for handwritten words in the visual word form area. *NeuroImage, 49*, 1786–1799.

Quiroga, R. Q., Mukamel, R., Isham, E. A., Malach, R., & Fried, I. (2008). Human single-neuron responses at the threshold of conscious recognition. *Proceedings of the National Academy of Sciences of the United States of America, 105*, 3599–3604.

Resulaj, A., Kiani, R., Wolpert, D. M., & Shadlen, M. N. (2009). Changes of mind in decision-making. *Nature, 461*, 263–266.

Rey, H. G., Fried, I., & Quian Quiroga, R. (2014). Timing of single-neuron and local field potential responses in the human medial temporal lobe. *Current Biology, 24*, 299–304.

Rounis, E., Maniscalco, B., Rothwell, J. C., Passingham, R., & Lau, H. (2010). Theta-burst transcranial magnetic stimulation to the prefrontal cortex impairs metacognitive visual awareness. *Cognitive Neuroscience, 1*, 165–175.

Sackur, J., & Dehaene, S. (2009). The cognitive architecture for chaining of two mental operations. *Cognition, 111*, 187–211.

Schurger, A., Sarigiannidis, I., Naccache, L., Sitt, J. D., & Dehaene, S. (2015). Cortical activity is more stable when sensory stimuli are consciously perceived. *Proceedings of the National Academy of Sciences of the United States of America, 112*, E2083–E2092.

Sergent, C., Baillet, S., & Dehaene, S. (2005). Timing of the brain events underlying access to consciousness during the attentional blink. *Nature Neuroscience, 8*, 1391–1400.

Sergent, C., & Dehaene, S. (2004). Is consciousness a gradual phenomenon? Evidence for an all-or-none bifurcation during the attentional blink. *Psychological Science, 15*, 720–728.

Sergent, C., et al. (2013). Cueing attention after the stimulus is gone can retrospectively trigger conscious perception. *Current Biology, 23*, 150–155.

Siegler, R. S. (1988). Strategy choice procedures and the development of multiplication skill. *Journal of Experimental Psychology General, 117*, 258–275.

Silver, D., et al. (2016). Mastering the game of go with deep neural networks and tree search. *Nature, 529*, 484–489.

Simons, J. S., Garrison, J. R., & Johnson, M. K. (2017). Brain mechanisms of reality monitoring. *Trends in Cognitive Sciences, 21*, 462–473.

Smith, J. D. (2009). The study of animal metacognition. *Trends in Cognitive Sciences, 13*, 389–396.

Tenenbaum, J. B., Kemp, C., Griffiths, T. L., & Goodman, N. D. (2011). How to grow a mind: Statistics, structure, and abstraction. *Science, 331*, 1279–1285.

Tononi, G., Boly, M., Massimini, M., & Koch, C. (2016). Integrated information theory: From consciousness to its physical substrate. *Nature Reviews Neuroscience, 17*, 450–461.

Trübutschek, D., et al. (2017). A theory of working memory without consciousness or sustained activity. *eLife, 6*, e23871. https://doi.org/10.7554/eLife.23871.

van Gaal, S., Lamme, V. A., Fahrenfort, J. J., & Ridderinkhof, K. R. (2010). Dissociable brain mechanisms underlying the conscious and unconscious control of behavior. *Journal of Cognitive Neuroscience, 23*, 91–105.

van Gaal, S., et al. (2014). Can the meaning of multiple words be integrated unconsciously? *Philosophical Transactions of the Royal Society of London. Series B, Biological Sciences, 369*, 20130212.

Vlassova, A., Donkin, C., & Pearson, J. (2014). Unconscious information changes decision accuracy but not confidence. *Proceedings of the National Academy of Sciences of the United States of America, 111*, 16,214–16,218.

Vorberg, D., Mattler, U., Heinecke, A., Schmidt, T., & Schwarzbach, J. (2003). Different time courses for visual perception and action priming. *Proceedings of the National Academy of Sciences of the United States of America, 100*, 6275–6280.

Vuilleumier, P., et al. (2001). Neural fate of seen and unseen faces in visuospatial neglect: A combined event-related functional MRI and event-related potential study. *Proceedings of the National Academy of Sciences of the United States of America, 98*, 3495–3500.

Vul, E., Hanus, D., & Kanwisher, N. (2009). Attention as inference: Selection is probabilistic; responses are all-or-none samples. *Journal of Experimental Psychology. General, 138*, 546–560.

Vul, E., Nieuwenstein, M., & Kanwisher, N. (2008). Temporal selection is suppressed, delayed, and diffused during the attentional blink. *Psychological Science, 19*, 55–61.

Wang, L., Uhrig, L., Jarraya, B., & Dehaene, S. (2015). Representation of numerical and sequential patterns in macaque and human brains. *Current Biology, 25*, 1966–1974.

Watanabe, K., & Funahashi, S. (2014). Neural mechanisms of dual-task interference and cognitive capacity limitation in the prefrontal cortex. *Nature Neuroscience, 17*, 601–611.

Could a Robot Be Conscious? Some Lessons from Philosophy

Markus Gabriel

Contents

Abstract

In this chapter, the question whether robots could be conscious is evaluated from a philosophical perspective. The position taken is that the human being is the indispensable locus of ethical discovery. Questions concerning what we ought to do as morally equipped agents subject to normative guidance largely depend on our synchronically and diachronically varying answers to the question of "who we are." It is argued here, that robots are not conscious and could not be conscious, where consciousness is understood as a systemic feature of the animal-environment relationship. It is suggested, that ethical reflection yields the result that we ought not to produce cerebral organoids implanted in a robotic "body."

M. Gabriel (✉)
Epistemology, Modern, and Contemporary Philosophy, Bonn University, Bonn, Germany
e-mail: gabrielm@uni-bonn.de

Keywords

Artificial intelligence · Robot · Consciousness · Philosophy · Existentialism · Ethics

Could a Robot be Conscious? The shortest answer to the question posed in my title is: "No." In what follows, I will lay out some reasons for why we should endorse the shortest answer. At the same time, I will argue that the right way of looking at the issues at stake has significant consequences for our relationship to the digital landscape we inhabit today.

Robots and A.I.-systems created by machine learning experts and research teams play a central role in our "infosphere" (Floridi 2014). Yet, in order to understand that role, it is crucial to update our conception of ourselves, the human being. For, as I will argue, the human being is the indispensable locus of ethical discovery. Questions concerning what we ought to do as morally equipped agents subject to normative guidance largely depend on our synchronically and diachronically varying answers to the question of who we are.

J. von Braun et al. (eds.), *Robotics, AI, and Humanity*, https://doi.org/10.1007/978-3-030-54173-6_5

My paper has two parts. In the first part, I argue that robots (1) are not conscious and (2) could not be conscious if consciousness is what I take it to be: a systemic feature of the animal-environment relationship.[1] In the second part, I will sketch an updated argument for the age-old idea (versions of which can be found in Plato, Aristotle, Kant, Hegel and beyond) that human sociality and, therefore, morality hinges on our capacity to think of ourselves as animals located in a context inhabited by humans and non-human animals alike. This context is grounded in inanimate nature, which presents us with necessary, but not sufficient conditions for consciousness.

Why There Could Not Be Any Conscious Robots

The Meaning of Existence

Ontology is the systematic investigation into the meaning of "existence". If successful, it leads to knowledge of existence, i.e. the property or properties constitutive of some object's being there. In a series of books, I have defended the view that to exist means "to appear in a field of sense" (Gabriel 2015a, b). To summarize the outcome of the arguments in this context: there is no system such that every system (except for itself) is a subsystem of that very system. It is impossible for there be a single, all-encompassing field of sense such that every object is part of it. This entails that, necessarily, every object belongs to a specific domain (or field of sense, as I call it), which conditions the field-relative properties that put it in touch with other objects in the same field.

For instance, the Vatican is a legal entity with an impressive history. The Vatican appears in the field of sense of history. There are other legal entities in that field: Europe, Italy, the University of Bonn, refugees, passports, taxes, airports etc. The number 5 appears in a different field of sense, such as the series of natural numbers or that of prime numbers etc. The number 5 does not belong to history; nor is history a mathematical object.

Any view of the form that there is just one overall domain of entities (such as the material-energetic layer of the universe or what have you) is incoherent, as it relies on the inconsistent and paradox-generating notion that there is an all of reality which encompasses everything there is.

Given that there cannot be an all-encompassing domain of objects subject to one set of laws or principles, we are entitled to reject "naturalism" in the sense of a view of the form that all knowledge is natural-scientific knowledge. Natural-scientific knowledge only deals with one domain of objects, namely the kinds of objects than are part of the material-energetic system of the universe with which we can causally interact.

As a short-cut to this result, one could, of course, simply point out that the very idea of such an all-encompassing domain of natural-scientific enquiry is quite evidently incompatible with mathematical logic and meta-mathematics. We actually know that there cannot be a scientific model (or any other model for that matter) such that absolutely everything (from the early universe to transfinite sets to Angela Merkel) falls within the range of its explanatory power. What is more, there is no such thing as "science" or "natural science" in the sense of a unified theoretical project whose singular terms (such as "boson", "dark matter", "molecule", "neuron", "glia cell", "galaxy", "the universe", "robot", "A.I." or what have you) each and all refer to well-defined entities in a single domain ("reality", "the universe", "being", "the world" or what have you).

Locating an entity in a domain of investigation presupposes a stable insight into what it is. What an entity is, depends on the field(s) of sense in which it makes an appearance. If we drop the field-parameter in our description of how certain entities relate to each other (such as humans and robots, minds and bodies, consciousness and brain-tissue, numbers and countable objects etc.), we will wind up with what Gilbert Ryle famously dubbed a category mistake.[2]

The idea of "conscious robots" potentially rests on a category mistake. The reason why it is easy to be misled is the following. Humans produce artifacts out of biological and non-biological matter. We build cars, tables, houses, guns, subway systems, smartphones, servers, high-performance computers, statues, etc. Throughout the recorded history of human behavior, we find that humans have produced artifacts, some of which resembled humans and other animals, including cave paintings, statues, etc.

What is equally remarkable is the fact that we find a long-standing desire to produce an artefact in our own image, i.e. something that resembles the feature that we still deem central to human beings: Logos.[3] The most recent and somewhat amplified version of this tendency is the idea that we

---

[1] The notion of consciousness is steeped in paradox and, therefore, highly questionable. For more on this see Gabriel (2019, 2017).

[2] What often goes unnoticed is that the paradigmatic category mistake according to Ryle is precisely a mistake in ontology in the sense deployed here: "It is perfectly proper to say, in one logical tone of voice, that there exist minds and to say, in another logical tone of voice, that there exist bodies. But these expressions do not indicate two different species of existence, for "existence" is not a generic word like "coloured" or "sexed." They indicate two different senses of "exist," somewhat as "rising" has different senses in "the tide is rising", "hopes are rising", and "the average age of death is rising". A man would be thought to be making a poor joke who said that three things are now rising, namely the tide, hopes and the average age of death. It would be just as good or bad a joke to say that there exist prime numbers and Wednesdays and public opinions and navies; or that there exist both minds and bodies" (Ryle 1949, p. 23).

[3] For a critique of this narrative see Smith (2019) and Gabriel (2020a).

might be able to produce intelligent agents, which resemble humans not only in intellectual capacity, but even in shape and movement.

Let us call these kinds of objects *anthropoids*. An *anthropoid* is a robot run by the kind of software nowadays subsumed under the heading of "A.I." Both robots and machine learning techniques are progressing at a rate that makes it possible for us to imagine robots moving in ways strikingly similar to humans (and other animals). To the extent to which they perform functions we classify as "intelligent" in humans (and other animals), we are likely to be prone to think of them as potential candidates for membership in the "kingdom of ends" (Kant 2016, AA IV, 439),[4] i.e. in the domain of autonomous and, therefore, moral agents.

However, this is a mistake, as I now want to argue. There is nothing we owe to our artefacts directly. What we owe to our robots is at most a function of what we owe to each other as proprietors of technology. If you destroy my garden robot, you harm me, but you cannot harm my robot in a morally relevant manner, just as you cannot harm a beach by picking up a handful of sand. A beach is a bunch of stones arranged in a certain way due to causal, geological pressures including the behavior of animals in its vicinity. A beach does not have the right kind of organization to be the direct object of moral concern.

Ontologically speaking, robots are like a beach and not like a human, a dog, a bee etc. Robots are just not alive at all: they might be at most (and in the distant future) zombies in the philosophical sense of entities hardly distinguishable from humans on the level of their observable behavior. Analogously, A.I. is not actually intelligent, but only seems to be intelligent in virtue of the projection of human intelligence onto the human-machine interface.

Here is a series of arguments for this view.

The Nature of Consciousness

Let us begin with the troublesome concept of *consciousness*. Consciousness is a process we can know only in virtue of having or rather being it. I know that I am a conscious thinker in virtue of being one. To be conscious is to be a state that is potentially self-transparent insofar as its existence is concerned.[5] This does not mean that we can know everything about consciousness simply in virtue of being conscious. This is obviously false and, by the way, has never been maintained

by anyone, not even Descartes who is often discredited in this regard for no good reason.[6] Both right now, as I am awake, and in certain dream states, I am aware of the fact that I am aware of something. This feature of self-awareness is consciousness of consciousness, i.e. self-consciousness. Some element of consciousness or other always goes unnoticed. As I am right now conscious of my consciousness, I can focus, for instance, on the structure of my subjective visual field only to realize that many processes in my environment are only subliminally available to conscious processing, which means that I am conscious of them without thereby being conscious of that very consciousness.

Trivially, not all consciousness is self-consciousness, as this leads into a vicious infinite regress. If all consciousness were self-consciousness, then either self-consciousness is consciousness or it is not. If self-consciousness is consciousness, there is a consciousness of self-consciousness and so on ad infinitum. We know from our own case that we are not conscious of anything without a suitable nervous system embedded in an organism. We know from neurobiology and human physiology that we are finite animals such that it is evidently impossible for us to be in infinitary states where each token of consciousness is infinitely many tokens of self-consciousness. There simply is not enough space in my organism for that many actual operations of self-reference.[7]

As a matter of fact, we do not know what, if anything, is the minimal neural correlate of consciousness (the MNCC). Recent contenders for a model designed to begin to answer this question despite the complexity involved in the neuroscientific endeavor to pinpoint such a correlate include "global workspace theory" (GWT) and "integrated information theory" (IIT).[8]

Whatever the right answer to the question concerning the MNCC will turn out to be (if it is even a meaningful

---

[4] Kant's famous phrase for the domain within which moral agents move. See AA IV, 439.

[5] On this well-known aspect of consciousness famously highlighted (but by no means first discovered) by Descartes, see the concept of "intrinsic existence" in the framework of Integrated Information Theory, one of the currently proposed neuroscientific models for the neural correlate of consciousness. See the exposition of the theory in Koch (2019).

[6] For a detailed historical argument according to which Descartes does not even have the concept of consciousness that he is often criticized for introducing, see Hennig (2007). For an outstanding philosophical reconstruction of Descartes' account of human mindedness that takes into account that his views are actually incompatible with the dualism typically ascribed to him see Rometsch (2018).

[7] The situation is different if we have a rational soul in the traditional sense of the term introduced by Plato and Aristotle and handed down to us via medieval philosophy. We should not naively discard this traditional option on the ground that we mistakenly pride ourselves for knowing that we are animals, because no one in the tradition denied this! The widespread idea that we began to realize that humans are animals in the wake of Darwin is unscientific and ignorant story-telling. Short proof: ζῷον λόγον ἔχον, *animal* rationale. It should be obvious that Plato and Aristotle, the inventors of logics, were able to accept the following syllogism: (P1) All humans are rational animals. (P2) All rational animals are animals. (C) All humans are animals.

[8] For a recent overview over the main contenders for the right model of the neural correlate of consciousness see Block (2019). For a critique of the concept of consciousness deployed by Dehaene et al. (2017), see Gabriel (2020b, §§9 and 15).

question), it has to respect the following *indispensability thesis*: the reality of the human standpoint is indispensable for any scientific account of consciousness.[9] There is no third-person point of view, no purely observational stance, such that we can distinguish between conscious and non-conscious entities/processes in the universe. Scientific knowledge-acquisition at some point or other presupposes a conscious knowledge-claim maintained and defended by a human thinker or group of human thinkers. For, scientific knowledge is a paradigm case of truth-apt justified belief. To make a (scientific) knowledge claim concerning a bit of reality means that one has reasons to believe that the bit under investigation has some of the central properties ascribed to it by a model. (Scientific) knowledge claims are not blind guesses. They are highly methodologically controlled outcomes of human activity which will always make use of some technology or other (including pencil and paper; conferences; fMRI; the LHC etc.). (Scientific) knowledge claims do not simply emerge from anonymous activity, they are high-level performances of human beings, often making use of non-human machinery.

Arguably, there are various potential philosophical confusions built into the very idea of searching for the MNCC, as there are many historically shifting meanings of the word "consciousness" (which differ, of course, between various natural languages dealing with the kind of phenomena grouped together by candidate meanings of "consciousness" in contemporary Anglophone philosophy and mind science).

To begin with we ought not to lose track of the distinction between "narrow" and "wide contents" of consciousness.[10] Consciousness has both an object and a content. The *object of consciousness* is that which it concerns, for instance, the Eiffel tower, if I look at it, or some of my internal states, when I feel hungry, say. Actually, whenever I am conscious of anything in my environment (such as the Eiffel tower), I am at the same time also conscious of some internal states of my organism. Typically, we are never conscious of just one thing alone.[11] Consciousness is of objects in a field.[12] Consciousness itself is a field which encompasses subfields. I can be conscious of a recently deceased friend, say, which means that I can have visual or olfactory memories of his presence. In this scenario, I am conscious of a past conscious episode relating to my friend and not directly of my friend. Memories are not perceptions despite the fact that they involve percepts.

The *content of consciousness* is the way in which the object appears to me. Sensory modality is, therefore, part of the content, as is perspective. In general, consciousness has an ego-centrical index: I am here and now conscious of a scenario (a dynamic field of processes and objects) in a bunch of sensory modalities (Burge 2013). Notice that I cannot be conscious of any scenario without myself being part of that scenario. I am here right now as the entire animal I am. It is impossible to perceive anything without literally being a part of the same field(s) as the objects, including the various physical fields whose interaction is required as a medium for the production of mental content.

Narrow content emerges in the context of internal self-awareness of my organism. It deals with states I am in. Pain, for instance, or the color sensations I experience when I close my eyes, have narrow content. They deal with internal states of my organism. Narrow content is a series of perspectives of the organism upon itself. Narrow content is, as it were, an internal window onto processes within the animal I am.

Wide content emerges in a systemic context which includes states of affairs beyond my ectodermic limitations. If I see a table, the table is nowhere to be found in my organism. My organism is evidently too small to encompass all objects of perception. St. Peter's Basilica is bigger than my organism. If I see it, it cannot be "in me". It is simply nonsensical to believe that there is no external world in which St. Peter's can be found on the dubious ground that it is allegedly impossible to directly perceive reality. Wide perceptual content deals directly with external reality. Yet, it does so in a species- and individual-relative way. St. Peter's looks different to you and

---

[9]This important methodological and ontological principle has recently been violated, for instance, by Hoffman (2019). If our perceptual system were constitutively out of touch with reality, how could we know this by deploying scientific methods which presuppose that our modes of information-processing in a laboratory are not only contingently reliable detectors of an unknowable thing in itself, but rather the correct instruments to figure out how things really are? The scientist who tries to make a case that all perception as such is a form of illusion cannot coherently establish this result, as she will make use of her perceptual apparatus in making her blatantly self-contradictory statement.

[10]On this distinction see the classical paper by Block (1986).

[11]Another important exception here are mystical experiences such as the unity with the One described by Plotinus or the *visio beatifica* known from the Christian tradition. Similar accounts can be found in any of the other major world religions. Again, we should not simply discard these accounts, which would make us incredibly ignorant vis-à-vis the very genealogy of the idea of "consciousness" which (like it or not) originates from that tradition.

[12]Consciousness is one field of sense among others. Not all fields of sense are conscious or related to consciousness. The consciousness-field and its objects are arguably entangled. In any event, there is a strong correlation between consciousness and its objects which does not entail without further premises that the objects of consciousness are necessarily correlated with consciousness. I reject the kind of premises that typically lead to the position that the objects of consciousness would not have existed, had we not been conscious of them. For a recent exchange on this issue, see Meillassoux (2006) and the highly sophisticated response from the point of view of philosophy (of quantum mechanics) by Bitbol (2019).

me and be it for the trivial reason that we will never strictly speaking occupy the same spatio-temporal location.[13]

Both wide and narrow content owe their specific structure in human (and non-human) conscious animals to, among other things, evolutionary parameters. This is a fact about animals. Consciousness is part of our adaption to our ecological niche, which in turn is causally shaped by our self-conscious adaption to that adaption guided by scientific and technological progress. If anything, we therefore have very strong evidence for the general hypothesis that consciousness is a biological phenomenon.

Kinds of Possibility

The next step in my argument consists in delivering the premise that we simply cannot reproduce the natural, necessary conditions for the existence of consciousness in the sense of anything like a full ego-centrical index. One reason for this is that we are astronomically far away from knowing the necessary (and jointly sufficient) conditions for human (and non-human) consciousness in the full sense, where "the full sense" encompasses both wide and narrow content. We do not know enough about the causal architecture of consciousness as a natural phenomenon in order to even begin constructing potentially conscious robots. Therefore, if any actually existing robot is anywhere near consciousness, this would be a sheer coincidence. It is as rational to believe that any actually existing robot or computer is conscious as that the Milky Way or a sandstorm in the Atacama Desert is conscious.[14]

To be sure, it is apparently at least logically possible that a given robot is conscious. But this does not entail that we have any actual reason to believe that a given robot is conscious (i.e. if it's being logically possible just means we cannot say we know it is not the case). It is, thus, irrational and unscientific to believe that currently existing robots are conscious.

At this point, someone might wonder if this train of thought rules out that robots could be conscious. So far, I seem not to have established that no robot *could* ever be conscious. At this stage, we need to be careful so that our modal "intuitions" do not idle. The central modalities are: actuality, possibility, contingency, and necessity. If we ask the question "could robots be conscious?" we are after a possibility. Is it possible that robots are conscious? So far, I have argued that

there is no reason to think that any currently existing robot is conscious. A robot is an artifact of human industry and so far, no relevant robot in that sense is conscious.[15] If we want to consider the possibility of robot consciousness, we therefore need either some evidence or argument that supports the rationality of the belief that *future* robots could meet the relevant threshold of consciousness. Otherwise we wind up with what I would like to call *extremely weak possibility.*

A possibility is extremely weak if (1) nothing we actually know supports it and (2) we have no reason to believe that the possibility amounts to a logical contradiction. Conscious robots are currently extremely weakly possible. What is more, I now want to argue that they are currently at most extremely weakly possible. The argument relies on the notion of "biological externalism," which I have proposed in my recent book *The Meaning of Thought* (Gabriel 2020a).[16]

In philosophy, *semantic externalism* is roughly the view that some terms refer to some natural kinds in such a way that a competent user of those terms need not know all essential feature of the kinds in order to count as a competent speaker.[17] For instance, I am a competent user of the term "fermion," but not an expert user. A professional nuclear physicist will use the term "fermion" in contexts where I would be likely to make nonsensical utterances. One strength of semantic externalism in general is that it gives an account of the fact that we often speak about things in a competent way without thereby knowing their essence. The standard example in philosophy is use of the term "water". When Thales or Aristotle referred to the stuff in the Aegean Sea as ὕδωρ, they were thereby referring to something that essentially involves H_2O molecules. However, if we asked them about ὕδωρ, they could not even entertain the thought that it essentially involves H_2O molecules, because there was no such thought available to them in their linguistic community. From the epistemic standpoint of the Ancient Greeks, it would havelooked possible that water could consist

---

[13]For a paradigmatic exposition of a direct realism which takes objective looks as relations between a perceiver and the perceived environment into account see the discussion in Campbell and Cassam (2014).

[14]On the recent resurgence of panpsychism in the sense of the view that basically everything is (or might be) conscious, see Harris (2019) and Goff (2019). I assume that it is reason to reject a given account of the meaning of "consciousness" if it either entails the truth of or significantly raises the probability of panpsychism.

[15]Of course, humans (and some non-human animals) are artifacts of human activity. As Aristotle nicely puts it: ἄνθρωπος ἄνθρωπον γεννᾷ (Aristotle, *Metaphysics*, VII 71032a 25 et passim). However, we do not classify humans as robots. If humans count as robots, machines, A.I.s or computers, it would be trivially true that robots, machines, A.I.s and computers are conscious because we are. The question "could robots be conscious?" deals exclusively with the concept of a robot as a non-biological (possibly anthropoid) artifact of human industry.

[16]This is a revised version of Gabriel (2018a). See also Gabriel (2020b) §§6–11, where I defend "mental realism," i.e. the view that mental terms (including: thinking, intelligence, consciousness) refer to processes which have necessary biological preconditions. Notice that the argument does not preclude the existence of future conscious robots controlled by cerebral organoids. Such hybrid entities are more likely to be possible than in silico conscious robots.

[17]For an overview see Rowlands (2014).

of various arrangements of elementary particles (atoms) in their sense. It would not have been irrational for Aristotle to assert that water consists of some Platonic atomic figure or other, as laid out in the *Timaios*. For him, that could be the case. Yet, Aristotle would have made a mistake, had he endorsed a view about water that rules out that water essentially involves H_2O molecules.

As far as consciousness is concerned, we are in a similar epistemic situation. We do not know which natural kinds, if any, are correlated with consciousness. However, we must not forget that the mainstream of contemporary scientifically minded theorists of consciousness tends to believe that consciousness has some necessary natural prerequisites or other. Unfortunately, this misleads far too many theorists into some version of "functionalism". It is certainly true that the biological prerequisites for conscious states significantly vary across species, across individuals within a species and even within an individual across time. In this sense, the physical underpinning of consciousness is multiply realizable in different structures. Yet, this factual variation in the (neural?) support of consciousness, of course, does not per se support the stronger claim that consciousness is realizable in inanimate matter or any matter that preserves the functional architecture correlated with consciousness in animals.

In this context, let *functionalism* be the view that consciousness is identical to a process which consists in the realization of a role in a system of sensory inputs and behavioral outputs. Most functionalists are committed to *multiple realizability*. According to this concept, the functional role of consciousness can be realized in different materials. We humans realize it in (neural?) tissue, other creatures might realize it in some other tissue. If consciousness is multiply realizable, it seems to be possible to produce it out of material other than biological tissue. This possibility is stronger than the extremely weak possibility that there could simply be robots. Functionalism to some extent supports the hypothesis that robots could be conscious. However, functionalism combined with multiple realizability is in serious trouble, as is well-known in the philosophical community, but often ignored by the interested bystander.[18] The major weakness is a consequence of the fact that we currently do not even know the MNNC. For all we know, if consciousness is identical to a functional role, this role could be performed by the universe as a whole or by some surprising subsystem of it (such as a galaxy cluster or a beach). This explains the presence of panpsychism in contemporary philosophy, i.e. the (misguided) notion that consciousness might be everywhere.[19]

Functionalism tends to lead to acceptance of the notion that consciousness could emerge out (or rather: be a property) of basically any system whose parts are arranged in such a way that we can describe their operations in terms of sensorial input and behavioral output.

According to biological externalism, "consciousness" and cognate terms in our mentalistic vocabulary refer to processes which have necessary biological prerequisites. Thus, there could be no conscious robot produced entirely out of inanimate matter. Support for biological externalism can be derived from our current knowledge base concerning consciousness. There should be no doubt that ever since humans have been in the business of thinking about their own mental states and those of other animals, they have been thinking about something with some biological underpinning. At least, this is not what is contentious between the functionalist and her opponent.[20] Thus, as far as we know, consciousness has essentially biological preconditions. This does not mean that consciousness is a purely biological product, as I will now argue.

Biological externalism alone does not suffice to rule out future robots controlled by cerebral organoids. However, I surmise that such robots are biologically impossible. This claim is empirical, but partially grounded in a further maneuver of philosophical theorizing. Above, I mentioned an indispensability thesis. As far as consciousness is concerned, the indispensability of consciousness consists in the fact that we cannot circumvent it in any complete account of human mindedness. Scientists are conscious and, therefore, consciously interested in consciousness for various reasons (including ethical considerations, because we believe that we ought to care for conscious creatures more than for entirely non-conscious matter). However, the level of indispensability is located on a scale which differs from that occupied by neural tissue alone. Any theorist we have encountered so far, has been a full-blooded human being with animal parts that include many cell types other than neurons. Human animals are not neural tissue implanted in an organism, as it were. My skin is not just a bag containing a central nervous system hidden from view by a more complex organization. Simply put: I am not a brain.[21] Human neural tissue is produced by a human organism out of stem cells in complex biological processes. It develops over the course of pregnancy in such

---

[18] See especially: Block (1978) and Searle (1992). An example of a cautionary tale of philosophical confusions concerning the alleged substrate-independence of life and intelligence is Tegmark (2017).

[19] Notice that IIT entails that panpsychism is false. See Tononi and Koch (2014). IIT's model for consciousness provides us with a defeasible criterion for the presence of consciousness in a system. In its current

---

state, it rules out that beaches could be conscious. It also rules out that computers and robots could be conscious.

[20] There is the additional difficulty that humans have been thinking about divine thinking for long stretches of our history harking back as far as the first written documents of humanity. This supports the notion that there could be divine non-biological thought and consciousness. However, it does not back up the idea of finite, conscious robots produced by human industry out of non-biological matter.

[21] For a detailed defense of a non-reductive account of human thinkers see Gabriel (2017).

a way that at some (as yet unknown) point a human fetus becomes conscious. No known neural tissue outside of an organism has the property of consciousness. This is probably not a coincidence, as consciousness is a product of processes that can be studied by evolutionary biology. All cell types came into existence in this way. Neural tissue comes into existence in causal contexts that produce organisms. Let us call the structure of this process *systemic organization*.[22] As far as I know, no sample of neural tissue that even comes close to being a candidate for a correlate of anything mental has been found outside of systemic organization.

The absolute majority of actual conscious states we know of has both objects and content. Consciousness is usually about something in some particular mode of presentation or other. Without integration into an organism, it is quite senseless to think of neural tissue as performing any function that correlates with consciousness, as we know it. Should it be possible to produce cerebral organoids that are in conscious states, those states would at most resemble a tiny fraction of a proper subset of our conscious states. No organized heap of neural tissue will perceive anything in its environment without proper sense organs. To perceive our environment is precisely not a kind of hallucination triggered by otherwise unknowable external causes. Such a view—which I attack under the heading of "constructivism"—is profoundly incoherent, as it amounts to the idea that we cannot ever really know anything about our environment as it really is. This makes it impossible to know that there is neural tissue sealed off from an external environment so that the view that we are literally "brains in a vat," i.e. neural tissue hidden under our skulls, is utterly incoherent (Putnam 1981). External reality as a whole cannot be a kind of hallucination or user illusion produced by a subsystem of the central nervous system. For, if it were, we could not know this alleged fact by studying the central nervous system, because the central nervous system itself belongs to external reality. Thus, the central nervous system is not a hallucination by the central nervous system. If we know anything about ourselves as animals capable of perception, we thereby know that we can know (parts of) external reality.

Here, we can use a famous thought-experiment by Donald Davidson as a conceptual magnifying glass. In a forthcoming paper I have co-authored with the cosmologist George F. R. Ellis, we use this thought-experiment in order to illustrate the concept of top-down causation[23] (Gabriel and Ellis 2020, forthcoming). Davidson asks us to imagine that lightning

strikes a dead tree in a swamp while Davidson is standing nearby. As Davidson's body dissolves due to the causal circumstances, the tree's molecules by coincidence turn into a replica of his body which begins to behave like Davidson, moves into his house, writes articles in his name, etc. (Davidson 1987). We maintain that Swampman is physically impossible. No molecule by molecule duplicate of a person could arise spontaneously from inanimate matter. The evolutionary pre-history and the adaptation of an organism to its causally local environment (its niche) are essential for the organism's existence. To the extent to which we could possibly recreate the conditions of survival for neural tissue complex enough to be a candidate for a token of the MNNC, our social activity of producing those conditions and artificially maintaining them in existence would at best replace the structure of organization. Thus, any robot actually capable of being run by "mental," i.e. actually conscious software would have to have the relevant biological hardware embedded in a context which plays the role of an organism.

The organism controls local patterns of causation in a top-down way. The organism is thus ontologically prior to the causal order of its elements (Noble 2016). If we randomly copied the order of an organism's elements, we would still not have copied the organism. To be more precise, we would have to copy the causal order of an organism's elements in the right way in order for a Swampman to be alive, which means that the contextual, social constraints on his production set the conditions for the lower-level elements to realize Swampman. Random physical structure is not enough for Swampman to be so much as alive for any amount of time. Hence, there could not be a Swampman replica of Davidson. Our use of the material of the thought experiment is supposed to illustrate that evolutionary causal history, including an organism's niche construction and its social contexts, is essential for the causal constitution of conscious life and thought. Even if we could replicate human organisms by some hitherto unavailable procedure, this would not be evidence for a bottom-up process, as the relevant causal context would, of course, include us and the technical apparatus needed in order to achieve the feat of bringing organic matter into shape.

Intelligent Robots?

One line of argument for the possibility of conscious robots draws on the notion of artificial consciousness and assimilates this discussion to that of AI. Yet, this is a red herring, as the term "intelligence" generates confusions similar to those associated with "consciousness".[24] In general, *intelligence*

---

[22]For an ecological account of consciousness see Fuchs (2018).

[23]M. Gabriel/G. Ellis, "Physical, Logical, and Mental Top-Down Effects," in: M. Gabriel/J. Voosholz (eds.), *Top Down Causation and Emergence*. 2020 (forthcoming). The thought experiment can be found in D. Davidson, "Knowing One's Own Mind", in: *Proceedings and Addresses of the American Philosophical Association* 60 (1987), pp. 441–458.

[24]On some of the confusions in artificial consciousness debates see Schneider (2019).

can be seen as the capacity to solve a given problem in a finite amount of time. Let us call this concept "undemanding intelligence". It can be measured by constraining the exercise of the capacity to time parameters. In this light, a system S* is more intelligent than a System S if it is more efficacious in solving a problem, i.e. if it finds a solution quicker. *Learning* is the process of replacing a given first-order *object-problem* by another higher-order *meta-problem* in such a way that finding solutions to the object-problem enhances the capacity of finding solutions to the meta-problem. A *standard artifact* is a non-biological product of human industry. Usually, a standard artifact is associated with a human goal-structure. In the case of modern technology, the human goal-structure is essentially tied to a division of labor. The division of labor of modern technology is too complex for any individual to know how each and every participating individual contributes to the production of the outcome. The management structure of our production of material goods (including the hardware required for any actually functioning robot) functions by producing meta-problems handed down in the form of object-problems to agents on the ground-floor of production. Without this socially immensely complex structure of the production of systems capable of justifying scientific knowledge-claims, there would be no robots.

More specifically, no AI-system has the property of intelligence outside of the top-down context realized at the human–machine-interface. No standard artifact (which includes software qua program-structure) has any degree of intelligence outside of a human use-context. AI essentially differs from human, animal intelligence for a simple reason: the parameters of our goal-structure are fundamentally set by our survival form. Intelligence first and foremost arises in the context of solving the central maintenance (survival) problem of human, animal organization. Animals have interests which in turn serve the goal of maintaining them in existence. This goal-structure came into being in the universe as a consequence of as yet not fully understood processes. We have no final answer to the question of the origin of life. Yet, whatever the actual causal context for the emergence of life, it is the breeding ground of intelligence.

The term "intelligence" derives from the Latin *intelligere/intelligentia* which means "understanding". We should distinguish intelligence in the traditional sense from undemanding intelligence. Accordingly, we can introduce the notion of *demanding intelligence* (D.I.). In his classic, *The Emperor's New Mind*, Sir Roger Penrose has shown that D.I. is not a matter of explicable rule-following (Penrose 1989). D.I. consists in finding a new solution space to an inherited problem by discovering an entirely new meta-problem. This requires the capacity for understanding oneself

as a creative thinker engaged in an activity of thinking that cannot be formalized at all. In this context, I have recently argued that our AI/machine learning programs amount at best to thought-models (Gabriel 2020a). Thought-models can be very powerful tools. Think of everyday modern technological products such as search engines, which serve the function of mining data by means of a formal representation of a mode of organizing potential material for thought. The internet significantly contributes to our cognitive enhancement in that it provides us with quick solutions to given problems so that we can use our mental time more efficiently. By deploying thought-models as instruments in our own struggle for survival and progress, we, humans, become more intelligent in that we create the potential for new modes of thinking. If anything, our digital technology produces conditions of emergence for intelligent human behavior. We can make intelligent use of our technology, and we should begin to realize that this does not at all entail that our technology is intelligent by itself.

D.I. is the capacity to change a problem space in virtue of an account of our activity of creating and changing problem spaces. In classical philosophical parlance, D.I. is self-consciousness or self-awareness: we, human beings, become aware of the fact that we are intelligent animals. In the context of exercises of that awareness we can produce thought-models designed to re-produce elements of our thought-activity in a simplified way. It is in the nature of a model of something to reduce the complexity of a target system. Models are modes of abstraction. They distinguish between an essential and an inessential feature of a target system relative to a goal-structure. A scientific model, such as the contemporary standard model of particle physics, is not a copy of physical reality, but a mode of abstracting away from levels of the universe we inhabit. It is crucial for the standard model that it does not mention the scientists who produced it in the quest for understanding the universe, precisely because scientists and their actual thoughts do not appear in the standard model. Scientists are not a bunch of elementary particles. The idea that scientists are ultimately reducible, i.e. logically replaceable by a bunch of elementary particles arranged in the right way, is a terrible confusion of model and reality. For more on this see Ellis (2016).

Analogously, the notion that human thinking is a rule-governed process exactly like that to be found in a Turing machine (or any other model of information-processing) conflates a model with the reality it is designed to make more intelligible to human thinkers. If we abstract away from the context we actually occupy as human thinkers, it should not be a surprise that we cannot recover our own minds from observing the behavior of our artifacts.

The Human Context

Neo-Existentialism

Human beings are sapient creatures. When Linnaeus suggested "homo sapiens" as the name for our species, he was fully aware of the fact that human beings fundamentally relate to themselves in a specific way. This is why he defines the human being in terms of our capacity for self-knowledge: "nosce te ipsum" (Linnaeus 1792). In this context, humans produce models of themselves. The German word for this is "Menschenbild", which means "image of humankind". A conception of man is an image, a model of what we think we are. Evidently, there is a variety of such images. Some believe that they have an immortal soul which is the locus of their humanity and dignity. Others think of themselves as sophisticated killer apes whose goal is to spread their genes. Whatever the right answer to the question of who or what we are as human beings, it must consider the remarkable fact that there is a range of answers to that question in the first place.

In this context, I propose a framework for the study of the human context I call "Neo-Existentialism" (Gabriel 2018b). Neo-Existentialism offers an account of what it is to be human, an account of humanity. *On this account, to be human is to instantiate the capacity to think of oneself as an agent of a certain kind and to (sometimes) act in light of that conception.* We can think of this as *higher-order anthropology.* The capacity to instantiate humanity in ways that differ synchronically and diachronically across individuals and populations does not itself differ synchronically and diachronically across individuals and populations.

Neo-Existentialism differs from many forms of classical existentialism in that it draws on a potentially unified conception of humans as both objects of natural science, medicine, etc. and subjects of truth-apt, historically variable self-conceptions that are the target of the humanities and social sciences. It thus bridges the perceived gap between the natural sciences and the humanities by locating scientific knowledge-acquisition in the human context.

Among other things, it has the advantage of offering a solution to the so-called mind-body problem that is designed to bring all academic disciplines to the same table, the one we all sit at in virtue of our humanity. In the philosophy of mind, neo-existentialism argues that there is no single phenomenon or reality corresponding to the ultimately very messy umbrella term "the mind". Rather, the phenomena typically grouped together under this heading are located on a spectrum ranging from the (by now) obviously physical to the non-existing. However, what does unify the various phenomena subsumed under the messy concept of "the mind" is that they result from the attempt of the human being to distinguish itself both from the purely physical universe and from the rest of the animal kingdom. In so doing, our self-portrait as specifically minded creatures evolved in light of our equally varying accounts of what it is for non-human being to exist.

Being a German-speaking philosopher, I suggest that we replace the messy term "mind" by what we call "Geist" in my neck of the woods. *Geist* is not a natural kind or complicated structure of natural kinds, but precisely something that does not exist independently of the specific descriptions used in order to point out phenomena whose very existence depends on mutual ascriptions of tokens of mental states such that their accuracy-conditions presuppose anchoring both in the external natural world and a linguistic division of labor. "Geist" is what you get under conditions of mutual action explanation in a context where you cannot delegate the vocabulary by which you conceive of yourself and others as human to a neutral, natural-scientific standpoint.

To look at reality from the standpoint of a human being (Geist), means that we produce thought-models in the context of the human life form. There is no way to circumvent this. This is why Neo-Existentialism rejects Daniel Dennett's influential distinction between the *physical*, the *design* and the *intentional* stance (Dennett 1987, 2017). There is no physical stance capable of dealing with the human being if this stance abstracts from the fact that the scientist is a human being endowed with a mind suitable for making knowledge-claims etc. The physical stance simply evaporates if we try to think of it entirely independently from the intentional stance. Dennett's mistake consists in thinking of the intentional stance as a kind of model or theory of human agency which serves the function of a useful illusion. According to contemporary philosophical classification systems, his view is a form of "mental fictionalism" according to which attributing mental states such as "consciousness" to an entity such as a human animal is literally false, but useful.[25]

The starting point of Neo-Existentialim's framework is the observation that human agents (sometimes) act in light of a concept of who/what they are. A striking example for this would be the difference between someone who does what she does in virtue of her belief that she has an immortal soul whose morality is tested during her earthly life by a transcendent God on the one hand and, on the other, someone who believes that she is a sophisticated killer ape without any homuncular control center; a complex biological cellular network whose goal is maintenance in the form of survival and the spreading of her genes via procreation. There are many other forms of actually being human, or realizing one's own humanity by acting in light of a shared conception of what humanity is. The humanities, such as anthropology, religious studies, the sociology of knowledge, etc. investigate such

---

[25] For a discussion of "mental fictionalism" see the various contributions in *The Monist* Vol. 96, no. 4 (2013).

ways of being human in their specific mode of institutional, historical, etc. manifestation.

In this context, Neo-Existentialism distinguishes between *two kinds of error*: error about a natural kind vs. error about oneself as an agent. From the standpoint of *the epistemic conception of reality,*[26] we can define a "natural kind" as a type of object that it is exactly what it is regardless of the truth or falsity of our attitude. Electrons or supernovae are candidates for natural kinds, as they have their properties no matter what anyone believes about them. At some level or other, the natural sciences discover properties of natural kinds even though they cannot be reduced to this feature, because they sometimes create new objects and, therefore, bring properties into being that are a function of their theories. This is what happens in material science or in a particle accelerator which is capable of generating new particles. Science is not just a list of natural kinds and their properties, a table of elements.

We can distinguish "natural kinds" from what Ian Hacking has helpfully labelled "interactive kinds" (Hacking 1999, 2002). Specifically, humans are interactive kinds in virtue of the fact that it matters for who we are how we conceive of ourselves. My favorite example is someone who thinks that he is a talented Tango dancer, but in reality, can hardly dance at all. This person might lead a deluded life to the extent to which his integration into a group can severely suffer from his wrong beliefs about himself. Wrong beliefs about myself—including paradigmatically: wrong beliefs about my Self—can change my properties. If I have wrong beliefs, I am in a different state than if I have true beliefs. Thus, beliefs matter for who and what we are. Our wrong beliefs can guide our actions. The deluded tango dancer acts in light of a misguided (partially false) conception of himself so that a feedback loop between wrong beliefs and action comes into existence.

Some proper subset or other of our mentalistic vocabulary is such that it comprises elements that do not refer to natural kinds. This does not rule out a priori that there is another proper subset of the same vocabulary that happens to pick out natural kinds. Vigilance, various urges we consciously experience, and maybe phenomenal consciousness (what-it-is-likeness) as a whole belong to this category. As things stand, our mentalistic vocabulary is differentiated both diachronically and synchronically over different natural languages and specialized idiolects. It is not unified in any specific way

beyond the fact that we typically invoke it in contexts where action explanation, including activities such as predicting or regulating future behavior, matters. But this is precisely a manifestation of "Geist". As long as humans interact in an institutionalized form of any kind, the game of mutual action-explanation and attitude adjustment to the observed and extrapolated actions of others will go on and produce new vocabularies and situations. Monarchies, right- and left-wing populism, neurosis, credit cards, fear of the Gods, love of wisdom, class struggle, ideology, moral righteousness, and indefinitely many other facets of human reality will never be replaced by a unified, centralized committee of neuroscientistic Newspeak headed by some eliminative materialist or other.

Neo-Existentialism is not a relapse into metaphysical dualism according to which there are exactly two kinds of objects in the universe: material and mental substances. That would only lead us back to the unsolvable mystery of their interaction. Mental causation is real in that tokens of the types picked out by the relevant proper subset of our mentalistic vocabulary that makes us *geistig*, are integrated into a meshwork of necessary and jointly sufficient conditions. This meshwork essentially involves natural conditions, such as nervous systems embedded in healthy organisms etc. There is no overall privileged grounding relation running through all token meshworks. Any actual token of the meshwork, any of its states, can take any form out of a huge, epistemically indefinite and historically open set of possible ways of being human. We continue to generate new ways of being human without there being any a priori catalogue. This is the sense in which humans do not have an essence: there is no surveyable totality of modes of realizing humanity.

Values and the Humanities

(Moral) values are grounded in the universal form of being human. The universal form of being human consists in our capacity to lead a life in light of a conception of the human being and its place in animate and inanimate nature. Our anthropological self-conception cannot be exhaustively studied by the natural sciences. The kinds of complexity involved in high-level human social systems, the dynamics of historical developments, the plurality and history of human languages, art forms, religion etc. cannot seriously be reduced to the models of explanation characteristic of natural-scientific knowledge-acquisition.

The humanities remind us of our own humanity. This is one of their crucial roles in modernity. Natural science will not survive the materialist attacks on the humanities that, among other things, are outcomes of a misguided scientific worldview according to which all (real) knowledge is natural-scientific knowledge. It is simply false that all there is the

---

[26]The epistemic conception of reality is the notion that to be real means to be the target of fallible belief. This conception of reality is broad enough to encompass immaterial objects, such as mathematical objects, (possibly) consciousness and laws of nature and to think of them as real. Reality cannot be reduced to the material-energetic layer of the physical universe. As a matter of fact, this is a lesson from modern physics itself, an insight we owe to quantum theory, which has ultimately superseded the naïve conception of "matter" and "causation" as a series of "micro-bangings" of "atoms in the void". For more on this see Ladyman and Ross (2007), Ellis (2016), Ismael (2016), and Falkenburg (2007, 2012).

material-energetic layer of the physical universe. We know that this is false from the various humanities and social sciences, which clearly study objects and processes that are by no means identical to objects and processes studied by any combination of actually existing disciplines from the range of the natural sciences.

The materialist version of the scientific worldview is an ideological distortion of scientific activity easily exploited by groups whose interest lies in pinning down an alleged specific essence of the human being, such as the false idea that the human self is identical to the brain or some other subsystem of the nervous system (Gabriel 2018c). If we are sophisticated killer apes, it does not make sense to resist Chinese or North-Korean style full-blown cyberdictatorship, say. There is no normativity inherent in the concept of a culturally sophisticated primate, let alone in that of a bunch of neural tissue to be found in an organism. If we were identical to one of those models of the human being, we would lose the very concept of human dignity underpinning the value system of the democratic rule of law.

Natural science as such is the value-free discovery of natural facts. This is why science has not only contributed to human health, security, and flourishing, but at the same time turned out to be the biggest killing machine humanity has ever witnessed. Millions of people were killed in the wake of scientific progress, and humanity is currently on the brink of self-extinction as a consequence of the misguided idea that human progress can be replaced by natural-scientific and technological progress. This ideology is quite literally a dead-end, based on flawed metaphysics.

Concluding Remarks

What we are currently witnessing on a global scale in our digital age is a struggle among different conceptions of the human. We rid ourselves of the very capacity to describe our situation, to make it transparent and, thereby, to defend the kinds of universal values that guarantee socio-economically mediated access to humanity's invariant core, if we march for science without marching for the humanities. That humanity should not destroy itself by ruining the only planet we will ever thrive on, cannot be deduced at all from natural scientific and technological knowledge. As long as we do not grant the humanities and all other academic disciplines equal epistemological standing, natural science too will be easy prey for those who do not care about the facts, but really are interested only in maximizing the reach of their will to power. Thinking that scientific knowledge is valuable is simply not a piece of natural-scientific knowledge. To think otherwise, is to be deluded and to fall in the target domain of the humanities and social sciences which, among other things, ought to study the delusions of the so-called scientific worldview.

For all the reasons sketched in my paper, we are entitled to reject the very idea of conscious robots. Let me conclude by pointing out that even if (*per impossibile*) there could be conscious robots, this very possibility does not entail the desirability of their actual existence. Rather, I suggest by way of a conclusion that ethical reflection yields the result that we ought not to produce cerebral organoids implanted in a robotic "body."

This is no argument against technological or medical progress. Rather, it is a reminder of the fact that scientific discovery is subject to the value-system of the human life form, the only system we can know of as humans. Whether or not it is somehow backed up by a transcendent God does not matter for ethics, as morality takes care of itself: what we ought to do cannot merely be a consequence of the fact that the Almighty dictates what we ought to do anyhow. The fact that a certain kind of action is good or evil, i.e. ethics, cannot be derived from the mere decree of any kind of will. If there is good and evil (morally recommended and morally prohibited action), God himself does not create it.[27]

References

Bitbol, M. (2019). *Maintenant la finitude. Peut-on penser l'absolu?* Paris: Flammarion.

Block, N. (1978). Troubles with functionalism. In C. W. Savage (Ed.), *Perception and cognition: Issues in the foundations of psychology* (Minnesota studies in the philosophy of science) (Vol. 9, pp. 261–325). Minneapolis: University of Minnesota Press.

Block, N. (1986). Advertisement for a semantics of psychology. *Midwest Studies in Philosophy, 10*(1), 615–678. https://doi.org/10.1111/j.1475-4975.1987.tb00558.x.

Block, N. (2019). What is wrong with the no-report paradigm and how to fix it. *Trends in Cognitive Sciences, 23*(12), 1003–1013. https://doi.org/10.1016/j.tics.2019.10.001.

Burge, T. (2013). Self and self-understanding: The Dewey lectures (2007, 2011). In T. Burge (Ed.), *Cognition through understanding* (Self-knowledge, interlocution, reasoning, reflection: Philosophical essays) (Vol. 3, pp. 140–226). Oxford: Oxford University Press.

Campbell, J., & Cassam, Q. (2014). *Berkeley's puzzle: What does experience teach us?* Oxford: Oxford University Press.

---

[27] In my book *Moralischer Fortschritt in dunklen Zeiten* (Berlin: Ullstein 2020) I defend a brand of "universal moral realism", according to which moral value (the good, the neutral, the bad/evil) is not in the eye of any beholder, but rather an objectively existing property of the action assessed by a morally trained participant in a morally relevant practice. Like all other forms of truth-apt judgment, moral judgment is fallible. Moral properties are relational: they essentially express relationships between human beings and, indirectly, between different organic life forms. Thus, there is neither a direct nor an indirect set of duties with respect to the inanimate layer of the physical universe. What we owe to our inanimate environment is always a function of what we owe to each other and to the rest of the animal kingdom. Conscious cerebral organoids implanted in a robot would be organisms to whom we owe something. In particular, we owe it to them not to produce them in the first place.

Davidson, D. (1987). Knowing one's own mind. *Proceedings and Addresses of the American Philosophical Association, 60*, 441–458. https://doi.org/10.2307/3131782.

Dehaene, S., Lau, H., & Kouider, S. (2017). What is consciousness, and could machines have it? *Science, 358*, 486–492. https://doi.org/10.1126/science.aan8871.

Dennett, D. C. (1987). *The intentional stance*. Cambridge: MIT Press.

Dennett, D. C. (2017). *From bacteria to Bach and back: The evolution of minds*. New York: W.W. Norton and Company.

Ellis, G. (2016). *How can physics underlie the mind? Top-down causation in the human context*. Berlin/Heidelberg: Springer.

Falkenburg, B. (2007). *Particle metaphysics: A critical account of subatomic reality*. Berlin/Heidelberg: Springer.

Falkenburg, B. (2012). *Mythos Determinismus: Wieviel erklärt uns die Hirnforschung?* Berlin/Heidelberg: Springer.

Floridi, L. (2014). *The fourth revolution: How the infosphere is reshaping human reality*. Oxford: Oxford University Press.

Fuchs, T. (2018). *Ecology of the brain: The phenomenology and biology of the human mind*. Oxford: Oxford University Press.

Gabriel, M. (2015a). *Why the world does not exist*. Cambridge: Polity.

Gabriel, M. (2015b). *Fields of sense: A new realist ontology*. Edinburgh: Edinburgh University Press.

Gabriel, M. (2017). *I am not a brain: Philosophy of mind for the 21st century*. Cambridge: Polity.

Gabriel, M. (2018a). *Der Sinn des Denkens*. Berlin: Ullstein Buchverlage.

Gabriel, M. (2018b). *Neo-existentialism*. Cambridge: Polity.

Gabriel, M. (2018c). Review of Owen Flanagan and Gregg Caruso, eds., Neuroexistentialism: Meaning, morals, and purpose in the age of neuroscience. Available via Notre dame philosophical reviews. Retrieved February 12, 2020, from https://ndpr.nd.edu/news/neuroexistentialism-meaning-morals-and-purpose-in-the-age-of-neuroscience/.

Gabriel, M. (2019). The paradox of self-consciousness: A conversation with Markus Gabriel. Available via edge. Retrieved February 12, 2020, from https://www.edge.org/conversation/markus_gabriel-the-paradox-of-self-consciousness.

Gabriel, M. (2020a). *The meaning of thought*. Cambridge: Polity.

Gabriel, M. (2020b). *Fiktionen*. Berlin: Suhrkamp.

Gabriel, M., & Ellis, G. (2020). Physical, logical, and mental top-down effects. In M. Gabriel & J. Voosholz (Eds.), *Top-down causation and emergence*. Dordrecht: Springer, Forthcoming.

Goff, P. (2019). *Galileo's error: Foundations for a new science of consciousness*. New York: Pantheon Books.

Hacking, I. (1999). *The social construction of what?* Cambridge/London: Harvard University Press.

Hacking, I. (2002). Historical ontology. In P. Gärdenfors, J. Wolenski, & K. Kijania-Placek (Eds.), *In the scope of logic, methodology and philosophy of science* (Vol. II, pp. 583–600). Dordrecht: Springer.

Harris, A. (2019). *Conscious: A brief guide to the fundamental mystery of the mind*. New York: HarperCollins.

Hennig, B. (2007). Cartesian conscientia. *British Journal for the History of Philosophy, 15*(3), 455–484. https://doi.org/10.1080/09608780701444915.

Hoffman, D. (2019). *The case against reality: Why evolution hid the truth from our eyes*. New York: W. W Norton & Company.

Ismael, J. T. (2016). *How physics makes us free*. Oxford: Oxford University Press.

Kant, I. (2016). *Grundlegung zur Metaphysik der Sitten*. Riga: Hartknoch.

Koch, C. (2019). *The feeling of life itself: Why consciousness is widespread but can't be computed*. Cambridge: MIT Press.

Ladyman, J., & Ross, D. (2007). *Every thing must go. Metaphysics naturalized*. Oxford/New York: Oxford University Press.

Linnaeus, C. (1792). *The animal kingdom or zoological system (trans: Kerr R)* (pp. 44–53). Edinburgh: Creech.

Meillassoux, Q. (2006). *Après la finitude. Essai sur la nécessité de la contingence*. Paris: Éditions du Seuil.

Noble, D. (2016). *Dance to the tune of life: Biological relativity*. Cambridge: Cambridge University Press.

Penrose, P. (1989). *The emperor's new mind. Concerning computers, minds, and the laws of physics*. Oxford: Oxford University Press.

Putnam, H. (1981). Brains in a vat. In H. Putnam (Ed.), *Reason, truth and history* (pp. 1–21). Cambridge: Cambridge University Press.

Rometsch, J. (2018). *Freiheit zur Wahrheit: Grundlagen der Erkenntnis am Beispiel von Descartes und Locke*. Frankfurt am Main: Klostermann.

Rowlands, M. (2014). *Externalism: Putting mind and world back together again*. New York/Oxford: Routledge.

Ryle, G. (1949). *The concept of mind*. Chicago: University of Chicago Press.

Schneider, S. (2019). *Artificial you: AI and the future of your mind*. Princeton: Princeton University Press.

Searle, J. R. (1992). *The rediscovery of the mind*. Cambridge: MIT Press.

Smith, B. C. (2019). *The promise of artificial intelligence: Reckoning and judgment*. Cambridge: MIT Press.

Tegmark, M. (2017). *Life 3.0: Being human in the age of artificial intelligence*. New York: Alfred A. Knopf.

Tononi, G., & Koch, C. (2014). Consciousness: Here, there but not everywhere. Available via Cornell University. Retrieved February 12, 2020, from https://arxiv.org/ftp/arxiv/papers/1405/1405.7089.pdf.

Part II

AI and Robotics Changing the Future of Society: Work, Farming, Services, and Poverty

Robotics and the Global Organisation of Production

Koen De Backer and Timothy DeStefano

Contents

Abstract

The growing investment in robotics is an important aspect of the increasing digitalisation of economy. Economic research has begun to consider the role of robotics in modern economies, but the empirical analysis remains overall limited. The empirical evidence of effects of robotics on employment is mixed, as shown in the review in this chapter. The effects of robots on economies go further than employment effects, as there are impacts for the organisation of production in global value chains. These change the division of labour between richer and poorer economies. Robotics may reduce offshoring of activities from developed economies towards emerging economies. Global spreading of automation with robotics can lead to faster de-industrialisation in the development process. Low-cost jobs in manufacturing may increasingly be conducted by robots such that fewer jobs than expected may be on offer for humans even if industries were to grow in emerging economies.

This contribution builds on the publication OECD (2018), "Industrial robotics and the global organisation of production", OECD Science, Technology and Industry Working Papers, No. 2018/03, OECD Publishing, Paris, https://doi.org/10.1787/dd98ff58-en. The opinions and arguments expressed herein do not necessarily reflect the official views of the OECD or its member countries. Koen De Backer is Head of Division and Tim DeStefano is Economist, both in the Structural and Industry Policy Division of the Directorate of Science, Technology and Innovation, OECD. The OECD, and not the Authors cited in the Contribution, retains all intellectual property in the Contribution.

K. De Backer (✉) · T. DeStefano
Division of Structural and Industry Policy, Science, Technology and Innovation (STI), OECD, Paris, France
e-mail: Koen.DEBACKER@oecd.org

© The Author(s) 2021
J. von Braun et al. (eds.), *Robotics, AI, and Humanity*, https://doi.org/10.1007/978-3-030-54173-6_6

Keywords

Robotics · Economy · Manufacturing industry ·
Employment · Outsourcing · Global value chains

Introduction

Industrial robotics have become an integral component in the
production processes of many firms around the world. The
growing investment in robotics is one aspect of the increasing
digitalisation of economy and society which is fundamentally
changing the outlook of manufacturing industries across the
globe. Governments in OECD and emerging economies are
betting heavily on robotics to safeguard the competitiveness
of their manufacturing industries, frequently supported by
(direct) government support for research and adoption of
robotics.

The rising prominence of robotics—and the digital
(r)evolution more broadly—is increasingly attracting the
attention of policy makers because of its possible effects
on national economies. While high on the policy agenda in
OECD economies, the impacts of new digital technologies
are however uncertain and not well understood in general.
The economic literature has recently begun to consider the
role of robotics in modern economies but the (empirical)
analysis remains overall limited. One exception is the rapidly
growing number of studies discussing the employment
impacts of robotics. Indeed, the discussion around robotics
has centred especially on the implications of labour
markets—especially in developed economies—as robots
are expected to have a major impact on employment.

There is a widespread concern that new technologies
might destroy a large number of jobs and cause "technological unemployment". Some economists believe that many of
the jobs today will be undertaken by robots in the coming
years (Brynjolfsson and McAfee 2014; Ford 2009).[1] Robots
are now capable of replacing a host of routine tasks performed within the firm and as their capabilities improve their
ability to carry out non-routine tasks will increase. As for
the types of jobs thought to be the most at risk to industrial
robots and automation, these include blue collar jobs and
routine occupations, while the groups of employees who are
most at risk of wage decline or job lose are low-skilled males
(Graetz and Michaels 2015; Acemoglu and Restrepo 2017;
and Frey and Osborne 2017). But the empirical literature
has produced mixed results until now. For example, while

Acemoglu and Restrepo (2017) do find a negative impact
of robots on employment and wages, Graetz and Michaels
(2015) find only marginal effects on hours worked. Also
OECD (2016) has concluded that the effects of robots on
employment may be significantly smaller than what others
have projected.

In addition to potential employment effects, some empirical work has analysed the effect of robotics on (labour)
productivity. Graetz and Michaels (2015) report for a panel
of 17 countries over the period 1993–2007 a positive impact of robots on labour productivity as well as total factor
productivity—and thus economic growth. In discussing the
future of robots, also BCG (Sirkin et al. 2015) expects a
strong rise of productivity as a result of the wider adoption
or robotics in manufacturing.

But the potential effects of robots on national economies
go further than employment and productivity effects. Increased robot use, fuelled by the continuous decline in the
price of robots and the increased dexterity of machines, can
be expected to increasingly impact existing/future production
technologies and the organisation of production within so-
called Global Value Chains (GVCs). In economic terms,
robots can be considered as a close substitute for lower-
skilled labour and a complement to higher-skilled labour.[2]
Significant investments in robotics will alter relative factor endowments and thus factor costs in countries and this
will/may change the location of production.

Robotics may limit the offshoring to emerging economies
and promote the reshoring of activities back to OECD
economies (De Backer et al. 2016; Dachs and Zanker 2015).[3]
Increased automation and robotics will overall decrease the
importance of labour costs in total production costs, hence
making the (re-) location of productive activities in OECD
economies (again) more attractive. This is exacerbated by
the fact that the gap in hourly wages between emerging and
developed economies is decreasing and robots continue
to become more economical. Differences in the cost of
production between developed and emerging markets may
thus narrow further, encouraging firms to restructure their
global activities (Finley et al. 2017; De Backer and Flaig
2017).

---

[1] Widely cited work by Frey and Osborne (2017) suggests that potentially 47% of US employment may be threatened by computerisation and
automation in the future. Comparable exercises have produced similar
results for other countries, all pointing to a significant to large impact of
robots and automation.

[2] The current systems of robotics replace primarily the "routine" activities of lower-skilled labour while higher-skilled labour is (still)
needed to handle, monitor and if necessary to intervene, the machines.
Future robots will be more autonomous and self-learning, potentially
diminishing the complementary character towards higher-skilled labour.
[3] De Backer et al. (2016) reported that the aggregate evidence on
reshoring is until now rather limited which stands somewhat in contrast
to the anecdotal and survey evidence on this new phenomenon. Also
the Eurofound report "European Monitor of Reshoring Annual Report
2016" notes that reshoring is a relevant phenomenon in the EU, not
decreasing in size but that further data and research are needed to
confirm whether it is growing.

This may be particularly pertinent for firms in developed countries who previously have offshored jobs to developing countries to benefit from lower labour costs (Lewis 2014; UNCTAD 2016). While mostly anecdotal, there are a host of examples of *botsourcing* (i.e. firms building new factories in the home country which are based on highly automated production plans) including Philips and Parkdale (Clifford 2013; Markoff 2012). Adidas recently opened a shoe factory in Germany called a Speedfactory with the objective of getting new shoe designs to consumers faster (Box 1).

This paper uses historic data on robots investments across industries and countries to deepen the analysis and to study the specific effects of robots on the location of production. In the following section, the paper provides a short overview of the changing location of production, which seems to pick up especially in recent years with GVCs drastically changing in nature. The following section provides empirical evidence on the growing importance of industrial robotics in today's manufacturing. The Impact of Robotics on the Global Location of Production section then analyses the links between these growing investments in robotics and the changes in production location across countries and industries; specific attention is paid to recent trends in offshoring and reshoring as well as the reallocation of resources within Multinational Enterprises (MNEs). One important aspect that has to be kept in mind is that the robotics revolution is (only) in its beginning stages, meaning that the potential impacts are just starting to emerge. The findings in this paper have to be interpreted in the light of this and most likely only signal more important effects to emerge once the robotics revolution is taking place.

Box 1 Speedfactory Adidas

Following the decision of one of the company's major competitors, Nike, to produce shoes through a robotised system in the United States, the world-known sport footwear German company Adidas decided to adopt a similar strategy by bringing production back from Asia to Germany. The first robotised plant has been opened in Ansbach in Southern Germany while the company also plans to establish a Speedfactory in the United States in 2017. Together, both factories are expected to produce a minimum of 1 million pairs of shoes every year. About 160 new jobs are expected to be created at the German plant, mostly highly skilled labour to maintain the robots. Bringing production back from China and Vietnam will help the company to offset long shipping times but also the rising cost of labour in some Asian countries. More importantly, it will also help Adidas to meet the demand for rapid innovation in designs and styles. Based on the current

supply chain model, the average industry time for shoes to reach consumers (from design to delivery) takes 18 months (unless transported by plane). Within a Speedfactory, however, the use of robots and 3D printers enables shoe components to be produced and assembled in-house, reducing the expected delivery time (from virtual design to a store shelf) to less than a week (The Economist 2017).

The Changing Location of Production

The Rapid Growth of GVCs in the 1990s and 2000s

The rapid growth of GVCs has been an important driver of globalisation during the past decades. Because of successive rounds of trade (and investment) liberalisation and rapid progress in Information and Communication Technologies (ICTs), it became easier—and profitable—for companies to offshore activities over large distances. Production processes have become more geographically dispersed as companies increasingly locate different production stages across different countries through a network of independent suppliers and their own affiliates. Within these GVCs, intermediate inputs such as parts and components are produced in one country and then exported to other countries for further production and/or assembly in final products. As a result, production processes have become internationally fragmented and products increasingly "made in the world".

After their explosive growth during the early 2000s, GVCs have gradually become the backbone of the global economy and dramatically changed its functioning. The result is that global production nowadays spans a growing number of companies, industries and countries and a number of emerging economies have become economic powerhouses because of GVCs. The large flows of goods, services, capital, people and technology moving across borders within these international production networks have resulted in a growing interconnectedness between countries (OECD 2013).

GVCs have become longer and more complex since their emergence in the 1980s. Production stages of a growing number of goods—more traditional products like textiles as well as more technology-intensive products like, e.g. electronics—and increasingly also services are spread out across a multiple of locations. This in turn has resulted in growing trade and transport flows over time. The organisation of production in long and complex GVCs to take advantage of optimal location factors for specific stages of production across the globe has shown its advantages for companies in terms of productivity, efficiency, scale economies, etc.

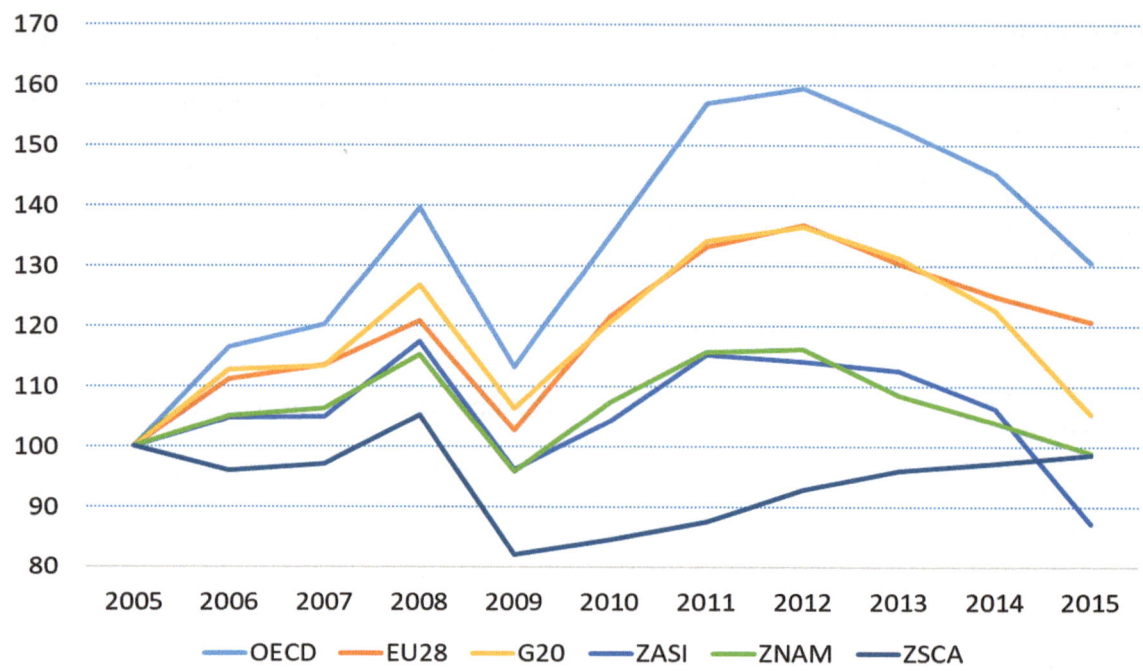

Fig. 1 Foreign value added of exports (index based on 2005). Source: De Backer et al. OECD—2018 TiVA database

The global dimension of economic integration has expanded rapidly as emerging countries such as the People's Republic of China, India and Brazil have become major players in the global economy, in part because of their increasing participation in GVCs. Numerous activities especially in manufacturing have been relocated to emerging economies as sourcing inputs from low-cost producers, either within or outside the boundaries of the firm, lowered production costs. The emergence of GVCs has allowed emerging economies to integrate in the global economy faster than in the past. Countries are able to specialise in a certain production activity according to their comparative advantage instead of developing the whole value chain itself (OECD 2013). Participation in GVCs is often viewed by governments as a fast track to industrialisation and strategies to increase the participation within GVCs are essentially part of economic development policies.

The emergence and growth of GVCs have been largely documented in recent years due to new evidence. Until recently, the empirical evidence on GVCs mainly consisted of case studies of specific products; early examples include the Barbie Doll (Tempest 1996; Feenstra 1998) and an average "American" car (WTO 1998). While these case studies offered interesting anecdotal evidence, more stylised analysis of the geographical distribution of costs, value added and profits has subsequently received a great deal of attention, in particular the well-known studies of Apple products (Linden et al. 2009; Dedrick et al. 2010). Afterwards, the OECD developed new aggregate—at the industry, national and global level—measures[4] and has documented the rapid growth of

these global production networks. OECD (2013) discussed this new empirical evidence in detail and analysed the important implications of these GVCs of different policy domains (including trade, investment, competiveness, etc.).

The End of GVCs in the Future?

In contrast to the ever-growing length and complexity of GVCs in the past, the international fragmentation of production appears to have lost momentum in recent years. A (limited) consolidation of GVCs had already been observed during the economic crisis in 2008/2009, but trade as well as GVC growth picked up again thereafter. But new OECD results show that the foreign value added of exports—which is largely considered to be one important indicator of GVCs, see OECD (2013)—shows a downward trend (Fig. 1). The recent trade and GVC slowdown is increasingly argued to stem from more structural determinants—in addition to more conjunctural factors—suggesting that a new normal of GVCs may be at the horizon. Illustrating this is the growing popularity of the concept of "peak trade".

As the continuous expansion of GVCs may (have) come to an end, concerns about future production, trade and economic growth are rapidly rising in countries. Trade has traditionally been an important driver of productivity and economic growth of economies. One question that is increasingly raised in discussions on (the future of) globalisation is whether the relationship between trade and GDP growth has been undergoing a structural shift in recent years. It is clear that a slowdown of trade within GVCs will rapidly have consequences for the global economy. Likewise, a new era of GVC

---

[4]See OECD's Trade in Value Added Database. A similar initiative is the World Input-Output Database.

dynamics will most likely result in a re-configuration of the international production landscape with significant shifts in competitiveness between regions and countries.

Different factors will (simultaneously) determine the future of GVCs as discussed in detail in De Backer et al. (2018) simulating how production and trade within GVCs may look like in 2030. On the one side, there are factors that have promoted the growth of GVCs in the past and these are expected to facilitate the future expansion of GVCs; thus, business as usual. New communication technologies (i.e. the "C" in ICTs) which allow for the coordination of activities across large distances, the integration of emerging economies because of their low (labour costs) in manufacturing, the growing middle class in emerging economies which gradually changes (consumer) demand and results in fast growing end-markets in these countries, the efficient provision of "modern" services (telecommunications, logistics, business services, etc.) which are the glue that ties GVCs together and the rapid growth of MNEs are all major reasons why GVCs have expanded significantly since the 2000s.

On the other side, there are other factors that push for "a new normal" of GVCs; these include old and new factors, i.e. factors which are known to negatively impact GVCs but also emerging factors of which the possible effects on GVCs are less known. These factors are expected to increasingly challenge the organisation of production in longer and complex GVCs and may shape the future evolution of GVCs differently. Strongly rising wage costs in (some) emerging economies and the growing digitalisation of production because of new information technologies (i.e. the "I" in ICTs: robotics, automation, artificial intelligence, etc.) are expected to restore the competitiveness of developed economies and discourage further offshoring to emerging economies. In addition, rising transport costs, the hidden and extra-costs of offshoring including the problems in protecting proprietary knowledge abroad, the growing need to balance cost efficiency with risk diversification which comes at a cost, will make the international dispersion of production more expensive. Also future extra costs arising from policy efforts to internalise the environmental costs of (international) transport may make the international trade of goods and services more costly and thus affect the further offshoring of activities within GVCs.

Further on, the current organisation of production in long and complex GVCs has made companies less responsive to changes in customer demand while at the same time product customisation is becoming essential for firms to maintain a competitive edge. Some have argued that a shift from mass production to mass customisation is happening, hence the need for companies to be able to quickly respond to market signals.

Using different scenarios that mirror the most likely evolution of the different factors, the results in De Backer et al. (2018) indicate that the future of GVCs may indeed look quite different from the past. Dramatic decreases in international sourcing and intermediate trade are observed for the future up to 2030 with GVCs regressing back to their 2005 levels. The growing digitalisation of production will be most likely the biggest game-changer in this process, reversing the importance and length of GVCs and reorienting global production and trade back towards OECD economies.

Reshoring Instead of Offshoring?

Within this changing topography of GVCs and global production, reshoring of activities is expected to become increasingly attractive especially when these activities can be highly automated through the increased use of robots. A growing number of (media) reports seem to indicate that manufacturing companies in OECD economies are increasingly transferring activities back to their home country (back-shoring) or to a neighbouring country (near-shoring). This stands in sharp contrast with the large offshoring of activities and jobs particularly in manufacturing away from developed economies over the past decades.

Policy makers in developed economies are banking on this and hope that reshoring will bring back jobs in OECD manufacturing. But within this ongoing debate on reshoring, considerable disagreement exists about how important this trend actually is and may become in the future. Some predict that reshoring will become a fundamental trend in the early twenty-first century, while more sceptical voices point to the overall small number of companies that have actually brought activities and jobs home. Indeed, while company surveys and anecdotal evidence suggest the growing importance of the reshoring trend, the more aggregate evidence indicates that the effects on national economies are (still) limited and only very recent (see for example De Backer et al. 2016). For example, claims that reshoring will result in a large number of extra jobs at home have not been received much empirical support. One reason is that reshored production is often highly automated through the use of robotics, meaning that only a limited number of additional jobs are created and that these jobs will increasingly be high-skilled.

The evidence also seems to indicate that the phenomenon of reshoring does not necessarily mean the end of offshoring nor that it will bring back all the activities that have been offshored during the past decades and restore manufacturing in OECD economies back to its level of the 1970s or 1980s. The evidence at the company and industry level demonstrates that offshoring is still taking place at the same time that reshoring is picking up. Further on, that same evidence tends to suggest that offshoring is still more important and larger today. Offshoring is still an attractive option since proximity to markets is a major reason for international investment:

the large size and the strong growth of emerging markets is an important explanation for the attractiveness of emerging economies.

But after years of large-scale offshoring and outsourcing, it becomes clear that companies increasingly opt for more diversified sourcing strategies including reshoring and consider more options in structuring their production processes. In addition to global hubs in GVCs, production is expected to become increasingly concentrated in regional/local hubs closer to end markets both in developed and emerging economies. For some products low (labour) costs and long value chains will continue to form important competitive advantages for some time, but for other goods and services production will become increasingly organised at the more regional level.

Growing Investment in Robotics

During the last decades, robots have become increasingly prominent in manufacturing industries with parts of—and in some cases, complete—production processes automated. The robotisation of manufacturing first took off in OECD economies as it helped to compensate for high and rising labour costs and safeguard international competitiveness (Fig. 2[5]). But in more recent years, strong robotics investment

can also be observed in several emerging economies, often supported by their governments, as part of their industrialisation and development strategies. The growing robotisation is part of the broader trend of the digitalisation of manufacturing, with new digital technologies expected to radically change the outlook of industries.[6] In combination with important advances in a number of complementary technologies, greater adoption of robotics is believed to be a key element in fostering "a next production revolution" (OECD 2017).

While costs of hardware and enabling software are expected to fall further, the performance of robotics systems will vastly improve. BCG (Sirkin et al. 2015) estimates that the cost of robots will decrease by 20% and their performance improve by around 5% annually over the next decade. Robots which are already widely used across manufacturing industries are rather suited for repetitive activities and very precisely defined environments. While some robots are equipped with on-board sensors, most of their movements are still preplanned and programmed. Machines are however expected to become more flexible due to the progress of artificial intelligence, self-learning and auto-correcting capabilities,

---

[5]The data on robots in this paper are sourced from the International Federation of Robotics (IFR) and relate to the number of robots (no information is available on the size and the quality of robots). IFR defines an industrial robot as "an automatically controlled, reprogrammable, multipurpose manipulator programmable in three or more axes, which can be fixed in place or mobile for use in industrial automation applications" (IFR 2016b).

[6]OECD (2017) has distinguished three broad technologies underpinning the digitalisation of production; the Internet of Things (IoT), which enables the interconnection of machines, inventories and good; big data and embedded software which allow for the analysis of the huge volumes of digital data generated by these objects; and cloud computing which provides the ubiquitous availability of computing power. The uptake and growth of (industrial) robots or autonomous machines within sectors will result from the conjunction of these different technologies and applications.

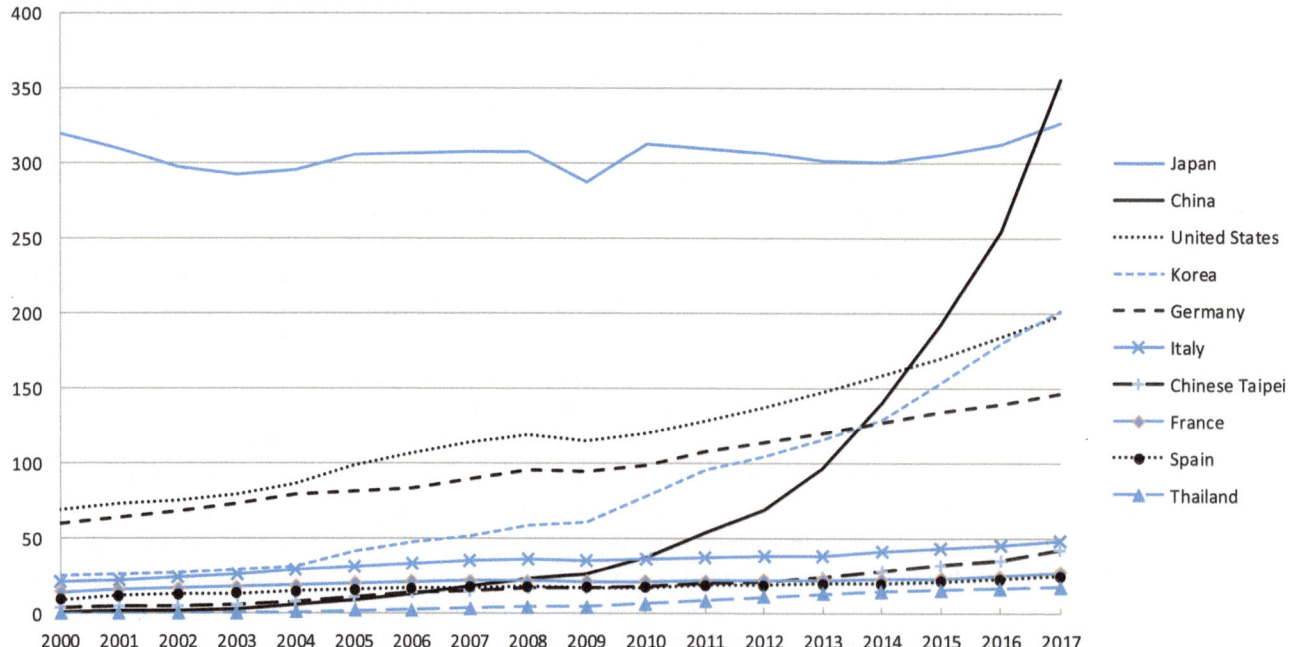

Fig. 2 Investments in industrial robots (units), by country, 2000–2017. Source: International Federation of Robotics (2016a)

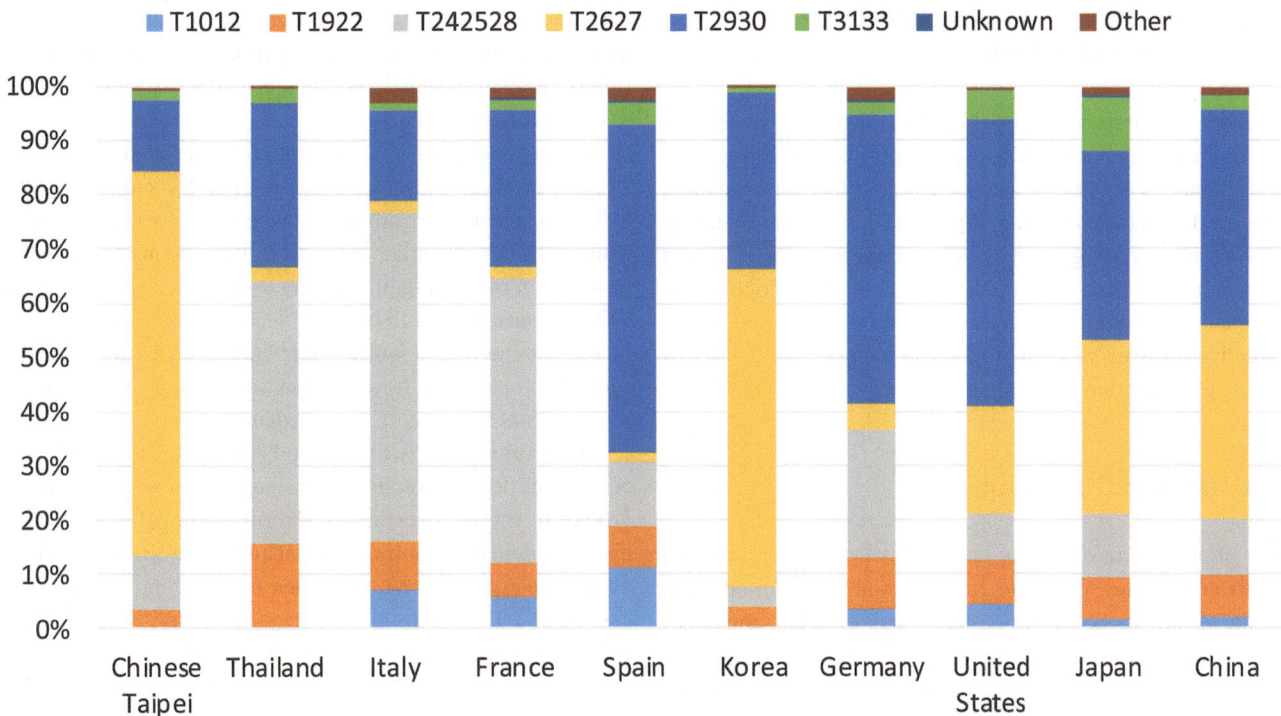

Fig. 3 Robot purchases by sector amongst highest users, 2017. *Note*: Sectors classified by ISIC version 4: T1012: food, beverage and Tabacco, T1922: petroleum products, chemicals, rubber and plastics, T242528: basic metals and machinery, T2627: computers and electrical machinery, T2930: motor vehicle and transport, T3133: other manufacturing and recycling, Other: to wood, paper and publishing non-metallic minerals and education and research. *Source*: Own calculations based on International Federation of Robotics (2016a)

allowing them to perform a wider range of complex activities. As a result, machines will be increasingly able to do many tasks more precisely, cheaper and faster.

Robots are highly concentrated in a limited number of industrial sectors (Fig. 3): the majority of robot use (roughly 70%) is concentrated within transport equipment, computers and electronics and chemical and mineral production and food and beverage production. For most economies, the transportation equipment sector is the largest user of industrial robots out of the economy. The deployment of robots in industries is generally dependent on a number of factors. First, technical requirements of the production process clearly determine the (further) usage of robots as some activities and jobs are easily automatable while others are not. Certain production tasks can only be replaced by very advanced robotics systems with a broad range of functions, which make them (too) expensive to implement. However, things are expected to change over time with higher performing robots—at a lower cost—being able to automate a growing number of activities and tasks.

Second, manufacturing industries in which labour costs account for a large(r) share in total production costs are more likely to invest in robotics because robots allow to save on labour and thus costs. But, third, location is another important determinant mediating this link between labour costs and wider robotics use. Industries located in emerging economies where labour costs are typically lower are less likely to adopt robots and automation compared to more developed (and thus higher labour cost) economies. The inflection point where robots become more cost efficient than human labour lies— cetris paribus—further in time for emerging economies. Interestingly however, some emerging economies are observed to heavily invest in robotics as a deliberate (government) strategy in order to compensate for their fast rising labour costs and/or to compete with the growing robotics manufacturing deployed in developed economies.

The high concentration of robots in sectors like transport equipment including automotive and electronics demonstrates the high stakes at play for emerging economies. The strong export position of these economies is largely based on their low labour costs in these industries. Strong investments in robots in developed economies may quickly result in the erosion of such a competitive advantage and make these activities exposed to reshoring of activities to developed economies. In other industries where emerging economies also benefit from their lower labour costs—e.g. garment and textiles—robots have not made a big inroad and are not expected to do so quickly.

The Impact of Robotics on the Global Location of Production

In order to address new opportunities and challenges following digitalisation, companies are reorganising their domestic and global production processes and adopting new business models. In what follows, the impact of robotics on offshoring as well as reshoring is separately analysed, as well as the broader reallocation of resources within MNEs.[7] Robots with a greater efficiency at lower costs may make it more cost-effective for firms to keep production in the home country or to move production back or close there.

Robotics and Offshoring

The offshoring to emerging economies in the past has been motivated by the search for lower labour costs, in addition to the desire to be closer to large and growing consumer markets. Attracted by the large labour force in these countries, companies in developed economies—typically characterised by higher labour costs—relocated (labour-intensive) activities during the past decades. Since robotics allow for labour cost savings, it can be hypothesised that this cost motive of offshoring from developed economies will become less pressing. In combination with the rising wage costs in (some) emerging economies[8] and persisting productivity differentials across countries, robotics are helping to lower the difference in the cost of production between developed and emerging economies. As robotic manufacturing becomes cheaper and offshoring more expensive, the cost/benefit analysis changes: instead of relocating activities away from home and sourcing inputs from abroad, production at home will increasingly become a viable/profitable alternative for companies.

Yet investments in robots are not only about saving on labour costs. Robotics are important tools to allow for more customised manufacturing, especially when artificially intelligent robots will increasingly become adaptable, programmable and autonomous. Industries in which market demand and consumer preferences change quickly have a lot to benefit from the usage of robots, compared to the alternative of offshoring—often far away—where suppliers do not al-

ways produce according to the right specifications, resulting in quality issues and long delivery times. The deployment of robots can therefore help companies get new products to the market much quicker.

The hypothesis thus is that larger robot usage increases the attractiveness of (developed) economies for manufacturing activities and as a result may reverse the past/current offshoring trends. In order to analyse the potential effects of robotics on offshoring, the widely used Feenstra and Hanson (1996) measure of offshoring[9,10] is related to investments in robotics across countries and industries over time. As such, the focus is on how robotics may change the extent of international sourcing of intermediates.

The results for the whole period 2005–2016 for developed economies do not directly seem to suggest a link between the growth in robots investments and offshoring (for a detailed discussion on the empirical model and variable construction, please see Appendix 1). But when focusing on the more recent years 2010–2016—i.e. a period characterised by rapidly rising investments in robotics—the results demonstrate a negative association of robotics investments (net of depreciation) with the growth of offshoring (Table 1). Industries in developed economies that increasingly invest in robotics witness a lower growth in offshoring, i.e. a decrease in the international sourcing of intermediates. In previous work, this negative association became larger as the labour intensity of industries increases, suggesting that robotics (help) hinder/stop the relocation of especially labour-intensive activities, as they help compensate for higher labour costs in developed economies. In this new analysis based on more recent data, this interaction effect disappears which may suggest that robotics are becoming more widespread across all manufacturing industries.

Robotics and Reshoring

The previous analysis suggests that robotics slow down—and in some cases, stop—offshoring and thus help to keep manufacturing activities in developed economies. A slightly different question is if investments in robots will lead to the actual reshoring of activities to developed economies, i.e.

---

[7]The focus of the analyses is on the offshoring from and reshoring to developed economies. The group of developed economies consists of the "high-income economies" identified by the World Bank. A high-income economy is defined by the World Bank as a country with a gross national income per capita US$12,236 or more in 2016.

[8]Nevertheless, rising wages have to be considered in combination with rising productivity. Further on, activities may be offshored from economies with rising wage costs to other emerging economies where wages are much lower and not rapidly increasing (for example from China to Cambodia).

[9]Feenstra and Hanson called this measure an indicator of "outsourcing" although the indicator actually measures offshoring since it is defined as companies' purchases of intermediate goods and services—excluding energy intermediate inputs—from foreign providers (at arms-length and from foreign affiliates).

[10]The indicator has been calculated on the basis of OECD TiVA data which are up-to-date until 2016. In addition, a number of control variables such as labour intensity, demand and absorptive capacity (measured by patent stock) are included. These data are sourced from UNIDO, TiVA and PATSTAT databases, respectively. After merging these datasets with the statistics on robotics, the sample includes roughly 40 countries over the period 2000–2016.

Table 1 Effects of growth in robotics on the growth of offshoring

	HDC	
Dependent var: Offshoring (annual growth)	2005–2016	2010–2016
Robot stock (annual growth)	−0.001	−0.013***
	0.00	0.00
Robot stock *Labour Intensity	−0.471	−0.233
	−0.35	−0.25
Labour Intensity	0.236	0.292*
	−0.27	−0.17
Patent Stock	−0.002	−0.003
	0.00	0.00
Demand	−0.007	−0.014
	−0.03	−0.02
Control Variables		
Year	✓	✓
Country*Industry	✓	✓
R-squared	0.245	0.214
Observations	4,635	2,897

Note: Robust standard errors in parenthesis. Level of significance
***$p < 0.01$, **$p < 0.05$, *$p < 0.1$
Source: Own calculations

bringing activities that were once offshored back home. De Backer et al. (2016) analysed the (re-)allocation of resources abroad and at home within groups of MNEs based on firm-level data and found some evidence of reshoring in terms of capital investments but not in terms of employment. One reason put forward for these trends was that robotics are very capital intensive investments, but at the same time labour-saving. This also explains why the employment impact of reshoring in developed economies is often rather limited, and does not lead to the manufacturing renaissance of (certain) OECD economies as some have advocated.

This paper extends the empirical firm-level analysis as in De Backer et al. (2016) and includes robotics investments at the country-industry level (please see Appendix 1 for a detailed discussion on the empirical model and variable construction). The idea is to check if within groups of MNEs a transfer of productive resources (i.e. fixed capital and employment) takes place from the affiliates abroad to the headquarters and affiliates at home[11] because of robotics investments. Interestingly, the results—now thus including more recent years—for the group of developed economies do show indications of backshoring in terms of employment over the most recent period 2010–2016 (Table 2). A negative change in aggregate employment abroad is associated with a positive employment growth in the home country within

the same business group, which thus give some support to the claims about the growing importance of backshoring in recent years.

Augmenting the model with robots investments—by interacting the negative/positive growth abroad with robots investments in order to see if there is more reshoring to home countries where robots investments are relatively more important[12]—does however not get support from the model. No extra effect is observed from robotics on the reshoring of productive resources to developed home countries, neither in terms of fixed capital or employment. Also the recent reshoring seems not be driven by investments in robotics, i.e. home countries investing strongly in robotics do not witness a stronger reshoring of jobs in 2010–2016. All in all, while robotics seems to have a negative effect on the pace of offshoring, the results suggest that robots do not (yet) trigger a reshoring of activities to developed economies.

Robotics and the Reallocation of Resources within MNEs

The reallocation of resources within networks of MNEs is not a two-way street between affiliates abroad and the head-quarters at home but instead happens between all affiliates mutually. By broadening the analysis beyond the reshoring of activities of MNEs to home countries, the potential effect of robotics on the total reallocation of resources across affiliates within groups of MNEs can be analysed. Because of their international networks, MNEs have a large (strategic and operational) flexibility in moving resources from one place to the other. The labour savings and increased production flexibility from robot use enable MNEs to shift production to other countries, for example, by locating production closer to the end customer so they can tailor and design goods based on the changing interests of the local market. Firms with plant facilities in multiple countries can quickly transfer product designs through CAD/CAM files between teams of robots making it easier for firms to shift production needs throughout the world. Greater advances in cloud computing and machine to machine communication will also facilitate real-time learning between robots in different locations and increase the responsiveness and efficiency of production (PWC 2014).

The objective of the next empirical analysis thus is to analyse if the (re-)allocation of productive resources within

---

[11]In order to check for this, the evolution of productive resources in affiliates abroad is split out in a positive and negative component (i.e. a negative coefficient for the negative growth abroad lends support for reshoring); see for more details De Backer et al. (2016).

[12]The robotics variable is constructed as the growth in robots stock of the home country relative to the average of the growth in robots stock in the countries where the group has affiliates; this in order to take into account the different geographical spread of MNE groups. Of course, this is only an indirect and rough proxy of the potential effect of robotics since the robotics is constructed on the country-industry level while the allocation of resources is based on firm-level data.

Table 2 Effects of growth in robotics on backshoring within MNEs

Dependent variable	2003–2016				2010–2016			
	Yearly growth rate				Yearly growth rate			
	Fixed assets		Employment		Fixed assets		Employment	
Growth abroad pos	0.00466	0.00543	−0.00210	−0.00841	0.00325	0.00392	0.00673	−0.000869
	(0.0131)	(0.0119)	(0.00874)	(0.00720)	(0.0130)	(0.0140)	(0.00700)	(0.00933)
Growth abroad pos* robot stock		−0.0293		0.00982		0.0208		0.0201
		(0.0240)		(0.0383)		(0.0667)		(0.0265)
Growth abroad neg	−0.00272	0.00203	−0.0145	0.0189	−0.00296	−0.00474	−0.0204**	−0.0234**
	(0.0364)	(0.0338)	(0.0154)	(0.0214)	(0.0162)	(0.0189)	(0.00850)	(0.00922)
Growth abroad neg* robot stock		0.182		0.0354		−0.0278		−0.0674
		(0.136)		(0.154)		(0.0795)		(0.0482)
Growth in robot stock		−0.00757		0.00114		−0.00955		−0.000614
		(0.00537)		(0.00363)		(0.00759)		(0.00494)
Ave growth group 2003–2015	0.701***	0.737***	0.830***	0.837***	0.723***	0.726***	0.806***	0.810***
	(0.123)	(0.107)	(0.0250)	(0.0262)	(0.125)	(0.125)	(0.0299)	(0.0314)
Control variables								
nace2-year FE	✓	✓	✓	✓	✓	✓	✓	✓
country FE	✓	✓	✓	✓	✓	✓	✓	✓
Observations	75,384	72,124	91,555	87,260	47,713	45,697	62,467	59,494
R-squared	0.029	0.031	0.029	0.029	0.030	0.031	0.028	0.028

Note: Robust standard errors in parenthesis. Level of significance ***$p < 0.01$, **$p < 0.05$, *$p < 0.1$
Source: Own calculations

MNEs shifts relatively more towards countries that invest (more) heavily in robotics. Similar to the previous analyses on offshoring and reshoring, the discussion focuses on how robotics may increase the attractiveness of countries for manufacturing activities, but now in particular for investment by (foreign and domestic) MNEs. The same firm-level information on fixed capital and employment as in the previous analysis is used but now at the level of individual affiliates of MNEs, again linked to robotics stock in the country of the affiliate[13] (see Appendix 1 for details on the empirical model and data construction).

The results in Table 3 lend some support for the hypothesis that the reallocation of resources across affiliates of MNEs is driven in part by investment in robotics. A positive correlation of robot investment on employment growth is observed for MNEs with headquarters in developed economies. In particular, the results suggest that affiliates located in economies with strong growth in robotics investments, relative to the average group growth of the MNEs, seem to attract larger resources in terms of jobs. However, there is no effect on capital investment, and the employment effect disappears when focusing on only the more recent years 2010–2016.

Conclusion and Policy Implications

The analyses in this paper demonstrate that robotics may impact the location of production within GVCs. The most important finding is the negative effect that robotics may have on the offshoring of activities from developed economies—i.e. robotics seem to decrease the need for relocating activities away from developed economies. Yet, while robotics may thus keep production activities in developed economies, these same investments in robots do not seem strong or large enough to bring back activities that have been offshored in the past. In addition, there is mixed evidence about robotics being a major factor in attracting international investment, when analysing the effect of robotics on the international reallocation of resources within MNEs.

There are a number of reasons that may help explain the rather limited evidence found in this paper. First and foremost, it may be too premature to observe the potentially disruptive effects of robotics on the location of production. Investments in robots have grown significantly indeed but

may have done so only recently and it can be expected that potential impacts will take some to materialise. If a robotics revolution is about to happen like some are arguing, one response this paper calls for is the need for further and follow-up research in the coming years.

Second, while information on robotics has become increasingly available across industries and countries including emerging economies, it should be taken into account that the available data only include information on the number (i.e. a count) of robots. Regretfully, no information is available on the size and especially the growing quality of robots—one can easily assume that new vintages of robot investments have a higher performance—but this is not reflected in the data.

Third, robots are only one part of the wider digital revolution that is currently taking place. Other developments including the Internet of Things, artificial intelligence, etc. will additionally contribute to the digital revolution and, consequently, it can be expected that companies will need to invest in complementary assets to fully benefit from their investment in robotics. The data in this paper do not include information on these other components of the digitalisation of manufacturing, which may mean that the effects of robots are somewhat underestimated.

The negative effect of robotics on offshoring that is found in this paper seems to be in line with the distinction made by Baldwin (2016) who argued about the differential effects of respectively communication and information technologies on the international fragmentation. The rapid progress in communication technologies has been one factor behind the rapid growth of GVCs in the past as these technologies allow for the monitoring and coordination of production activities across large distances. Information technologies, including robots investment, instead may curb the further international dispersion of activities and may make developed economies more attractive (again) for manufacturing activities. One reason is that information technologies reduce the share of labour costs in total production costs. A second reason is that information technologies allow companies to customise their products better and much faster, which is increasingly becoming important in a number of industries.

The effect of robotics on offshoring in developed economies, based on historical data, also supports the results of De Backer and Flaig (2017), who simulated the future of GVCs based on a number of scenarios for the future. They reported that one specific scenario, namely the rapid progress in information technologies including robotics, will increase the attractiveness of OECD economies for manufacturing activities. This would be reflected in a decreased sourcing of intermediates from abroad, lower levels of offshoring and a stronger export position of OECD manufacturing in global markets.

[13] The robotics variable is now constructed as the growth in robots stock of the affiliate country relative to the average of the growth in robots stock in the countries where the group has affiliates; as in the previous analysis on backshoring, this is done in order to take into account the different geographical spread of groups of MNEs. Again, this is only an indirect and rough proxy of the potential effect of robotics (country-industry level of robotics versus firm-level performance).

Table 3 Effects of growth in robotics on the reallocation of resources within MNEs

	2003–2016		2010–2016	
	Employment growth	Fixed asset growth	Employment growth	Fixed asset growth
Change in affliate stock/ave change affiliates	0.006**	−0.005	0.000	−0.01
	0.00	−0.01	0.00	−0.01
Employment growth of group	0.117***		0.107***	
	−0.01		−0.01	
FA growth of group		0.134***		0.134***
		−0.01		−0.01
Control variables				
Sector*year				
Country*year				
R-squared	0.038	0.013	0.029	0.011
Observations	373,287	336,542	224,312	192,516

Note: Robust standard errors in parenthesis. Level of significance ***$p < 0.01$, **$p < 0.05$, *$p < 0.1$
Source: Own calculations

The evidence in this paper albeit preliminary—and the need for follow-up analysis using more recent data cannot be emphasised enough—seems to indicate that robotics may contribute to manufacturing competitiveness. Most OECD countries see digital production as a new source of competitiveness in restoring their position in manufacturing industries. This is reflected in the growing number of policy measures that several OECD countries have implemented or are in the process of implementing.

Faced with these economic challenges, also governments in emerging economies have identified robotics as a key determinant of their future economic success. The growth model of emerging economies which is largely based on low labour costs will come under increasing pressure if developed economies increasingly automate their manufacturing and are successful in keeping/attracting production activities at home. In a number of emerging economies, labour costs have been rising quickly which makes the option of automation and larger robot usage—at home or abroad—increasingly attractive, especially for companies who have offshored activities to these countries in search of low wages.

The likely outcome will be different across industries and countries. But, as robotics are getting cheaper and becoming more performant also in the number of activities they can be applied to, it can be expected that the so-called inflection point for investment in robots (BCG estimates this inflection point to be reached when using robots becomes 15% cheaper per hour than employing human staff) will come closer in a growing number of industries and countries. This process of growing automation, in developed as well as emerging economies, may further stimulate the process of premature de-industrialisation that has been observed recently in a num-

ber of economies. While the traditional model of economic development involves a shift from agriculture over manufacturing to services, Rodrik (2015) reported that a number of emerging economies are de-industrialising quicker than expected. The fact that automation—in developed economies but increasingly also in emerging economies—may increasingly replace low-cost jobs risks that manufacturing will not be able to offer large numbers of jobs for the large labour supply in these countries.

Appendix 1: Empirical Strategies and Variable Descriptions

Robot Stock and Offshoring

The model used to estimate the effects of robot stock on offshoring is illustrated in Eq. (A.1):

$$offshore_{ict} = robot_{ict} + robot * labint_{ict} + labint_{ict}$$
$$+ Absorb_{ict} + Demand_{ict} + \vartheta_i + \varepsilon_{ict}$$
$$(A.1)$$

offshore reflects the annual growth in offshoring across sector i in country c in time t.[14] Offshoring is defined as the share of imported intermediate goods and services—excluding energy intermediate inputs over the sum of intermediate goods and services—excluding energy intermediate inputs (Feenstra and Hanson 1996). Our variables of interest

---

[14]Growth is calculated as the log difference of the variable over the specified period.

robot and robot * labint capture the effects of the growth in robot stock and robots conditional on labour intensity on offshoring. We also include a number of control variables believed to be related to offshoring including growth in absorptive capacity (measured by patent stock) Absorb, and growth in Demand of the country and sector over time. Country*sector and year fixed effects are also included to control for economy industry as well as year variation over time.

Robot Stock and Backshoring

In Eq. (A.2) employment or fixed assets growth rate at home of company i part of group g in year t is regressed over the aggregate growth rate of the same variable of affiliates abroad. The growth abroad variable abd is split in two variables abd _ p and _n, depending on whether it takes a positive or negative value, respectively. Doing so allows the change in growth rate at home to be different in sign and magnitude depending on whether affiliates abroad are expanding or contracting, respectively.

If there is backshoring, the relationship is expected to be negative—i.e. an increase in employment or investment at home is associated with a decrease in employment or investments abroad. However, a negative association may actually be also a symptom of offshoring, i.e. a decrease in employment or investment at home is associated with an increase in employment or investments abroad. Therefore in this kind of analysis it is important to distinguish positive from negative changes in employment and fixed capital assets.

Interacting the growth abroad variables with relative robot stock of the group over the average affiliate ($abd_p * robot$ and $abd_n * robot$) allows us to assess the extent to which robots are contributing to the backshoring or resources within the MNE. If robotics contributes to backshoring one should expect to find a negative coefficient for the $abd \_ n_{gt} * robot$ variable. In order to examine whether robot use in the home country is related to backshoring here our robot measure represents the log different in robot stock of the country and sector of the headquarters over the average log difference of the robot stock for the sectors and countries of the affiliates abroad.

$$home_{igt} = abd\,p_{igt} + abdn_{igt} + abd\,p_{igt} * robot$$
$$+abdn_{igt} * robot + grpave_{igt} + \delta_{igt} + \varepsilon_{it}$$
$$(A.2)$$

Robot Stock and Reallocation

Equation (A.3) uses the same firm-level information on fixed capital and employment as in Eq. (A.2) for the dependent variable but now at the level of individual affiliates of MNEs

signified by aff _ factor. This is regressed on robot stock of the affiliate country and sector. Control variables are also added, including growth in demand within the country c sector s and year t of the affiliate, the average growth of the group (either employment or fixed assets) signified as factor and sector*year country*year fixed effects represented by ϑ.

$$aff\,factor_{asct} = robot_{cst} + demand_{sct} + factor_{gsct}$$
$$+\vartheta_t + \varepsilon_{asct}$$
$$(A.3)$$

References

Acemoglu, D., & Restrepo, P. (2017). Robots and jobs: Evidence from US labor markets. *NBER working paper 23285*. doi: https://doi.org/10.3386/w23285

Baldwin, R. (2016). *The great convergence: Information technology and the new globalization*. Cambridge: Harvard University Press.

Brynjolfsson, B., & McAfee, A. (2014). *The second machine age: Work, progress and prosperity in a time of brilliant technologies*. New York/London: W.W. Norton and Company.

Clifford, S. (2013). *U.S. textile plants return, with floors largely empty of people. Available via The New York Times*. Retrieved February 13, 2020, from https://www.nytimes.com/2013/09/20/business/us-textile-factories-return.html.

Dachs, B., & Zanker, C. (2015). Backshoring of production activities in European manufacturing. *MPRA paper 63868*. Available via MPRA. Retrieved February 13, 2020, from https://mpra.ub.uni-muenchen.de/63868/1/MPRA_paper_63867.pdf.

De Backer, K., Desnoyers-James, I., Menon, C., & Moussiegt, L. (2016). Reshoring: Myth or reality? *OECD Science, Technology and Industry Policy Papers 27*. doi: https://doi.org/10.1787/5jm56frbm38s-en.

De Backer, K., Destefano, T., Menon, T., & Suhn, J. R. (2018) Industrial robotics and the global organisation of production. *OECD Science, Technology and Industry Policy Papers 2018/03*. doi: https://doi.org/10.1787/dd98ff58-en.

De Backer, K., & Flaig, D. (2017) The future of global value chains: Business as usual or "a new normal"? *OECD Science, Technology and Industry Policy Papers 41*. doi: https://doi.org/10.1787/d8da8760-en

Dedrick, J., Kraemer, K. L., & Linden, G. (2010). Who profits from innovation in global value chains?: A study of the iPod and notebook PCs. *Industrial and Corporate Change, 19*(1), 81–116. https://doi.org/10.1093/icc/dtp032.

Feenstra, R., & Hanson, G. (1996). Globalization, outsourcing, and wage inequality. *The American Economic Review, 86*(2), 240–245. https://doi.org/10.3386/w5424.

Feenstra, R. C. (1998). Integration of trade and disintegration of production in the global economy. *The Journal of Economic Perspectives, 12*(4), 31–50. https://doi.org/10.1257/jep.12.4.31.

Finley, F., Bergbaum, A., Csicsila, A., Blaeser, J., Lim, L. H., Sun, Y., & Bastin, Z. (2017). *Homeward bound: Nearshoring continues, labour becomes a limiting factor, and automation takes root*. Available via AlixPartners. Retrieved February 13, 2020, from https://emarketing.alixpartners.com/rs/emsimages/2017/pubs/EI/AP_Strategic_Manufacturing_Sourcing_Homeward_Bound_Jan_2017.pdf.

Ford, M. (2009). *The lights in the tunnel: Automation, accelerating technology and the economy of the future*. United States: Acculant Publishing.

Frey, C., & Osborne, M. (2017). The future of employment: How susceptible are jobs to computerisation? *Technological Forecasting and Social Change, 114*(2017), 254–280. https://doi.org/10.1016/j.techfore.2016.08.019.

Graetz, G., & Michaels, G. (2015). Robots at work. *CEP discussion paper 1335.*

International Federation of Robotics (IFR). (2016a). *Wold robotics report: Industrial robots.* Frankfurt.

International Federation of Robotics (IFR). (2016b). *World robotics report 2016: IFR press release.* Frankfurt.

Lewis, C. (2014). *Robots are starting to make offshoring less attractive.* Available via Harvard Business Review. Retrieved February 13, 2020, from https://hbr.org/2014/05/robots-are-starting-to-make-offshoring-less-attractive.

Linden, G., Kraemer, K. L., & Dedrick, J. (2009). Who captures value in a global innovation network? The case of apple's iPod. *Communications of the ACM, 52*(3), 140–144. https://doi.org/10.1145/1467247.1467280.

Markoff, J. (2012). *Skilled work, without the worker.* Available via The New York Times. Retrieved February 13, 2020, from https://www.nytimes.com/2012/08/19/business/new-wave-of-adept-robots-is-changing-global-industry.html.

OECD. (2013). *Interconnected economies: Benefitting from global value chains.* Paris: OECD Publishing. https://doi.org/10.1787/9789264189560-en.

OECD. (2016). The risk of automation for jobs in OECD countries: A comparative analysis. *OECD Social, Employment and Migration Working Papers 189.* doi: https://doi.org/10.1787/5jlz9h56dvq7-en.

OECD. (2017). Benefits and challenges of digitalising production. In: OECD (ed) The next production revolution: Implications for governments and business. OECD Publishing, Paris. doi: https://doi.org/10.1787/9789264271036-6-en.

Pricewaterhouse Coopers (PWC) (2014). *The new hire: How a new generation of robots is transforming manufacturing.* Available via PWC. Retrieved February 13, 2020, from https://www.pwc.com/us/en/industrial-products/assets/industrial-robot-trends-in-manufacturing-report.pdf.

Rodrik, D. (2015). *Economics rules: Why economics works, when it fails, and how to tell the difference.* Oxford: Oxford University Press.

Sirkin, H. L., Zinser, M., & Rose, J. R. (2015). *The robotics revolution: The next great leap in manufacturing.* Available via CIRCABC. Retrieved February 13, 2020, from https://circabc.europa.eu/sd/a/b3067f4e-ea5e-4864-9693-0645e5cbc053/BCG_The_Robotics_Revolution_Sep_2015_tcm80-197133.pdf.

Tempest, R. (1996). *Barbie and the world economy.* Available via Los Angeles Times. Retrieved February 13, 2020, from https://www.latimes.com/archives/la-xpm-1996-09-22-mn-46610-story.html.

The Economist. (2017). *Adidas's high-tech factory brings production back to Germany.* Available via The Economist. Retrieved February 13, 2020, from https://www.economist.com/business/2017/01/14/adidass-high-tech-factory-brings-production-back-to-germany.

UNCTAD. (2016). *Robots and industrialization in developing countries. UNCTAD policy brief 50.* Available via UNCTAD. Retrieved February 13, 2020, from https://unctad.org/en/PublicationsLibrary/presspb2016d6_en.pdf.

WTO. (1998). *Annual report.* Geneva: WTO.

AI/Robotics and the Poor

Joachim von Braun and Heike Baumüller

Contents

Abstract

Artificial intelligence and robotics (AI/R) have the potential to greatly change livelihoods. Information on how AI/R may affect the poor is scarce. This chapter aims to address this gap in research. A framework is established that depicts poverty and marginality conditions of health, education, public services, work, small businesses, including farming, as well as the voice and empowerment of the poor. This framework identifies points of entry of AI/R, and is complemented by a more detailed discussion of the way in which changes through AI/R in these areas may relate positively or negatively to the livelihood of the poor. Context will play an important role determining the AI/R consequences for the diverse populations in poverty and marginalized populations at risk. This chapter calls for empirical scenarios and modelling analyses to better understand the different components in the emerging technological and institutional AI/R innovations and to identify how they will shape the livelihoods of poor households and communities.

Keywords

Poverty · Inequality · Artificial intelligence · Robotics · Research and development · Marginalization

J. von Braun (✉) · H. Baumüller
Center for Development Research (ZEF) Bonn University, Bonn, Germany
e-mail: jvonbraun@uni-bonn.de; hbaumueller@uni-bonn.de

Introduction[1]

Artificial intelligence based on the utilization of big data, machine learning, and applications in robot technologies will have far-reaching implications for economies, the fabric of society and culture. It can be expected that artificial intelligence and robotics (AI/R) offer opportunities but also have adverse effects for the poor segments of societies and that the effects will differ for specific applications of AI/R. The implications of AI/R for poverty and marginalization are, to date, not much studied, but are important for ethical considerations and AI/R related policies. It actually seems that attention to robots' rights is overrated whereas attention to implications of robotics and AI for the poorer segments of societies are underrated. Opportunities and risks of AI/robotics for sustainable development and for welfare of people need more attention in research and in policy (Birhane and van Dijk 2020).

In order to derive potential implications for the poor, a theoretical framework is needed that captures the structural and dynamic factors shaping incomes and livelihood capabilities of the poor without AI/R and that identifies points of entry of AI/R as well as their impact on poor households and communities. For this reason, a framework of opportunities and risks for conceptualizing this review is outlined in A Framework of AI/Robotics Impacts on the Poor and Marginalized section.

There are several caveats to such framing. The field of AI/R is developing quickly, bringing about changes of limited predictability. Uncertainties also surround the potentially changing regulatory regimes and policies that will frame AI/R innovations and applications.[2] It is important to also stress that "the poor" are not at all a uniform set of people but that this descriptor denotes a highly diverse group whose composition changes over time. Therefore, the impacts of AI/R on diverse groups of the poor will differ. Poverty and inequality effects will also be context-specific across countries, partly depending on infrastructures. AI/R may theoretically reduce poverty while increasing inequality or vice versa.

Key areas particularly relevant for poor people's livelihoods that are likely to be significantly impacted by AI/R in the medium and long term are *education, health, financial and public services, employment, small businesses incl. farming, natural resources management, voice and empowerment*. These key areas form the structure of this review. In the final section, some policy measures will be discussed that might contribute to poverty reducing and inclusive AI/R innovations and applications.

A Framework of AI/Robotics Impacts on the Poor and Marginalized

There has been significant progress in the reduction of poverty in the developing world over the past few decades. In 1990, 36% of the world's people lived in poverty (income of less than US$1.90 a day in 2011 purchasing power parity (PPP)). By 2015, that share had declined to 10%. The number of people living in extreme poverty stood at 736 million in 2015, indicating a decrease from nearly 2 billion in 1990 (World Bank 2018). This progress is the result of various factors, including economic growth reaching the poor and, in many countries, an increase in attention to social protection policies.

While poverty levels are on the decline at the global level, they remain high in some regions, notably in Sub-Saharan Africa and South Asia where 85% of the extreme poor live. The World Bank (2018) also warns that the decline may be slowing down. In the case of Sub-Saharan Africa, the absolute numbers are in fact increasing and the region is predicted to host 87% of the extreme poor by 2030 (World Bank 2018). Therefore, this review of potential AI/R impacts on the poor pays particular attention to Sub-Saharan Africa. At the same time, Sub-Saharan Africa's growing population also brings opportunities. The region already has the world's youngest population, being home to over 250 million young people (aged between 15 and 24 years)[3] who can learn quickly and adapt to new technologies such as AI/R.

Being poor is not defined just by lack of income. Access to education, to basic utilities, health care, nutrition, and security are also critical for well-being.[4] In Sub-Saharan

---

[1]This paper is based on Joachim von Braun's contribution to the conference on "Robotics, AI, and Humanity: Science, Ethics, and Policy", jointly organized by the Pontifical Academy of Sciences and the Pontifical Academy of Social Sciences, at the Vatican on May 16–17, 2019. von Braun's contribution was also published as a ZEF Working Paper (von Braun 2019) and later expanded upon for the purposes of this present publication.

[2]It is important to highlight that access to ICTs supporting infrastructure for the poor (incl. network coverage, speed, costs) is very unequal. The Pontifical Academy of Sciences addressed this issue in a conference: *Connectivity as a Human Right*, 10 October 2017, Scripta Varia 140, Vatican City, 2018. Further information on costs and connectivity are, for instance, available regarding the network speed by country (SPEEDTEST n.d.), the expansion of the 4G coverage (McKetta 2018), the Price Index (ITU 2019), and further figures on global digitalization (Kemp 2018). Moreover, gender bias has to be kept in mind when reflecting on benefits and challenges of AI/R for the poor and marginalized. Research found that women have less access to ICTs, especially in areas with high levels of poverty. Compared to men, women in South Asia are 26% less likely to own a mobile phone and 70% less likely to use mobile internet. In Sub-Saharan Africa, the shares are 14% and 34%, respectively (GSMA 2018).

[3]Data from UN Department of Economic and Social Affairs, downloaded on 10 January 2020. https://population.un.org/wpp/Download/Standard/Population/

[4]Such a multidimensional view of poverty has been adopted by the World Bank and it is found that, "[a]t the global level, the share of poor according to a multidimensional definition that includes consumption, education, and access to basic infrastructure is approximately 50 percent

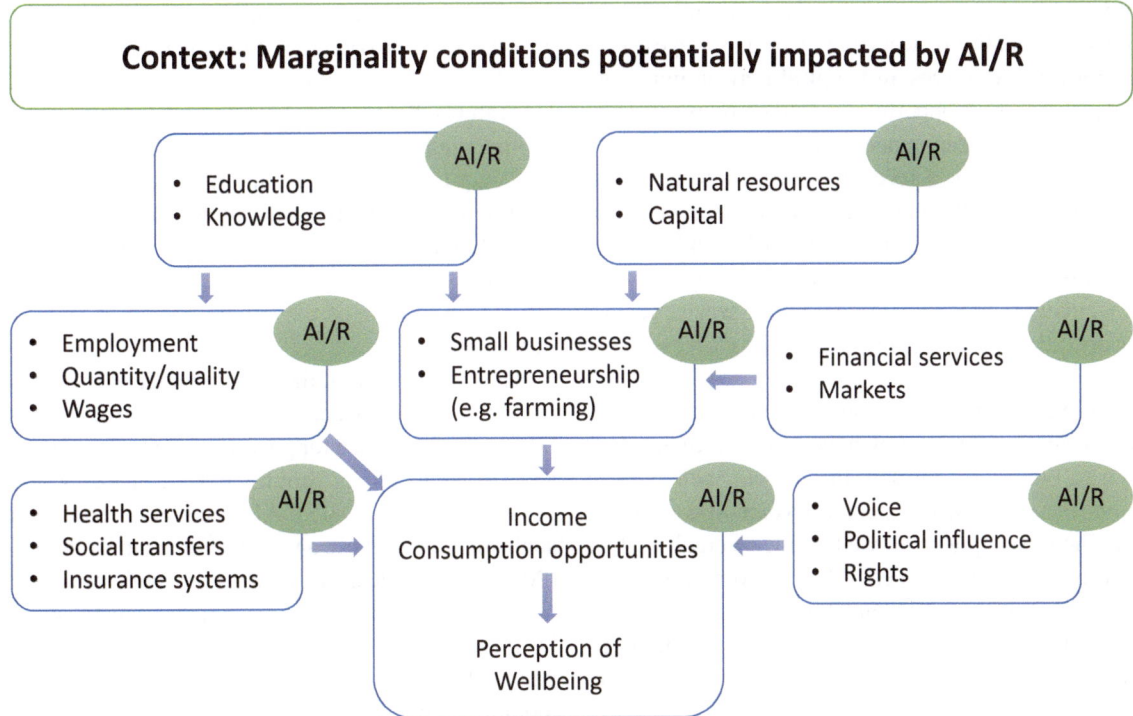

Fig. 1 Framework for assessing the impacts of AI/R on poor people. Source: designed by authors

Africa, low levels of consumption and high levels of food insecurity are often accompanied by challenges in nonmonetary dimensions. Poverty is to a great extent a matter of structural forces such as exclusion, discrimination and the deprivation of rights, governance deficiencies and corruption, and the forces of ecological change that increase the vulnerability of the poor, as many of them depend on natural resources living at the margins in rural areas or in high-risk margins of urban areas. These aspects are captured by the concept of marginality.[5] Marginality encompasses broader approaches such as relative deprivation and social exclusion or the capabilities approach.

Relative deprivation builds on the idea that the value of objective circumstances depends on subjective comparisons (Stark and Bloom 1985). The relative deprivation concept adds an important dimension to absolute poverty concepts because it involves comparisons with other people. AI/R may change the patterns of comparing one group of people with "others". Such comparisons may include the comparison of people aided by robotics with others not having access to a specific robotics aid, or even a direct comparison of people

---

higher than when relying solely on monetary poverty" (World Bank 2018, p. 9).

[5]Marginality can be defined as an involuntary position and condition of an individual or group at the edge of social, economic, and ecological systems, preventing the access to resources, assets, services, restraining freedom of choice, preventing the development of capabilities, and causing extreme poverty (von Braun and Gatzweiler 2014).

and robots, for instance when robots take on tasks which had previously been carried out by low-income workers. Certainly, AI/R will change patterns of relative deprivation.

A Framework

Following the above-mentioned framing, Fig. 1 outlines a set of factors influencing poor households and their members' wellbeing. Each of these factors, as well as the forces determining marginality, may be impacted by AI/R as depicted within the blocks of Fig. 1. There are many potential points of entry for AI/R on the different dimensions of this framework. Naturally, these emerging, disruptive technologies bring both opportunities and risks for poor households.

Walking through the framework shown in Fig. 1, starting in the middle, the influence of **employment** and **small businesses and entrepreneurship**, respectively, on the **income and consumption opportunities** are of significance. Through the introduction of AI/R, mobility costs in general may be reduced and opportunities for mobility may also improve for the poor. Labour tends to be the most important asset of the poor. Thus, employment and small enterprise opportunities have a key influence on income and consumption of poor households.

As a prerequisite for improved access to remunerative employment and business opportunities, **education and knowledge** are essential for impoverished communities. Improving households' education and knowledge base not only affects the quantity and quality of work but also their wages. Furthermore, **natural resources and capital** play an important role in shaping business opportunities and entrepreneurship. AI/R is likely to have deep impacts on these linkages.

Improving poor people's **access to services and markets**, such as financial, health, and insurance services or social transfers, is another driving factor in increasing and stabilizing poor households' income and consumption opportunities and their wellbeing. AI/R may influence these service sectors and their operations, for instance, by lowering access barriers.

As mentioned above, how (poor) people perceive their situation is heavily influenced by how they view themselves in relation to others, i.e. by how they assess their relative deprivation. By strengthening people's **voice and political influence**, their perceptions as well as their actual wellbeing can be positively influenced. AI/R contributions could be immense in improving accountability and transparency of governments for their citizens, thus ensuring improved service delivery and empowering local development actors and policy makers.

Opportunities of Data and Information Systems about Poverty

The lack of reliable data in developing countries is a major obstacle to planning and investment in sustainable development, food security, and disaster relief. Poverty data, for example, is typically scarce, sparse in coverage, and labour-intensive to obtain. Remote-sensing data such as high-resolution satellite imagery, on the other hand, is becoming increasingly available and inexpensive. In addition, the growing use of digital technologies is yielding a wide range of incidentally collected data e.g. on the movement of people or spending on airtime. Importantly, many of these analytical tools can provide real-time alerts to take timely actions, including in emergency situations.

An important contribution of AI to poverty reduction is the enhanced capability to **identify the poor**. Data and innovative machine learning-based identification strategies are already advancing to provide a more up-to-date and well-defined information base compared to traditional household surveys or indicator systems (e.g., Mirza 2018; Zurutuza 2018). For instance, night time lights are substantially correlated with economic wealth. In addition, automated processes can be used to identify the roof material, source of light, and access to water sources—based on satellite images— to derive households' economic situation.

Jean et al. (2016), for instance, demonstrate the usefulness of AI for estimating consumption expenditure and asset wealth from high-resolution satellite imagery. They suggest a learning approach in which night time light intensity is used as a data-rich proxy and a model is trained to predict night time lights from daytime imagery simultaneously with learning features that are useful for poverty prediction. The model applies filters that identify different terrains and structures, including roads, buildings, and farmlands. Using survey and satellite data from five African countries—Nigeria, Tanzania, Uganda, Malawi, and Rwanda—they show that neural networks can be trained to identify image features that can explain up to 75% of the variation in local-level economic outcomes.

The rapid spread of the mobile phone across the developing world is offering access to a wide range of new data which has already been used for poverty mapping (e.g. Blumenstock and Cadamuro 2015; Smith-Clarke et al. 2014; Soto et al. 2011; Steele et al. 2017). AI-based tools could improve the sophistication of these analyses. For example, Sundsøy et al. (2016) use deep learning to analyse mobile phone datasets in order to classify the socio-economic status of phone users. Data used in the analysis include basic phone usage, top-up pattern, handset type, social network structure and individual mobility. Any use of personal data will have to go hand-in-hand with well-developed data protection rules, awareness campaigns and digital literacy training, in particular for the poor and marginalized who may be less informed about associated risks.

Accurately mapping populations through AI can inform **pro-poor targeting** of e.g. infrastructure, social security, health care or emergency relief. Facebook, for instance, is developing detailed population density maps by using satellite imagery, machine learning algorithms, and population statistics to ensure that connectivity and other infrastructure reaches the entire population even in remote, sparsely populated areas.[6] The maps can also be used e.g. to facilitate the provision of humanitarian aid or plan health care interventions. The map of Malawi, for instance, was used by local partners to better target a measles and rubella campaign (Bonafilia et al. 2019). Other examples of using AI-assisted data outputs to improve the provision of services to the poor are presented in AI-Assisted Financial Services section.

Education and Knowledge Links with AI/Robotics

AI/R may be of great benefit to the poor to the extent that these technologies enhance their access to education and knowledge, upskill their capabilities in the labour market, and

---

[6]Maps for Africa have already been completed (available at https://data.humdata.org/dataset/highresolutionpopulationdensitymaps) and maps for the rest of the world are expected to be added in the near future (Singh 2019).

increase their income earning capabilities in small businesses and farming. However, education systems tend to directly and indirectly exclude or marginalize the poor: directly through school fees or low quality education in poor communities, and indirectly through a lack of time on the part of the children and other constraints on the part of the parents in poor communities. Time constraints arise for millions of children that work in addition to attending school; when excessive time is spent at work, education suffers and life-time earnings are reduced as Mussa (2018) shows for cohorts of rural child labourers in Ethiopia. The inequalities tend to grow with levels of education, i.e. secondary and tertiary levels. Girls remain underrepresented in these levels leading to gender gaps (von Braun 2018). Digital literacy constraints may additionally inhibit the acquisition of skills by use of interactive distance learning systems empowered by AI.

There are new opportunities through AI/R in education which foster the inclusion of the poor. Du Boulay et al. (2018) argue that AI technologies can assist both educators and learners by providing personalized support for students, offering tools for teachers to increase their awareness, and freeing up time to provide nuanced, individualized support. Nye (2015) reviews intelligent tutoring systems (ITS) targeting the developing world. An interesting example from Kenya is the "Kio Kit" (BRCK n.d.), a fully integrated education platform designed in Kenya which turns a school room into a digital classroom using tablet computers, wireless tablet charging and securing the tablets in a hardened, water-resistant, lockable case. Such education platforms empowered by AI could make a big difference in the future, provided it is made accessible to the poor in marginal rural areas. However, barriers to using ITS need to be addressed, including students' basic computing skills and factors like hardware sharing, mobile-dominant computing, data costs, electrical reliability, internet infrastructure, language, and culture (Nye 2015).

AI-assisted translation and speech recognition may hold great potential for the poor by overcoming barriers of language and illiteracy. Google's speech-to-text service, for instance, uses neural network models to convert audio to text, including several languages spoken by the poor, such as Swahili, Zulu, Cantonese, Mandarin, and several Indian languages (Google Cloud 2019). These technologies could substantially increase access to information and services for disadvantages groups. For now, the data requirements to train machine learning algorithms or neural networks still pose an obstacle to including less commonly spoken languages. However, as more and more written and spoken material is being digitalized and made accessible online, big data analytics will become increasingly efficient and further improve over time.

Health Services for the Poor Facilitated by AI/Robotics

The poor have difficulties to claim their rights to public services. The **health sector** is particularly interesting for AI/R applications, given the ongoing digitalization of all types of health data and health information. The long-term potential for advancing the field of digital health and precision and personalized medicine can be immense when AI/R supported medical and public health decision-making reduces costs (Ciechanover 2019). Recognizing this potential, the International Telecommunication Union (ITU 2018) has established a new Focus Group on "Artificial Intelligence for Health" (FG-AI4H) in partnership with the World Health Organization (WHO) to develop standardized assessment frameworks for AI-based health solutions, such as AI-based diagnosis, triage or treatment decisions. Thereby, the group aims at assuring the quality of solutions, fostering their adoption and improving global health. As these developments bring down costs, they may enhance poor people's access to health services. To date, however, these innovations still tend to be far from reaching the poor.

There are clear advantages for **worker health** from AI/R, for instance when harmful work tasks can be handed over to robots. Pesticide spraying in fields is an important example, including the drone-based detection and mapping of insect infestations at micro level for optimal targeting. A similar safety-enhancing example is the employment of AI technology to improve the safety of mining workers by reducing their intensity of work and exposure to high-risk assignments that could be taken over by robots.

Another important area of application with particular relevance to the poor is the detection and prevention of **malnutrition**, which is one of the leading causes of infant mortality in developing countries. Khare et al. (2017) designed a prediction model for malnutrition based on a machine learning approach, using the available features in the Indian Demographic and Health Survey (IDHS) dataset. Their findings suggest that this approach identifies some important features that had not been detected by the existing literature. Another example is the Child Growth Monitor app developed by the German NGO Welthungerhilfe, which identifies malnutrition in children using inbuilt infrared cameras for scanning and machine learning to correlate the scan with anthropometric measures (Welthungerhilfe 2018). The app was pre-tested in India and the results are promising. It may replace the manual measures of weight and height, which are costly, slow, and often inaccurate. Apps also offer targeted nutritional advice directly to households. Numerous nutrition apps already exist which monitor food intake and provide dietary recommendations to its users. One available feature of these apps, which is enabled through deep learning, is the tracking of macronutrient content by taking a photo of the meal that is about to be consumed. In the future, these apps may also be used for personalized dietary advice and tailored health messages in developing countries. In this way, smartphone apps can be a low-cost intervention for improving dietary quality and health even in remote areas with limited access to dieticians or health workers.

AI-enabled **diagnostic tools** could improve health services in remote areas (Wood et al. 2019). These technologies could support self-diagnosis or assist health workers. For example, researchers at Stanford University have trained an algorithm to diagnose skin cancer from images (Kubota 2017). The Israeli company Zebra Medical Vision is partnering with Indian authorities to roll out its AI-based medical imaging technology which can diagnose tuberculosis using chest X-rays (Zebra Medical Vision Team 2019). AI-technologies also enable a faster detection of emerging epidemics such as Ebola or Zika (Wood et al. 2019). It will be important to ensure that AI technologies are adapted to a wide range of people's characteristics. Research has shown that using training data from specific population groups may exclude some social groups (Gershgorn 2018). Machine learning requires large amounts of data which may not be easily available from the poor and marginalized who currently do not have access to health services. Moreover, well-trained human intermediaries would still be required to explain and build trust in diagnostic results. Importantly, diagnostic tools will only be useful if they are complemented with improved access to treatment for the poor.

Robots may help to improve access to and quality of **surgical care**. The Lancet Commission estimates that 9 out of 10 people in low- and middle-income countries do not have access to safe, affordable surgical and anaesthesia care (Meara et al. 2015). Only 6% of the 313 million surgeries performed annually take place in the poorest countries (Meara et al. 2015). Robots are already being used to assist and thereby enhance the capability of surgeons to perform complex procedures. In the long-run, they could even be used to undertake surgery remotely. While it is still early days for such treatments, rapid advances in technological capacities will bring them closer within reach. China, for instance, is already experimenting with remotely operated robots to carry out orthopaedic surgery (Demaitre 2019). However, given that such procedures are costly, still rely on trained personnel on-site and require 5G networks, it may be a long time until they reach rural areas that still struggle to access even 3G networks.

AI may also assist in **humanitarian health crises** by extracting valuable information for decision-making. Fernandez-Luque and Imran (2018) find that while there are many examples of the use of artificial intelligence in health crises, they are largely limited to outbreak detection while their application in low-income countries is under-researched. Capacities for data sharing and real-time analysis, combined with human and organizational aspects, are the main constraints to the use of AI in this area.

AI-Assisted Financial Services

A large share of the poor lack access to formal financial services such as banking, credit or insurance. In 2017, only one third of adults in rural Sub-Saharan Africa owned a bank account and only 5% held a loan from a formal financial institution (World Bank 2019). Access to insurance for small and medium sized enterprises in Africa, which could help them better manage risks and improve their risk profile vis-à-vis lenders, also remains limited. Similarly, only few smallholders in the region held crop insurance in 2016 (Hess and Hazell 2016). AI could in particular assist in facilitating access to credit and insurance for the poor, but there are also risks for the poor when AI is used to support discriminatory screening in credit approval procedures or predatory operations of "fintech" organizations (Hill and Kozup 2007; Variyar and Vignesh 2017).

Banks are often reluctant to provide **loans** to low-income households because they are unable to assess the risk of loan default. AI-based systems can assist in credit-scoring by using data collected by mobile phones, such as call-detail records, social network analytics techniques or credit and debit account information of customers (Óskarsdóttir et al. 2019), and combining them with agronomic, environmental, economic, and satellite data. The Kenyan company Apollo Agriculture,[7] for instance, uses machine learning to analyse data about farmers collected by field agents and satellite imagery to create a profile of the farmers and evaluate their credit-worthiness. Loans are then disbursed via agro-dealers and also bundled with insurance in case of yield losses.

AI could also assist in the provision of **insurance** to a large number of dispersed, small-scale farmers and businesses by providing high-quality data, minimizing uncovered basis risk and lowering cost delivery mechanisms. In view of climate change and the related increase in weather risks, important areas are weather services and weather risk insurances building upon AI-enhanced weather observation and monitoring at pixel levels.[8] For instance, index-based insurance using remote-sensing data has shown to be promising as a means of providing crop insurance in areas and to customers that were previously out of reach (De Leeuw et al. 2014; IFAD 2017). AI can also help to monitor or develop

---

[7] https://www.apolloagriculture.com/

[8] Aerobotics (South Africa) is an example for machine learning on aerial imagery to identify problems in crop yields (World Wide Web Foundation 2017).

efficient index-based insurance for pastoralists by predicting livestock mortality sufficiently in advance to ensure ex-ante asset protection insurance.

The provision of such services are greatly facilitated by **mobile money** services. The most prominent example in Africa is M-Pesa (M is for mobile, "pesa" is Swahili for "money") which started in Kenya in 2007. M-Pesa allows people from all around the country, even in the most remote areas, to transfer money directly, saving considerable amounts of time and money. In 2018, just over 80% of the adult population of Kenya was registered (CA 2018). While M-Pesa is not AI-based, it facilitates recording even small business and farm performance and financial transactions including productivity, expenses, sales and revenues, thereby providing valuable information to build credit profiles. The data can also be used to inform other decisions, such as investments in roads, storage facilities for farmers or financial infrastructure.

AI and Social Transfers

In recent years, **social protection programs** including cash transfers to the poor, some of them implemented with certain conditions such as children's school participation, and labour-intensive public works programs (PWPs), have expanded in many countries. By combining different datasets, machine learning can contribute to increasing the efficiency of targeting social interventions and channeling humanitarian responses in the course of crises.

For instance, PWPs' benefits are sometimes obscured to such an extent that PWPs become deficient, e.g. when not targeting the intended beneficiaries or due to leakages related to corruption. The main objective of PWPs is to provide social protection to the working-age poor by transferring cash or in-kind to beneficiaries to protect households' consumption while at the same time promoting savings and investments in productive assets, the generation of public goods, and the provision of training to support rural transformation. AI-based tools can enable real-time monitoring by facilitating the collection, processing, management, validation, and dissemination of data for operations, accountability, and policy decision making. In the case of vast PWPs in India, the development of a management information system (MIS) to support program processes and structures ensured more reliable and on-time management of big data that comes from multiple sites and levels of program implementation, thus minimizing errors, frauds, and corruption (Subbarao et al. 2013).

AI-enabled welfare programmes have also attracted criticism (Alston 2019; Eubanks 2018). The UN Special Rapporteur on extreme poverty and human rights, Philip Alston, warns in a report to the UN General Assembly that algorithms used for targeting and enforcing social protection programs may lead to wrong decisions or discrimination, especially in the absence of data for certain societal groups. Moreover, the poor and marginalized with limited digital and technological capacities may find it difficult to interact with digital systems. Issues of data privacy may also arise, given that such systems rely on the collection and integration of large sets of personal data.

AI/Robotics Effects for the Poor in Employment, Small Business, and Smallholder Farming

Employment generation is a key policy priority for many African governments to create job and income opportunities for the 300 million young people who are projected to enter the labour market by 2030 (AfDB 2019). The impact of AI/R on employment through the introduction of industrial automation and labour saving technologies is perhaps the most widely debated issue when it comes to distributional effects of these technologies.[9] The main concern relates to the loss of jobs for low-skilled labour, which is of particular relevance to the poor. The capacity to work is the key asset of the poor, be it in the labour market or as small entrepreneurs in the service sector or small-scale farming.

Theoretically, this concern may be derived from the classical model designed by Arthur Lewis (1954; cf. Polany Levitt 2008). Lewis' model assumes an unlimited supply of unskilled labour while the opportunity cost of labour determines wages in the modern sector. More specifically, the model supposes a traditional working environment of peasants, artisanal producers, and domestic servants, which is subjected to population pressures. This situation in the traditional sector grants the modern sector "unlimited supplies" of labour, with wages exceeding the subsistence level only by a small margin. As the modern sector expands, employment and output, as well as the share of profits in national income, rise. At a certain point, the surplus labour from the traditional sector is exhausted, which leads to the wage rate increasing as productivity further rises.

Within the context of AI/R technologies, however, this so-called Lewis inflection point may no longer be reached in emerging economies if there were to evolve an unlimited "robot reserve army" which competes with the labour force of the poor (Schlogl and Sumner 2018). In adaption of the Lewis model of economic development, Schlogl and Sumner (2018) use a framework in which the potential for automation creates "unlimited supplies of artificial labor" particularly in the agri-

---

[9]See, for instance, Acemoglu and Restrepo (2017), Chiacchio et al. (2018), Frank et al. (2019), Korinek (2019), Korinek and Stiglitz (2017), also published in Agrawal et al. (2019), and Marin (2018).

cultural and industrial sectors due to technological feasibility. This is likely to push labour into the service sector, leading to a further expansion of the already large and low-productive service-sector employment. Gries and Naudé (2018) assess these issues by incorporating AI-facilitated automation with modelling to allow for demand-side constraints and thereby finding less adverse effects for employment.

Acemoglu and Restrepo (2017) distinguish two potential employment effects of automation. First, robots may displace and thereby reduce the demand for labour (displacement effect). Through linkages within global value chains, workers in low-income countries may also be affected by robots in higher-income markets, which could reduce the need for outsourcing such jobs to low-wage countries overseas. As a result, low-skill, labour-intensive industrialization, as observed in many East Asian countries, may no longer be a promising development model. Second, robot use could increase the demand for labour either by reducing the cost of production which leads to industry expansion (price-productivity effect) or by increasing total output overall (scale-productivity effect).

The key question from a poverty and distributional perspective will be which jobs are replaced and which are generated. Policies should aim at providing social security measures for affected workers while investing in the development of the necessary skills to take advantage of newly created jobs. This is easier said than done, however, especially for the poor who often start from a very low level of education.

Small businesses can grow rapidly thanks to digital opportunities, expanding their boundaries and reshaping traditional production patterns. The rise of digital platform firms means that technological effects reach more people faster than ever before. Technology is changing the skills that employers seek. Ernst et al. (2018) point out that opportunities in terms of increases in productivity, including for developing countries, can ensue, given the vastly reduced costs of capital that some applications have demonstrated and the potential for productivity increases, especially among low-skilled workers. Makridakis (2017) argues that significant competitive advantages will accrue to those utilizing the Internet widely and willing to take entrepreneurial risks in order to turn innovative products/services into commercial success stories.

A large proportion of the poor live on **small farms**, particularly in Africa and South and East Asia. Digital technologies, services, and tools can offer many opportunities for smallholder farmers to make more informed decisions, and to increase productivity and incomes. Key benefits of digitalization include greater access to information and other services including finance, links to markets, a sustainable increase in productivity, and better informed policies.[10] Examples are:

- *Land ownership certification:* In many developing countries, land rights are unclear and contested. Smartphones, cameras or drones, which can capture geospatial and topographical data—using, for example, Global Positioning Systems and global navigation satellite systems—are therefore useful tools for land mapping and land tenure programs. For instance, machine learning can be used to predict the boundaries of parcels based on common property boundary features. If this geospatial data and land transaction history is saved through blockchain technology, land registries could become extremely robust.
- *Precision technologies:* including big data, satellite imagery, sensors, robotics, drones, etc.[11] For instance, in South Africa, the drone start-up Aerobotics is using drone technology and AI to assist farmers in optimizing their yields, thereby greatly contributing to cost reduction. In Kenya, AI and big data analytics provide useful information about farming trends and productivity based on data that is generated through soil analysis. Another AI-assisted application is image recognition to diagnose plant diseases, offer information about treatments and to monitor the spread of diseases based on photos of the plant taken with a smartphone.
- *Farm machinery:* Tractors mounted with sensors and connected to mobile platforms allow users to remotely capture farm information related to soil, water, and crop conditions, to calibrate usage of inputs accordingly, and to monitor progress. Innovative models such as "uberization" of tractors and farm machinery, which make farm machinery available on rent and make mechanization more effective and affordable for the farmers, are gaining popularity in developing countries.
- *Innovations in irrigation and energy:* Solar-driven micro irrigation systems can help reduce energy costs and, when connected to the grid, allow farmers to benefit from selling generated surplus energy. AI/R complex system-based algorithms help to build efficient water systems for communal irrigation or optimization of water resources, distribution, and infrastructure planning, for instance.
- *Access to information:* AI-enabled digital services can enhance farmers' access to personalized information related to weather, soil health, market prices, and finance that is specifically adapted to their local context and information needs. This enables farmers to plan their farming activities, project the potential output and hence bargain for better prices. Through AI-based speech recognition, information can become accessible for farmers with low levels of literacy.

---

[10] A study on 81 countries revealed significant positive effects of an increase in ICTs adoption on agricultural productivity already at the early stages of digitization (Torero and von Braun 2006; Lio and Liu 2006). Substantial progress has been made in this regard in the past decade.

[11] This section draws on Ganguly et al. (2017).

- *ICTs platforms connecting buyers and sellers:* AI can improve the functioning of virtual market platform, e.g. to better match buyers and sellers, facilitate transactions or respond to customer queries. In addition, the data collected by e-commerce platforms can be analysed using AI tools in order to inform decision-making, for instance on which crops to plant depending on demand or price trends or on government investments in logistics infrastructure.

The challenge lies in the effectiveness of these technologies in addressing the field level issues and making these applications user friendly for poor farmers. Hence, the focus needs to be on integrating several of these applications in user-friendly platforms designed with in-built data interpretation and predetermined actions to equip users with end-to-end solutions.

AI/Robotics Links to Voice and Empowerment

To the extent that AI/R take over functions that are currently provided by poor and marginalized people, these technologies may have adverse effects on their rights and entitlements, and - by extrapolation - their capacities to influence social and economic policies may suffer. However, if new AI/R functions were usable by or even come under the direct control of marginalized and poor people, these people might become more empowered. The latter *direct* effect will be more likely if collective actions by the poor are facilitating access at scale to specific AI/R applications, such as AI-facilitated land rights or R-facilitated drone technologies reducing hazardous work on farms. Yet, so far, there remain huge digital gaps between the rich and the poor, and it may remain the same in the field of AI/R applications, thus limiting direct benefits of AI/R for the poor.

More impactful may be *indirect effects* of AI on voice and empowerment, given the weight of poor populations in political decision making, including elections or protest movements. In this regard, AI-based tools can both **empower and disempower** the poor (Polonski 2017; Savaget et al. 2019). Automated translation tools, for instance, can empower people by helping them to vote or be more informed about political processes. AI can also increase the transparency of government by analysing increasingly large volumes of open data which may otherwise be difficult to interpret, in particular for less educated readers. It can also be used to verify the accuracy of information and highlight articles or messages that contain misinformation. However, the same technologies can also be used to generate and spread false information and thereby manipulate public opinion (Shorey and Howard 2015).

At the same time, AI-enabled technologies risk **reinforcing existing inequality**. The World Wide Web Foundation (2017) emphasizes that "... AI present a tremendous set of opportunities and challenges to human well-being [but] there is also concern that AI programs and decision-making systems supported by AI may include human biases, leading to further discrimination and marginalization" (p. 3). Indeed, algorithms and neural networks are not neutral, but are shaped by the data used for training. Caliskan et al. (2017) show that machine learning can take on biases, for example with regard to race or gender when trained with standard textual data from the web. There are initiatives that aim to change this, such as the "Whose Knowledge?" initiative which defines itself as a "... global campaign to center the knowledge of marginalized communities ... on the internet."[12] They work particularly with women, people of colour, communities like Dalits in India, and others to create and improve related content.

Regarding privacy and access to information rights, the poor are particularly threatened because of their current lack of power and voice. **New forms of regulating the digital economy** are called for that ensure proper data protection and privacy, and help share the benefits of productivity growth through a combination of profit sharing, (digital) capital taxation, and a reduction in working time. Investments in skills need to be enhanced, and the enabling business environment strengthened to include the poorer and more vulnerable groups (World Bank 2017).

Another important dimension to consider is the impact of AI on **human autonomy** or agency as information provision and decision-making is handed over to seemingly intelligent machines (Pew Research Center 2018). This may be a particular concern for the poor and disadvantaged who often have limited access to alternative information sources.

Policy Conclusions

A broadly based policy agenda will be needed to include the poor and marginalized in opportunities of AI/R and to shield them from discrimination and indirect adverse effects. Options for policy actions are presented in Fig. 2. For the areas of influence that were identified in Fig. 1, examples of possible actions to harness the potential of AI/R for the poor are listed, including targeted AI/R applications and supporting framework conditions. A number of cross-cutting policy actions are proposed:

Strengthen skills to develop and use AI/R technologies and enable the poor to take advantage of related employment opportunities. AI/R risks for employment and the poor call for investing in human capital in order to build the skills in

---

[12]https://whoseknowledge.org/

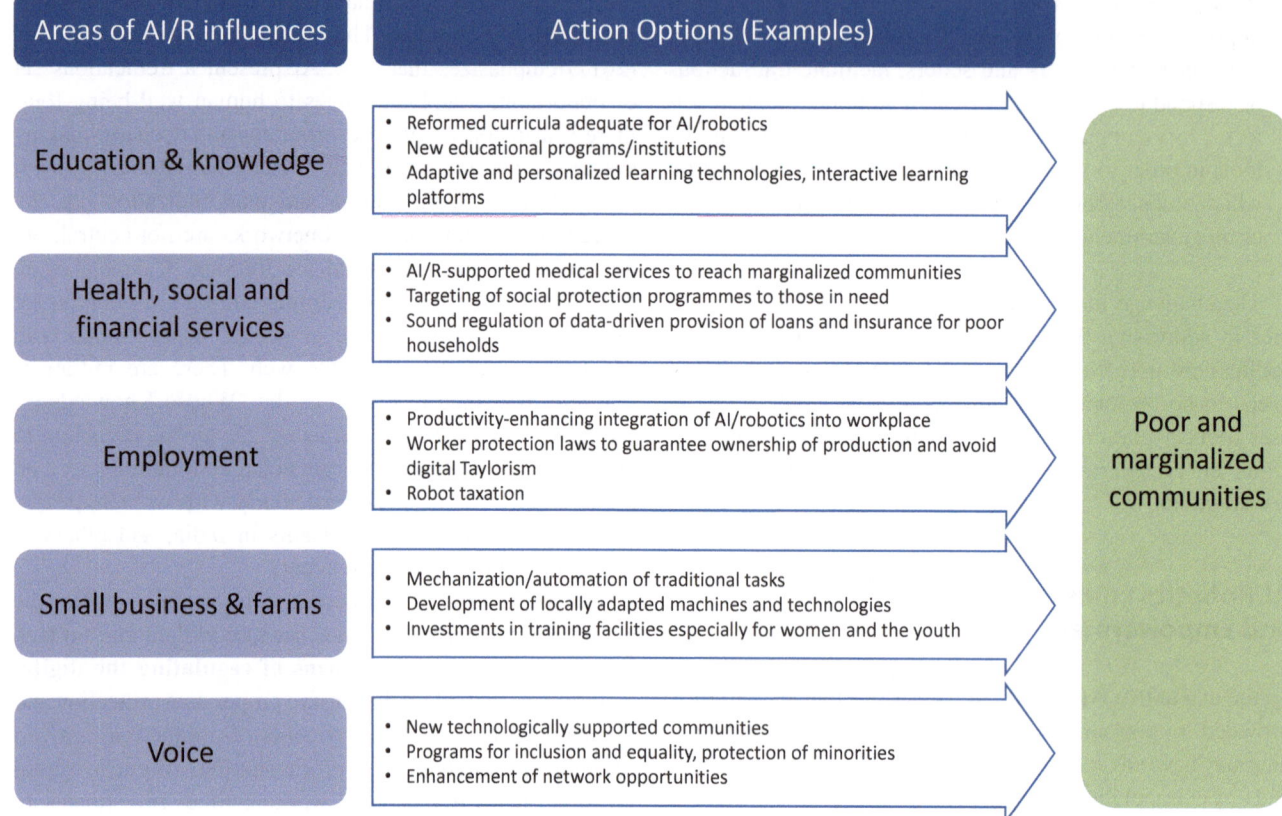

Areas of AI/R influences	Action Options (Examples)
Education & knowledge	• Reformed curricula adequate for AI/robotics • New educational programs/institutions • Adaptive and personalized learning technologies, interactive learning platforms
Health, social and financial services	• AI/R-supported medical services to reach marginalized communities • Targeting of social protection programmes to those in need • Sound regulation of data-driven provision of loans and insurance for poor households
Employment	• Productivity-enhancing integration of AI/robotics into workplace • Worker protection laws to guarantee ownership of production and avoid digital Taylorism • Robot taxation
Small business & farms	• Mechanization/automation of traditional tasks • Development of locally adapted machines and technologies • Investments in training facilities especially for women and the youth
Voice	• New technologically supported communities • Programs for inclusion and equality, protection of minorities • Enhancement of network opportunities

Fig. 2 Policy action options for pro-poor AI/R. Source: designed by authors

demand in the labour market. Indeed, skill development is perhaps the most important area of intervention to enhance benefits from AI/R. This needs to happen at various levels. Existing curricula need to be adapted, starting from school all the way to tertiary education. In addition, new education programs and institutions will be needed. One set of skills includes basic digital literacy of end-users, including intermediaries such as extension agents, agro-dealers or community leaders who could help cover the last mile to reach the poor and marginalized. Another set of skills include more advanced technical skills for the local development of AI/R solution and the use of more sophisticated applications in business operations. These skills would allow local communities to take advantage of direct job and income opportunities related to AI/R applications. These are unlikely to benefit the poor and marginalized, however. It is therefore imperative to ensure that these groups have the necessary skills to benefit from jobs or income opportunities that may be indirectly created where AI/R raises productivity and industrial output e.g. increased demand for raw materials or transport services.

Adapt AI/R solutions to the local context of the poor and marginalized. To ensure that AI/R solutions also benefit the poor, their usability will be crucial. Interfaces need to be adapted so as to be accessible for users with low literacy levels and also match the technological capacities

of devices commonly used by these groups. Dissemination strategies should take advantage of intermediaries to reach and assist these users. Similarly, the functions and content of related solutions need to be adapted to the needs of the poor. This includes the use of locally adapted models for AI applications that employ data from all societal groups to train algorithms, preventing any discriminations. Significant investments and efforts need to be made to collect such data.

Introduce safeguards to mitigate negative employment effects resulting from the introduction of AI/R. Realistically, the poor will find it difficult to benefit directly from many of the employment opportunities that could be generated by AI/R. Therefore, governments need to enhance social protection and extend it to all people in society, irrespective of the terms on which they work. This will enable the redistribution of the benefits of AI/R and digitalization more generally. Taxation policy may provide a suitable avenue to address the distributional effects. Several countries are already taxing certain digital activities, among them some lower and lower-middle income economies such as India, Kenya, Pakistan and Zimbabwe, while others are considering to follow suit (KPMG 2019). In addition, a "robot tax" (Delaney 2017), could be considered. Bruun and Duka (2018) suggest that the resulting tax revenue could be used to mitigate future technological unemployment through the introduction of a basic income scheme, accompanied by reforms in school curricula and retraining programs.

Implement supporting measures to increase the effectiveness of AI/R for the poor. On the one hand, more investments will be required to ensure that AI/R applications can be used by and are useful for the poor. Several areas of intervention were already mentioned above, such as digital literacy, usability and data collection. Significant investments in the digital infrastructure will also be needed, e.g. to expand fast mobile networks into rural areas. In addition, complementary measures will be needed that support the implementation of AI/R solutions, such as health workers that can assist in the use of AI/R-enabled health services or investments in the agriculture sector to support smallholders farmers e.g. by improving access to digital farming, financial services or markets. On the other hand, measures are required to support local start-ups in the development of pro-poor AI/R innovations, such as innovation hubs, usability labs, education or access to finance. The development of AI/R technologies compatible to the needs of the poor needs to be on the agenda of social businesses and NGOs. In addition, more collaboration of AI/R specialists with social scientists is required to arrive at pro-poor AI/R.

To conclude, there are opportunities and risks of AI/R for the poor. AI/R are not neutral in terms of their poverty and distributional effects. Research on the distributional implications of AI/R is currently not being paid sufficient attention. The AI/R opportunities are mainly expected to benefit richer segments of society but as this situation is changing, for instance with jobs lost in employment in the banking sector, the attention to the distributional effects of AI/R will increase. Empirical scenarios and modelling analyses are needed in the different national and societal contexts to better understand the components in the emerging technological and institutional AI/R innovations and identify how they may shape the livelihoods of poor households and communities. Fascination with AI/R must not divert attention from poverty and marginalization. It is important to explicitly pay attention to the outcomes for the poor, and focus on inclusive artificial intelligence and robotics.

References

Acemoglu, D., & Restrepo, P. (2017). *Robots and jobs: Evidence from US labor markets. NBER working paper no. 23285.* Cambridge: National Bureau of Economic Research.

AfDB. (2019). *African economic outlook 2019.* Abidjan: African Development Bank.

Agrawal, A., Gans, J., & Goldfarb, A. (2019). *The economics of artificial intelligence.* Cambridge and Chicago: National Bureau of Economic Research and University of Chicago Press.

Alston, P. (2019). *Report of the special rapporteur on extreme poverty and human rights. A/74/493.* New York: UN General Assembly.

Blumenstock, J., & Cadamuro, G. (2015). Predicting poverty and wealth from mobile phone metadata. *Science, 350,* 1073–1076.

Bonafilia, D., Gill, J., Kirsanov, D., & Sundram, J. (2019). *Mapping for humanitarian aid and development with weakly and semi-supervised learning.* Facebook Artificial Intelligence Blog, 9 April. Retrieved November 11, 2019, from https://ai.facebook.com/blog/mapping-the-world-to-help-aid-workers-with-weakly-semi-supervised-learning/.

BRCK. (n.d.). *Hardware, software and connectivity tools to enable learning.* Retrieved November 15, 2019, from https://www.brck.com/education/.

Bruun, E. P. G., & Duka, A. (2018). Artificial intelligence, jobs and the future of work: Racing with the machines. *Basic Income Studies, 13*(2). https://doi.org/10.1515/bis-2018-0018.

Birhane, A., & van Dijk J. (2020). Robot Rights? Let's Talk about Human Welfare Instead. *Proceedings of the AAAI/ACM Conference on AI, Ethics, and Society.* New York: Association for Computing Machinery, 207–213. https://doi.org/10.1145/3375627.3375855

CA. (2018). *Second quarter sector statistics report for the financial year 2018/2019.* Nairobi: Communications Authority of Kenya.

Caliskan, A., Bryson, J. J., & Narayanan, A. (2017). Semantics derived automatically from language corpora contain human-like biases. *Science, 356*(6334), 183–186. https://doi.org/10.1126/science.aal4230.

Chiacchio, F., Petropoulos, G., & Pichler, D. (2018). *The impact of industrial robots on EU employment and wages: A local labour market approach. Working paper 25186(2).* Brussels: Bruegl.

Ciechanover, A. (2019). The revolution of personalized medicine—Are we going to cure all diseases and at what price? Concept Note of the Pontifical Academy of Sciences-summit on the revolution of personalized medicine, Vatican, 8–9 April 2019.

De Leeuw, J., Vrieling, A., Shee, A., Atzberger, C., Hadgu, K. M., Birandar, C. M., Keah, H., & Turvey, C. (2014). The potential and uptake of remote sensing in insurance: A review. *Remote Sensing, 6*(11), 10888–10912. https://doi.org/10.3390/rs61110888.

Delaney, K. J. (2017). *The robot that takes your job should pay taxes, says Bill Gates.* Quartz, 17 February. Retrieved December 18, 2019, from https://qz.com/911968/bill-gates-the-robot-that-takes-your-job-should-pay-taxes/.

Demaitre, E. (2019). *Remote surgery using robots advances with 5G tests in China.* The Robot Report, 9 September. Retrieved January 6, 2020, from https://www.therobotreport.com/remote-surgery-via-robots-advances-china-5g-tests/.

Du Boulay, B., Poulovassilis, A., Holmes, W., & Mavrikis, M. (2018). Artificial intelligence and big data to close the achievement gap. In R. Luckin (Ed.), *Enhancing learning and teaching with technology: What the research says* (pp. 256–285). London: UCL IoE Press.

Ernst, E., Merola, R., & Samaan, D. (2018). *The economics of artificial intelligence: Implications for the future of work. ILO future of work research paper series.* Geneva: International Labor Organization.

Eubanks, V. (2018). *Automating inequality: How high-tech tools profile, police, and punish the poor.* New York: St Martin's Press.

Fernandez-Luque, L., & Imran, M. (2018). Humanitarian health computing using artificial intelligence and social media: A narrative literature review. *International Journal of Medical Informatics, 114,* 136–142. https://doi.org/10.1016/j.ijmedinf.2018.01.015.

Frank, M. R., et al. (2019). Toward understanding the impact of artificial intelligence on labor. *Proceedings of the National Academy of Sciences, 116*(14), 6531–6539. https://doi.org/10.1073/pnas.1900949116.

Ganguly, K., Gulati, A., & von Braun, J. (2017). *Innovations spearheading the next transformations in India's agriculture. ZEF working paper 159.* Bonn: Center for Development Research, University of Bonn. https://doi.org/10.22004/ag.econ.259006.

Gershgorn, D. (2018). *If AI is going to be the world's doctor, it needs better textbooks.* Available via prescription AI: The algorithm will see you now. Quartz, 6 September. Retrieved November 11, 2019, fromhttps://qz.com/1367177/if-ai-is-going-to-be-the-worlds-doctor-it-needs-better-textbooks/.

Google Cloud. (2019). *AI & machine learning products: Cloud speech-to-text: Language support.* Available via Google Cloud. Retrieved November 15, 2019, from https://cloud.google.com/speech-to-text/docs/languages.

Gries, T., & Naudé, W. (2018). *Artificial intelligence, jobs, inequality and productivity: Does aggregate demand matter? IZA DP no. 12005.* Bonn: Institute of Labor Economics.

GSMA. (2018). *The mobile gender gap report 2018.* London: GSM Association.

Hess, U., & Hazell, P. (2016). *Innovations and emerging trends in agricultural insurance: How can we transfer natural risks out of rural livelihoods to empower and protect people?* Bonn and Eschborn: Deutsche Gesellschaft für Internationale Zusammenarbeit (GIZ).

Hill, R. P., & Kozup, J. C. (2007). Consumer experiences with predatory lending practices. *Journal of Consumer Affairs, 41*(1), 29–46.

IFAD. (2017). *Remote sensing for index insurance: An overview of findings and lessons learned for smallholder agriculture.* Rome: International Fund of Agricultural Development.

ITU. (2018). *Focus group on "Artificial Intelligence for Health".* Retrieved November 15, 2019, from https://www.itu.int/en/ITU-T/focusgroups/ai4h/Pages/default.aspx.

ITU. (2019). *ICT Statistics home page.* Retrieved November 15, 2019, from https://www.itu.int/en/ITU-D/Statistics/Pages/default.aspx.

Jean, N., Burke, M., Xie, M., Davis, W. M., Lobell, D. B., & Ermon, S. (2016). Combining satellite imagery and machine learning to predict poverty. *Science, 353*(6301), 790–794. https://doi.org/10.1126/science.aaf7894.

Kemp, S. (2018). *Global digital report 2018.* Retrieved November 15, 2019, from https://wearesocial.com/blog/2018/01/global-digital-report-2018.

Khare, S., Kavyashree, S., Gupta, D., & Jyotishi, A. (2017). Investigation of nutritional status of children based on machine learning techniques using Indian demographic and health survey data. *Procedia Computer Science, 115*, 338–349. https://doi.org/10.1016/j.procs.2017.09.087.

Korinek, A. (2019). *Labor in the age of automation and artificial antelligence.* Econfip Research Brief, Economists for Exclusive Prosperity.

Korinek, A., & Stiglitz, J. E. (2017). *Artificial intelligence and its implications for income distribution and unemployment. NBER working paper 24174.* Cambridge: National Bureau of Economic Research.

KPMG. (2019). *Taxation of the digitalized economy.* Birmingham: KPMG LLP.

Kubota, T. (2017). *Artificial intelligence used to identify skin cancer.* Standford News, 3 May. Retrieved November 15, 2019, from https://news.stanford.edu/2017/01/25/artificial-intelligence-used-identify-skin-cancer/.

Lewis, W. A. (1954). Economic development with unlimited supplies of labour. *The Manchester School of Economic and Social Studies, 22*(2), 139–191. https://doi.org/10.1111/j.1467-9957.1954.tb00021.x.

Lio, M., & Liu, M. (2006). ICT and agricultural productivity: Evidence from cross-country data. *Agricultural Economics, 34*(3), 221–228. https://doi.org/10.1111/j.1574-0864.2006.00120.x.

Makridakis, S. (2017). The forthcoming artificial intelligence (AI) revolution: Its impact on society and firms. *Futures, 90*, 46–60. https://doi.org/10.1016/j.futures.2017.03.006.

Marin, D. (2018). Global value chains, the rise of the robots and human capital. *Wirtschaftsdienst, 98*(1), 46–49. https://doi.org/10.1007/s10273-018-2276-9.

McKetta, I. (2018). *The world's Internet in 2018: Faster, modernizing and always on, Speedtest,* 10 December. Retrieved November 15, 2019, from https://www.speedtest.net/insights/blog/2018-internet-speeds-global.

Meara, J. G., Leather, A. J. M., Hagander, L., Alkire, B. C., Alonso, N., Ameh, E. A., Bickler, S. W., Conteh, L., Dare, A. J., Davies, J., Mérisier, E. D., El-Halabi, S., Farmer, P. E., Gawande, A., Gillies, R., Greenberg, S. L. M., Grimes, C. E., Gruen, R. L., Ismail, E. A., Kamara, T. B., Lavy, C., Lundeg, G., Mkandawire, N. C., Raykar, N. P., Riesel, J. N., Rodas, E., Rose, J., Roy, N., Shrime, M. G., Sullivan, R., Verguet, S., Watters, D., Weiser, T. G., Wilson, I. H., Yamey, G., & Yip, W. (2015). Global surgery 2030: Evidence and solutions for achieving health, welfare, and economic development. *Lancet, 386*, 569–624. https://doi.org/10.1016/S0140-6736(15)60160-X.

Mirza, A. H. (2018). Poverty data model as decision tools in planning policy development. *Sci J Informatics, 5*(1), 39. https://doi.org/10.15294/sji.v5i1.14022.

Mussa, E. (2018). *Long-term effects of childhood work on human capital formation, migration decisions, and earnings in rural Ethiopia.* Doctoral Thesis, University of Bonn, Bonn.

Nye, B. D. (2015). Intelligent tutoring systems by and for the developing world: A review of trends and approaches for educational technology in a global context. *International Journal of Artificial Intelligence in Education, 25*(2), 177–203. https://doi.org/10.1007/s40593-014-0028-6.

Óskarsdóttir, M., Bravo, C., Sarraute, C., Vanthienen, J., & Baesens, B. (2019). The value of big data for credit scoring: Enhancing financial inclusion using mobile phone data and social network analytics. *Applied Soft Computing, 74*, 26–39. https://doi.org/10.1016/j.asoc.2018.10.004.

Pew Research Center. (2018). *Artificial intelligence and the future of humans.* Washington, DC: Pew Research Center.

Polany Levitt, K. (2008). *W. Arthur Lewis: Pioneer of development economics.* UN Chronicle, XLV(1). Retrieved November 15, 2019, from https://unchronicle.un.org/article/w-arthur-lewis-pioneer-development-economics.

Polonski, V. (2017). *How artificial intelligence conquered democracy.* The Conversation, 8 August, Retrieved December 18, 2019, from https://theconversation.com/how-artificial-intelligence-conquered-democracy-77675.

Savaget, P., Chiarini, R., & Evans, S. (2019). Empowering political participation through artificial intelligence. *Science and Public Policy, 46*(3), 369–380. https://doi.org/10.1093/scipol/scy064.

Schlogl, L., & Sumner, A. (2018). *The rise of the robot reserve army: Automation and the future of economic development, work, and wages in developing countries. Working paper 487.* London: Center for Global Development.

Shorey, S., & Howard, P. N. (2015). Automation, big data, and politics: A research review. *International Journal of Communication, 10*, 5032–5055.

Singh, I. (2019). *Facebook's highly-accurate Africa population maps are available for free download.* GEO NEWS, 23 April. Retrieved November 11, 2019, from https://geoawesomeness.com/facebooks-highly-accurate-africa-population-maps-are-available-for-free-download/.

Smith-Clarke, C., Mashhadi, A., & Capra, L. (2014). Poverty on the cheap: Estimating poverty maps using aggregated mobile communication networks. In R. Grinter, T. Rodden, & P. Aoki (Eds.), *Proc. of the SIGCHI conference on human factors in computing systems* (pp. 511–520). New York: ACM.

Soto, V., Frias-Martinez, V., Virseda, J., & Frias-Martinez, E. (2011). Prediction of socioeconomic levels using cell phone records. In J. A. Konstan, R. Conejo, J. L. Marzo, & N. Oliver (Eds.), *User modeling, adaption and personalization* (pp. 377–388). Heidelberg: Springer.

SPEEDTEST. (n.d.). *Speedtest Global Index.* Retrieved November 15, 2019, from https://www.speedtest.net/global-index.

Stark, O., & Bloom, D. E. (1985). The new economics of labor migration. *The American Economic Review, 75*(2), 173–178. http://www.jstor.org/stable/1805591.

Steele, J. E., Sundsøy, P. R., Pezzulo, C., et al. (2017). Mapping poverty using mobile phone and satellite data. *Journal of the Royal Society Interface, 14*(127), 20160690. https://doi.org/10.1098/rsif.2016.0690.

Subbarao, K., del Ninno, C., Andrews, C., & Rodríguez-Alas, C. (2013). *Public works as a safety net: Design, evidence, and implementation.* Washington D.C: The World Bank.

Sundsøy, P., Bjelland, J., Reme, B. A., et al. (2016). Deep learning applied to mobile phone data for individual income classification. In A. Petrillo, A. Nikkhah, & E. P. Edward (Eds.), *Proceedings of the 2016 international conference on artificial intelligence: Technologies and applications.* Amsterdam: Atlantic Press.

Torero, M., & von Braun, J. (Eds.). (2006). *Information and communication technology for development and poverty reduction: The potential of telecommunications.* Baltimore: Johns Hopkins University Press.

Variyar, M., & Vignesh, J. (2017). *The new lending game, post-demonetisation,* ET Tech, 6 January. Retrieved November 12, 2019, from https://tech.economictimes.indiatimes.com/news/technology/the-new-lending-game-post-demonetisation/56367457.

von Braun, J. (2018). *Rural girls: Equal partners on the path to employment.* The Chicago Council. Retrieved November 11, 2019, from https://digital.thechicagocouncil.org/girls-rural-economies-2018-report/von-braun-151YF-25398G.html.

von Braun, J. (2019). *AI and robotics implications for the poor. ZEF working paper 188.* Bonn: Center for Development Research, University of Bonn. https://doi.org/10.2139/ssrn.3497591.

von Braun, J., & Gatzweiler, F. (Eds.). (2014). *Marginality—Addressing the nexus of poverty, exclusion and ecology.* Dordrecht: Springer. https://doi.org/10.1007/978-94-007-7061-4.

Welthungerhilfe. (2018). *Child growth monitor app detects hunger.* Retrieved November 11, 2019, from https://childgrowthmonitor.org/ and Welthungerhilfe 20.03.2018 | Blog at https://www.welthungerhilfe.org/news/latest-articles/child-growth-monitor-mobile-app-detects-hunger/.

Wood, C. S., Thomas, M. R., Budd, J., et al. (2019). Taking connected mobile-health diagnostics of infectious diseases to the field. *Nature, 566,* 467–474. https://doi.org/10.1038/s41586-019-0956-2.

World Bank. (2017). *G20 NOTE: Technology and jobs in the developing world.* Retrieved November 11, 2019, from https://www.ilo.org/wcmsp5/groups/public/%2D%2D-europe/%2D%2D-ro-geneva/%2D%2D-ilo-berlin/documents/genericdocument/wcms_556987.pdf.

World Bank. (2018). *Poverty and shared prosperity 2018: Piecing together the poverty puzzle.* http://hdl.handle.net/10986/30418

World Bank. (2019). *The Global Findex Database 2017: Measuring financial inclusion and the Fintech Revolution,* doi: https://doi.org/10.1596/978-1-4648-1259-0.

World Wide Web Foundation. (2017). *Artificial intelligence: Starting the policy dialogue in Africa.* Retrieved February 25, 2019, from https://webfoundation.org/research/artificial-intelligence-starting-the-policy-dialogue-in-africa/.

Zebra Medical Vision Team. (2019). *From a dream to reality. Transforming India's healthcare.* Zebra Medical Vision Blog, 12 September.https://zebramedblog.wordpress.com/2019/09/12/transforming-indias-healthcare.

Zurutuza, N. (2018). *Information poverty and algorithmic equity: Bringing advances in AI to the most vulnerable.* ITU News, 2 May. Retrieved November 15, 2019, from https://news.itu.int/information-poverty-and-algorithmic-equity-bringing-advances-in-ai-to-the-most-vulnerable/.

Robotics and AI in Food Security and Innovation: Why They Matter and How to Harness Their Power

Maximo Torero

Contents

Abstract

From strawberry-picking robots to satellite remote sensing and GIS techniques that forecast crop yields, the integration of robotics and AI in agriculture will play a key role in sustainably meeting the growing food demand of the future. But it also carries the risk of alienating a certain population, such as smallholder farmers and rural households, as digital technologies tend to be biased toward those with higher-level skills. To ensure that digital technologies are inclusive and become a driver for development, countries should make technology affordable and invest in institutions and human capital, so that everyone can participate in the new digital economy. Digital agriculture also represents an opportunity for young people as agriculture value chains can be developed to create new service jobs in rural areas, making agriculture an attractive sector for youth.

M. Torero (✉)
Chief Economist, Food and Agriculture Organization of the United Nations, Rome, Italy
e-mail: maximo.torerocullen@fao.org

Keywords

Digital farming · Field robots · Agricultural economics · Innovation · Food security

Challenge: The Great Balancing Act

Feeding nearly 10 billion people by 2050, while maintaining economic growth and protecting the environment is an urgent, unprecedented challenge. It is a challenge of managing trade-offs between the immediate need to produce food and the long-term goal of conserving the ecosystem. Robotics and artificial intelligence will play a crucial role in this. However, as digital technologies revolutionize, the risks of unequal access and digital exclusion loom large. Countries must make investments in human capital and put policies and regulations in place to minimize such risks and ensure that everyone, especially smallholders who produce the majority of the world's food, can participate in a new digital economy.

© The Author(s) 2021
J. von Braun et al. (eds.), *Robotics, AI, and Humanity*, https://doi.org/10.1007/978-3-030-54173-6_8

To achieve sustainable food future, countries must meet the following needs simultaneously (Searchinger et al. 2013):

1. Close the gap between the amount of food available today and the amount that would be required in 2050. In order to do this, the world needs to increase food calories by 50% (FAO 2018).
2. Boost agricultural development, so that it can contribute to inclusive economic growth and social development. The agriculture sectors employ 28% of the global population and are still one of the most important pathways to help people escape poverty and hunger (World Bank 2019a).
3. Reduce environmental impact. Food production is the largest cause of environmental change globally. Agriculture contributes up to 29% of global greenhouse gas emissions (Vermeulen et al. 2012).

Additionally, countries must save water and protect land. Agriculture is the culprit of water scarcity, accounting for 70% of all water withdrawals (FAO 2019). About 40% of the world's land area is used for agriculture (Owen 2005). And the expansion of croplands and pastures undermines the capacity of ecosystems to sustain food production.

Climate variability will exacerbate food security in emerging economies (Torero 2016). Climate variability goes beyond extreme weather events and may include shifts in meteorological patterns. This could result in large-scale changes in agricultural sectors, causing major uncertainties and volatility. Even though this is integral information for food security and sustainable agriculture, policymakers have been slow to integrate this into long-term agricultural policy development.

What Is Happening: Robotic Farming

Mechanization has played a significant role in agriculture in the past. As shown in Fig. 1 below, mechanization was characterized by large farms (150–240 ha) focusing on increasing yields. The result of this process was a vicious cycle in which large farms adopted mechanization, increasing the competition for input and output markets, which then resulted in a more concentrated agricultural sector in terms of not only land but also input and output markets (see Fig. 1 below). This was identified earlier (Schmitz and Seckler 1970), but the social costs of past mechanization were never properly tackled.

Robotics is increasingly becoming a part of agriculture, but it is starting a different evolution process because of its potential to be scale neutral. This development can be broadly understood by studying what is happening at Amazon (Corke 2015) and other companies like Alibaba. Amazon's auto-

mated fulfillment centers employ a large number of small robots that operate in parallel, optimizing the allocation of resources and automation processes. The robotics and technology in agriculture is moving in the same direction. In the future, a farm could be populated by a large number of small, autonomous robots operating in different fields at the same time, performing the equivalent work of one large machine.

As robots become scale neutral and cost effective, they could equally benefit large-scale and small-scale farmers. Smallholders in rural areas could use apps such as Uber-for-tractor to book and hire expensive agricultural equipment, which would have been beyond their reach previously (Vota 2017). They might also find ways to share the fixed costs of small robots with other smallholder farmers through associative institutional designs.

Since most of the rural land is farmed by smallholders, it is essential to figure out how to adopt and incorporate technology to ensure that they can benefit from it. The key is to lower costs to make it sustainable. Access to numerous cheap and smaller robots that can work 24/7 could solve big agricultural production problems. For example, "AgBots" could reduce the cost of weed operations by 90% by cutting pesticide use (Corke 2015).

Farming is becoming a complex information space. From risk management of farm operations to special analytics, farming technology is advancing rapidly. Examples:

- *Evidence-based agriculture decision support.* There are key factors that might affect crop yields, including soil temperature, moisture levels, crop stress, pests, and diseases. IBM's Watson Decision Platform for Agriculture comprises artificial intelligence and cloud solutions that generate evidence-based insights into these key factors (Wiggers 2018). Watson can identify pests based on drone imagery. Its smartphone app can identify crop disease. It can also identify the best practices for irrigation, planting, and fertilization, and indicate the ideal time of the year to sell a given crop.
- *Potato: yield, pest, and diseases.* Satellite remote sensing and GIS techniques have been employed to forecast potato yields and detect blight disease and whiteflies. A smartphone app developed by Penn State University's PlantVillage and the International Institute of Tropical Agriculture gives farmers real-time diagnoses of crop diseases using AI algorithms trained with images of diseased crops (Kreuze 2019). The PlantVillage app detects diseases in sweet potato and monitors the spread of caterpillar in around 70 countries.
- *Automatic fruit grading and counting.* Cognitive image processing and computer vision techniques can support harvest by providing automated fruit grading, ripeness detection, fruit counting, and yield prediction—time-consuming tasks that are prone to human error (Mavridou

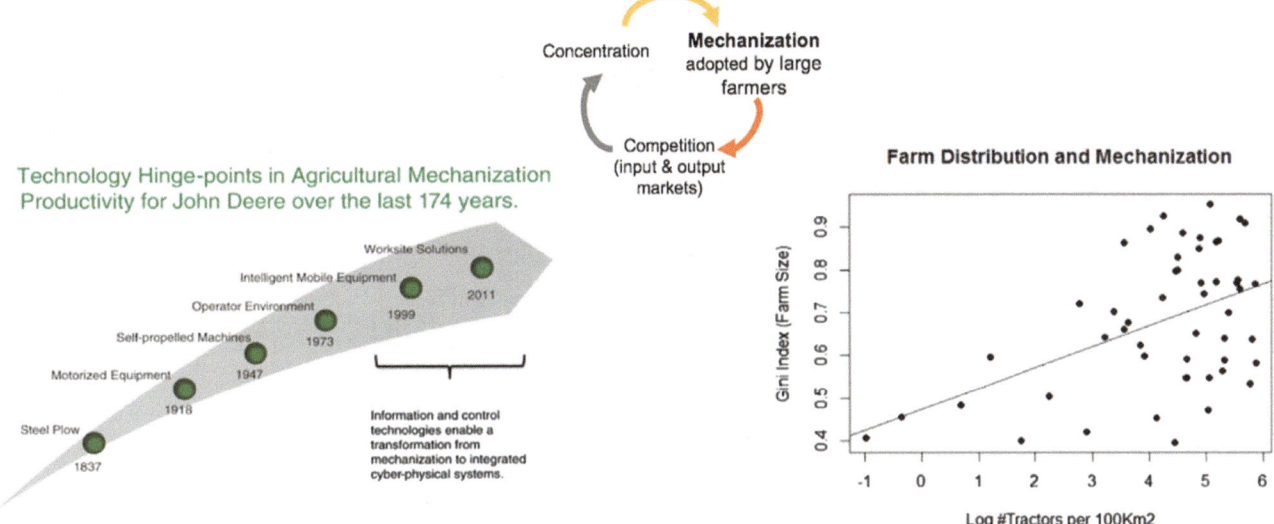

Fig. 1 Past mechanization resulted in concentrated agriculture sector. Source: Reid (2011), reprinted with permission from the National Academy of Engineering

et al. 2019). Machine vision can also protect plant health by detecting deficiencies and diseases using clustering based on color features.

- *Crop-harvesting robots.* The automatic harvester from Harvest CROO Robotics mimics human pickers, without changing how farmers grow their crop (Harvest CROO 2020). The use of multiple robotic components to separate the picking functions—leaf gathering, visual inspection, picking, and then packing—makes it possible to build a simple robot, which can operate faster. It takes a robot harvester 8 s to pick a single strawberry plant; 1.5 s to move to the next plant. A single robot harvester can work eight acres a day, which is equivalent to the work of 30 human laborers.

- *Soil and water management.* Traditionally farmers have used chemical analysis to test the quality of their soil, a complex and time-consuming method. It has created a barrier for smallholder farmers. In many rural areas of Africa, Asia, and Latin America, farmers are accruing debts to purchase packages of soil, but it does not help them achieve higher yields, because they do not know the condition of the soil they are purchasing. IBM's AgroPad, a smartphone app, is a paper-based analysis of soil that could provide the kind of information smallholder farmers need (Steiner 2018). Using sensors and AI algorithms, the app lets farmers understand the chemical make-up of their soil and also monitor the health of soil and water. It is low cost, and the paper-based device can be mass produced and deployed at a large scale through mobile and cloud technologies.

- *Crop price prediction.* FAO's Agricultural Market Information System (FAO-AMIS) uses machine learning to perform sentiment-based market prediction and crop forecasting to predict prices of wheat, maize, rice, and soybeans (Agricultural Market Information System 2020). Based on supply-and-demand position data, as well as satellite information calibration and testing on the ground, and their probable short-term development and outliers, the system predicts prices. It does so by combining the intensity of positive and negative forces, which means that when the density of positive forces is higher, it helps to predict positive price increases. To assess the market forces variables, several subject scopes are considered: monetary policy, market conditions, finance and investment, elections, negotiations, polices, regulations, energy, environment, and agriculture infrastructure are also included. The platform bolsters the preparedness and resilience to external shocks like food price surges (Fig. 2).

Mechanisms to Promote Development

The integration of robotics and AI in agriculture has the potential risk of alienating certain population, as digital technologies tend to be skill-biased, productivity-biased, and voice-biased. There will be challenges of achieving inclusion, efficiency, and innovation. Countries must expand access to technologies, lower costs, and scale up innovations to ensure smallholder farmers and rural households, too, become beneficiaries (Fig. 3).

Create Complements to Minimize the Risks

Rapid technological progress will make it possible for poor people to afford many digital services. But for them to do

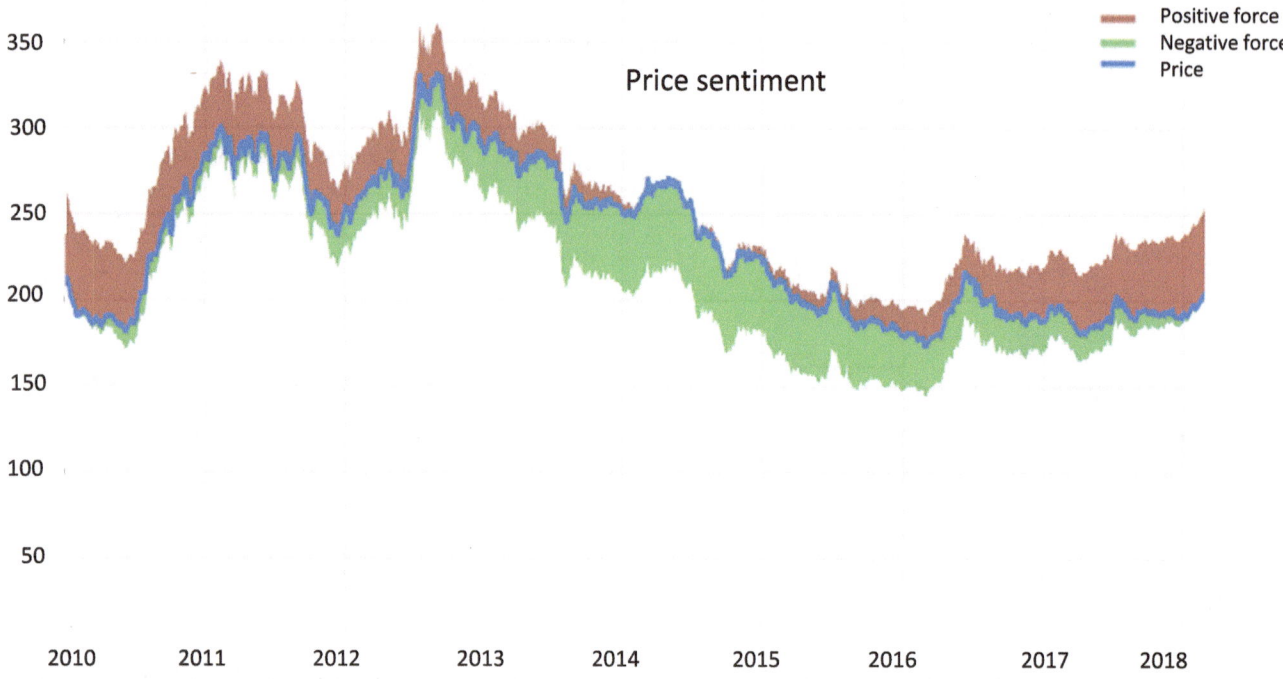

Fig. 2 Agriculture Market Information System performs sentiment-based market prediction. Source: FAO-AMIS

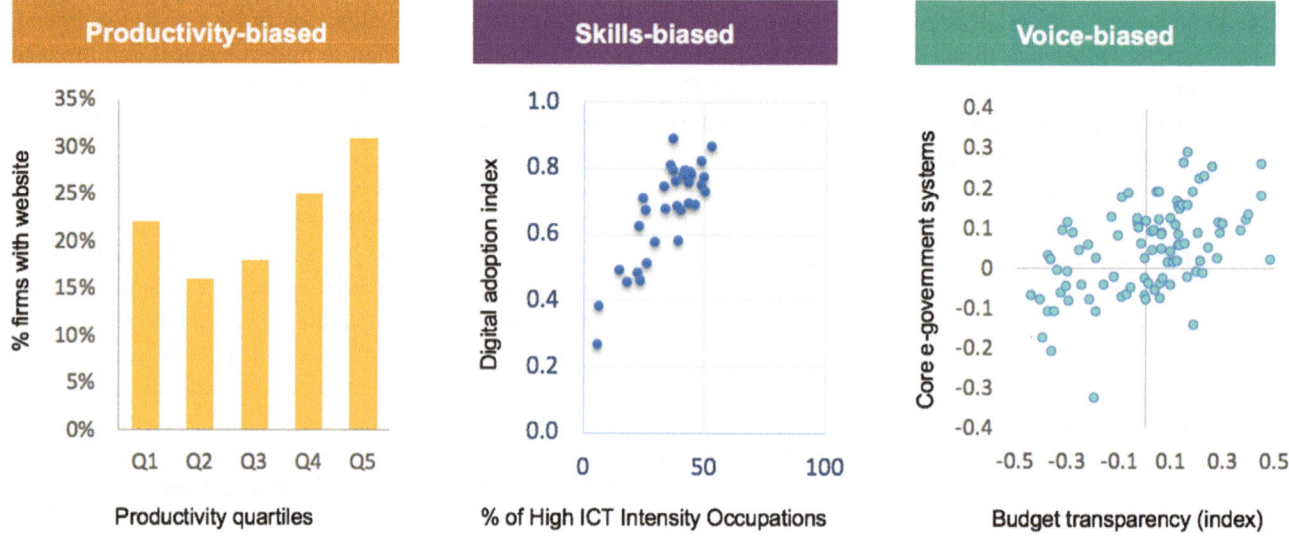

Fig. 3 Digital technologies tend to be biased. Source: World Bank Group (2016) and World Development Report Team (2016)

so, governments must invest in creating an environment that can avoid concentration of benefits and minimize inequality. Unfortunately, complements—such as regulations promoting entry and competition, skills enabling workers to access and leverage the new economy, and institutions delivering efficient mobile phone-based services and an e-government—have not kept pace (Fig. 4).

The importance of providing the complements, or laying down the "analog foundations," for a new digital economy as early as possible cannot be overstated. Specifically, governments must provide an environment that ensures

accessibility, adaptability, and appropriateness. This means the cost of accessing technologies should be affordable for all users. Technologies become adaptable when the content (knowledge or information) they are used to access is beneficial for users. Finally, technologies should be appropriate, which means that users must have the capability to use them (Fig. 5).

Countries should formulate digital development strategies that are broader than the current information and communication technology strategies. For example, when Ethiopia was given a loan of $500 million from the World Bank Group to

Fig. 4 Benefits and risks of digital technologies. Source: World Bank Group (2016) and World Development Report Team (2016)

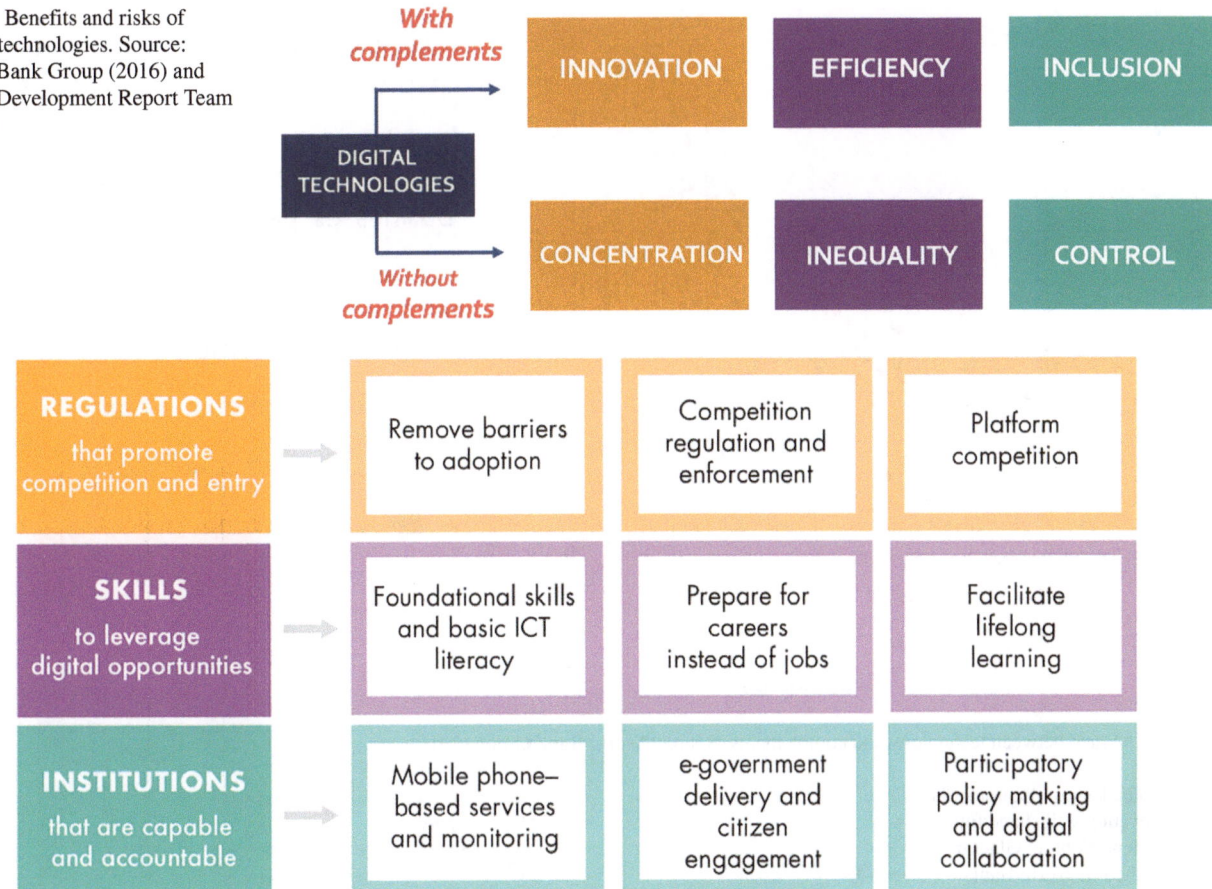

Fig. 5 National priorities: analog foundations for a digital economy. Source: World Bank Group (2016)

privatize all of its telecommunications sector, its investment plan had no regulatory mechanisms to create complements that could reduce inequality. Countries should create a policy and institutional environment for technology that fosters the greatest benefits for everyone. The quality of complements and technology both rise with incomes (Fig. 6).

Illiteracy is a potential constraint that could limit the benefits of digital technologies, as any technology will require the user to have a certain level of literacy. One way of overcoming this constraint is to work with children, as they can read and write, are more likely to be familiar with information and communication technology, and can be the bridge transferring knowledge, information and capability to their parents.

Why It Matters: Robotics in Agriculture

Robots Will Address Labor Shortage

One of the defining features of agriculture is a decrease in the available agriculture workforce, especially in Africa, Asia, and Latin America. Robotics and AI will have a significant effect on the labor markets, as they address farm labor shortage.

As stated earlier, the traditional mode of mechanization used to mean combining one worker with a large machinery to maximize the areas covered. This is no longer the case. Capabilities of these machines include being able to operate 24/7, which means they will increase productivity. Subsequently, labor-intensive agriculture in low-income countries will not be economically sustainable (Fig. 7).

- *New skills development.* As robots address labor scarcity on the farm, it will inevitably stoke fear. But the old jobs have already been lost. Now governments must focus on increasing the skill levels that farm workers have, so that they can be integrated into the agricultural value chains. Without preparation, many people would not benefit from the advancement of digital technologies and be left behind.

- *Positive social outcomes.* Labor-saving technologies could free up time for women smallholder farmers. With more time, they could undertake more productive work, earn more income, and also participate in decision-making, which would help improve their livelihoods and

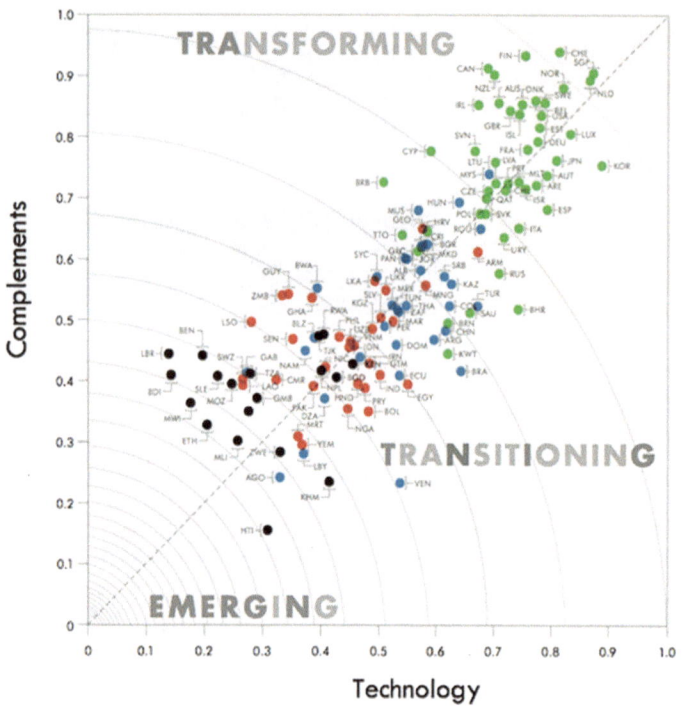

Fig. 6 The race between technology and complements. Source: World Bank Group (2016)

Fig. 7 The forces of automation and innovation will shape future employment. Source: Glaesar (2018), cited in World Bank Group (2016)

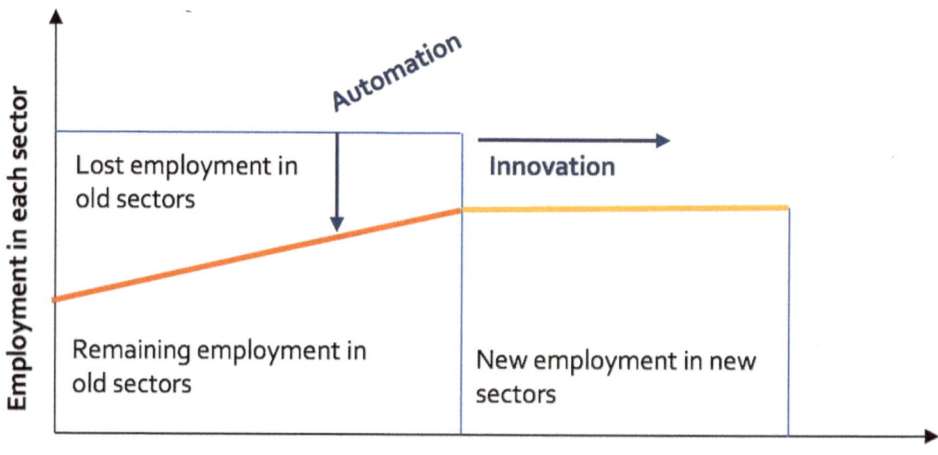

social status. Labor-saving technologies could also free up children from having to perform repetitive tasks and go back to school to receive high-quality education.[1] Such technologies provide an opportunity to increase human capital.

- *Youth employment.* Digital agriculture represents an opportunity for young people. Farming used to mean arduous physical labor, exposure to harsh weather conditions, and repetitive tasks. Suddenly, it has become a sector with robots that appeals to young people. There would be new service jobs in rural areas for medium- or high-

skilled young people who are IT specialists, engineers, or mechanics. For example, the digital advances would require a new class of skilled workforce that can handle the machines as managers and technicians. It would help reduce rural exodus.

Capital and Technologies Are Opportunities for Smallholders

New technologies that are affordable represent opportunities for smallholder households.

[1] About 59% of all children in hazardous work aged 5–17 work in agriculture (ILO 2020).

- *Technical opportunities.* Small-size mechanization means new possibilities for remote areas, steep slopes, or soft soil areas. Previously marginal areas could be productive again. Precision farming could be introduced to farmers that have little capital. It will allow farmers to adopt climate-smart practices. Farmers can be providers and consumers of data, as they link to cloud technologies using their smartphones, connecting to risk management instruments and track crop damage in real time.
- *Economic opportunities.* Buying new machinery no longer means getting oneself into debt thanks to low-risk premium, better access to credit and leasing options, and more liquid renting markets. There is a reduced risk of concentration due to capital constraint. The reduced efficient scale of production means higher profitability for smallholders.

Robots in the field also represent opportunities for income diversification for farmers and their family members. There will be no need to use family labor for low productivity tasks, and time can be allocated for more profit-generating activities. Additionally, robots can operate 24/7, allowing more precision on timing of harvest, especially for high-value commodities like grapes or strawberries.

Quantifying Robots' Impact on Poverty

There are sunk cost and entry cost into any market (Sutton 1991). Evidence shows that sunk costs may contribute to either an increase or a decrease in industry concentration, depending on how the cost components are distributed. If entering into a market with certain technologies requires significant investment, it could be a factor that discourages market entry, resulting in exclusion. Governments have to find innovative ways to make market entry easier.

In an ongoing study, researchers ran a simulation exercise to quantify the impact of technology on poverty.[2] Using a generic weed-picking robot, they applied the effects of labor-saving technologies to crop producers. Such a robot costs $10,000 or an average of annual cost of $3000 for a five-year life expectancy or an hourly rental of $0.07. Potentially, the weed-picking robot could work 61,320 h annually. Labor requirement for manual weeding would have taken 377 people per hectare per hour.

Using 2013 as a baseline, the researchers ran a dynamic multi-country, multi-sector general equilibrium model, MIRAGRODEP (Agrodep 2013).[3] The researchers simulated a labor-saving technological shock (not a total factor pro-

ductivity shock) on all crop producers looking at increases in services costs (renting of the robot), decreasing required amount of hired labor, reducing the amount of household labor, and how this affects cost structure, labor demand, and services demand. Finally, they assessed the effects on wages, input and output prices, and poverty impacts using the $1.90 Purchasing Power Parity poverty line.

A preliminary simulation assuming a 20% labor productivity-saving technology was run for two scenarios. Scenario 1 was "exclusive" technologies, which meant the productivity shock affected the whole agricultural sector, but not the smallholders. Therefore, poor people would benefit from higher wages (potentially) and lower food prices. Scenario 2 was "inclusive" technologies, and the productivity shock could include the smallholders. In this case, smallholders would benefit from the productivity gains for their own production.

As shown in Fig. 8, there were important reductions in the number of poor people in both scenarios, because of the labor productivity shock, resulting from the use of automatization. But the reductions were significantly higher for inclusive technologies. Notwithstanding important assumptions behind these simulations, the results clearly indicate that there is a lot of potential for these new labor-saving technologies, and that it is extremely important they reach smallholders. Unlike in the past, it is possible to be inclusive since there is a significant opportunity for "adoption at scale."

Challenges

It is worth reiterating a number of challenges that need to be overcome to ensure digital technologies are innovative, efficient, and inclusive.

- *Innovation for smallholders.* It is essential to provide useful and relevant information and knowledge to users. Providing access to the technology is of course important, but it would be rendered useless, if the content cannot be adapted by smallholders.
- *Energy supply.* Countries must focus on sourcing energy supply for robotics at scale. Renewable energy, especially solar energy with its consistently declining prices, have a vital role to play.
- *Value chain.* As jobs are created and eliminated with the advancement of digital technologies, agriculture value chains must be developed to harness their benefits and create higher-wage jobs, as previous laborers become managers of machinery. It provides an ideal opportunity to minimize inequalities and develop human capital.
- *Connectivity.* Robotics innovation requires the right intellectual property system. Governments must also put regulations in place to reduce the risk of oligopoly. New

[2]Laborde and Torero (in progress). "The Effects of Digitalization."

[3]MIRAGRODEP is a new version of the MIRAGE model of the world economy.

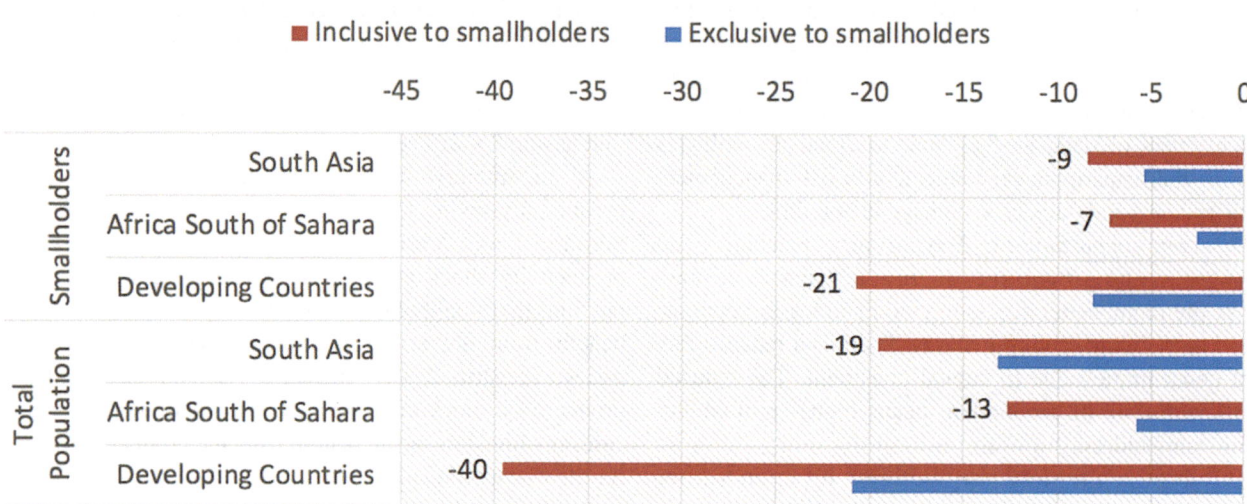

Fig. 8 Labor-saving technologies can help reduce the number of poor people. Source: Laborde and Torero (work in progress)

connectivity should protect data collected on the farm. Cyber-security concerns for farm robots should be addressed.

- *Complements*. Governments must invest in developing human capital, policy and accountable institutions to lay the foundation for smallholders and rural households to participate in a new digital economy.

- *Capabilities*. Fighting illiteracy is a key strategy in the effort to promote inclusive digital revolution. Without investing in human capital, especially quality education and integration of women in the workforce, it is not possible to minimize exclusion. Upward inter-generational transfer of knowledge is an option that has been tested successfully in different regions.

References

Agricultural Market Information System. (2020). Retrieved February 3, 2020, from http://www.amis-outlook.org/.

Agrodep. (2013). http://www.agrodep.org/model/miragrodep-model

Corke, P. (2015). *Robots in agriculture*. https://www.youtube.com/watch?v=Q69RVHy-Xbw&t=2s

FAO. (2018). *The future of food and agriculture: Alternative pathways to 2050*. http://www.fao.org/3/I8429EN/i8429en.pdf

FAO. (2019). *Water scarcity–One of our greatest challenges of all time*. Available via FAO. http://www.fao.org/fao-stories/article/en/c/1185405/

FAOSTAT. (2020). Retrieved February 3, 2020, from http://www.fao.org/faostat/en/#data.

Harvest CROO Robotics. (2020). https://harvestcroo.com/about/#technology-highlights

ILO. (2020). *Child labour in agriculture*. Available via ILO. https://www.ilo.org/ipec/areas/Agriculture/lang%2D%2Den/index.htm

Kreuze, J. (2019). *The bots are here and they're protecting our crops*. Available via Scientific American. https://

blogs.scientificamerican.com/observations/the-bots-are-here-and-theyre-protecting-our-crops/

Laborde, D., & Torero, M. (in progress). The effects of digitalization.

Lowder, S. K., Skoet, J., & Raney, T. (2016). The number, distribution of farms, smallholder farms, and family farms worldwide. *World Development, 87*, 16–29. https://doi.org/10.1016/j.worlddev.2015.10.041.

Mavridou, E., Vrochidou, E., Papakostas, G. A., Pachidis, T., & Kaburlasos, V. G. (2019). Machine vision systems in precision agriculture for crop farming. *Journal of Imaging, 5*(12), 89. https://doi.org/10.3390/jimaging5120089.

Owen, J. (2005). *Farming claims almost half Earth's land, new maps show*. Available via National Geographic. https://www.nationalgeographic.com/news/2005/12/agriculture-food-crops-land/

Reid, J. F. (2011). The impact of mechanization on agriculture. *National Academy of Engineering, 41*(3), 22–29. Available via National Academy of Engineering. https://www.nae.edu/52645/The-Impact-of-Mechanization-on-Agriculture.

Schmitz, A., & Seckler, D. (1970). Mechanized agriculture and social welfare: The case of the tomato harvester. *American Journal of Agricultural Economics, 52*(4), 569–577. https://doi.org/10.2307/1237264.

Searchinger, T., Hanson, C., Ranganathan, J., Lipinski, B., Waite, R., Winterbottom, R., Dinshaw, A., & Heimlich, R. (2013). *The great balancing act: Creating a sustainable food future, installment one*. Available via World Resource Institute. https://www.wri.org/publication/great-balancing-act

Steiner, M. (2018). *AI-powered technology will help farmers' health-check soil and water*. Available via IBM https://www.ibm.com/blogs/research/2018/09/agropad/

Sutton, J. (1991). *Sunk costs and market structure*. Cambridge: MIT Press.

Torero, M. (2016). *Scenarios on climate change impacts for developing APEC economies*. https://maximotorero.files.wordpress.com/2020/02/scenarios-on-climate-change-impacts-for-developing-apec-economies.pdf

Vermeulen, S. J., Campbell, B. M., & Ingram, J. S. I. (2012). Climate change and food systems. *The Annual Review of Environment and Resources, 37*, 195–222. https://doi.org/10.1146/annurev-environ-020411-130608.

Vota, W. (2017). *Uber for tractors is really a thing in developing countries*. Avalabe via ICTworks. https://www.ictworks.org/uber-for-tractors-is-really-a-thing-in-developing-countries/#.XkvPx0BFyUl

Wiggers, K. (2018). *IBM's Watson agriculture platform predicts crop prices, combats pests and more*. Available via Venturebeat. https://venturebeat.com/2018/09/24/ibms-new-ai-platform-for-agriculture-predicts-crop-prices-combats-pests-and-more/

World Bank. (2019a). *Agriculture and food*. Available via World Bank. https://www.worldbank.org/en/topic/agriculture/overview

World Bank. (2019b). *World development report 2019: The changing nature of work*. Washington, DC: World Bank. https://openknowledge.worldbank.org/handle/10986/30435.

World Bank Group. (2016). *World development report 2016: Digital dividends*. Washington, DC: World Bank. https://openknowledge.worldbank.org/handle/10986/23347.

World Development Report Team. (2016). *Digital dividends presentation*. Washington, DC: World Bank. Available via http://pubdocs.worldbank.org/en/668301452797743927/WDR-2016-Overview-Presentation-DC-launch.pdf.

Robotics in the Classroom: Hopes or Threats?

Pierre Léna

Contents

Abstract

Artificial intelligence implemented in a great diversity of systems, such as smartphones, computers, or robots, is progressively invading almost all aspects of life. Education is already concerned by this revolution, as are medicine or care for elderly people. Education is indeed a special case, because it is fundamentally based on the relationship, involving love and emotions as well as knowledge, between a fragile child and an adult. But teachers are becoming rare and education expensive: The Earth demography is here an economical challenge. We examine some of the various modalities of teacher substitution, companionship or computer-resources which are already experimented, and discuss their ethical aspects. We conclude on the positive aspects of computer-aided education, which does not substitute the teacher, but may help and provide continued professional development.

Keywords

Education · Emotions · Children · Adults · Love · Knowledge relationship · Robots

P. Léna
Observatoire de Paris & Université de Paris, Paris, France
e-mail: pierre.lena@obspm.fr

Pontifical Academy of Sciences & Fondation La main à la pâte, Paris, France

Introduction

In 2017, Sir Anthony Seldon, specialist of education, vice-Chancellor of the University of Buckingham (UK), prophesized that within a decade: "...intelligent machines that adapt to suit the learning styles of individual children will

J. von Braun et al. (eds.), *Robotics, AI, and Humanity*, https://doi.org/10.1007/978-3-030-54173-6_9

soon render traditional academic teaching all but redundant [. . .]. They will learn to read the brains and facial expressions of pupils, adapting the method of communication to what works best for them" (Bodkin 2017). As the digital world, supported by the seemingly endless developments of artificial intelligence, data collection and mining, progressively invades all sectors of private and public life, will education resist to this invasion, or benefit from it? Indeed, since over two millennia, schooling has been based on a face-to-face relation between a teacher and the student. Will robotics make outdated this traditional vision, at an epoch when the amount of mass education is required on Earth at an unprecedented scale? Is this perspective a fantasy, a likely nightmare or an interesting evolution?

First, we discuss in general terms the hopes and changes which these perspectives could offer, while having been explored since barely a decade. Second, we address the simplest issue, dealing with the use of robots as pedagogical tools, with the specific goal to introduce the pupils to computer science. Third, we enter into the hot question of "robot teachers," analyzing the diversity of situations and current experiments and research which can be considered under such a broad and somewhat provocative expression associating "teachers" and "robots": it may span from a simple machine, designed to help a human teacher, to a full humanoid substitute of the teacher. At this point, it is necessary to introduce an ethical discussion, since one must consider the fragility of the child, exposed for instance to the possibility for a robot "to read the child's brain and face." Finally, we try to focus on the most promising and probably most realistic contribution of artificial intelligence to education, namely the computer-aided education, understood in a less revolutionary sense than the existence of humanoid robot teachers.

We shall observe that actual implementations to-date are only beginning, and research on their impacts very limited. Hence, our conclusions will be careful and probably fragile.

Emerging Needs, Hopes, and Threats

Since over two millennia, education at school is based on a face-to-face relation between the teacher and the student. The person-to-person dialog between Socrates and Menon happens today in classes at a different scale, with tens, hundreds, or even more pupils in the class, but a "vertical" transmission of knowledge remains the general rule of primary and secondary education. Worldwide, teachers are trained for this purpose and implement their pedagogy in this context. Is this the most efficient way to transmit knowledge and to prepare the youth to read the present world and be ready for its future? Various attempts to explore alternate ways of improvement have been made, based on more or less empirical hypothesis on the learning process. Since several

decades and following John Dewey's ideas (1859–1962), an *inquiry* pedagogy, more "horizontal" and making the students more active, has developed. Neurosciences developments with Stanislas Dehaene are supporting the early intuitions of Maria Montessori (1870–1952), Lev Vygotski (1896–1934), and Jean Piaget (1896–1980), for a better respect of the stages which characterize the cognitive and physical development of the child (Dehaene 2018). Our own effort since 1996 on early science education with *La main à la pâte* has been inspired by these pedagogies (Charpak et al. 2005).[1] Recently, the scheme of "flipped (or inversed) classroom" (*classe inversée*) has become popular and begins to be implemented in various countries,[2] especially in higher education. There, the autonomy of the learner is stimulated, while the teacher is considered as a support, answering the questions and leading discussions.

Breaking the passivity of the "purely listening" (and often bored!) student is considered essential for an efficient learning of a foreign language: hence language laboratories have been among the first to replace, for some activities, the teacher by a machine. More recently, computers and tablets have emerged in the classrooms as teaching aids, and specific software becomes available to teachers of all disciplines. Geometry benefits from visual tools, geography from Google Earth, language from orthographic correction and voice helpers, etc.

With the advent of the digital revolution, progressively impacting all sectors of human activities and professional practices, an unescapable question emerges: will teachers disappear, or have to adapt to completely new schemes of professional activity? The physician profession is interesting to observe. In its practice, the personal relation with the patient, as for teachers, has always been considered essential. Yet, the profession is currently confronted to this same question and already encounters significant evolutions, such as telemedicine, robotics helpers . . . Similarly, the magnitude of aging population in China or Japan already leads to some care-taking robots for old people, a case which is not without some similarities with the issue of teaching.

Some relatively significant impacts on classroom practices are already perceivable: students have an unlimited access to information through Internet; collaborative work between schools across a country, or even worldwide, becomes common practice, especially on universal issues like sustainable development or climate education;[3] special education for children with dyspraxia draws on digital resources,[4] etc.

---

[1] See also (in English) www.fondation-lamap.org

[2] https://en.wikipedia.org/wiki/Flipped_classroom

[3] For example the networks organized by various organisations such as *La main à la pâte*, *Scholas Occurentes*, *Eco-schools*, etc. (see Battro et al. 2017).

[4] In France, inspired by S. Dehaene, the 'Cartable fantastique' (*The fantastic schoolbag*) https://www.cartablefantastique.fr/

Looking further into the future, several factors may indicate that a massive evolution of the classical schooling methods may come or will even be forced to happen. The cost of education may be the dominant factor. In a developed country such as France, the current offer of primary and secondary education—up to ages 16–18—to the whole of an age class represents over 10% in the budget of the nation. The goal of *"equal access to quality education"* is included as one of the 17 Sustainable Development Goals of the United Nations,[5] promulgated in 2015. Yet, attaining worldwide this goal seems entirely out of prospect during the next decades. It would need adding 20.1 million primary and secondary school teachers to the workforce, while also finding replacements for the 48.6 million expected to leave until 2030, because of their retirement, or the end of a temporary contract, or the desire to pursue a different profession with better pay or better working conditions (UIS 2016).[6] On top of the costs themselves, which yet maintain mediocre salaries for teachers, the supply and quality of these teachers remains a central problem, almost everywhere. In France, the traditional status of public teachers as selected "expert" civil servants is no longer sustainable for mathematics or English teachers in secondary schools, and other schemes of recruitment, with a lesser guarantee on quality, are being implemented. In Africa especially, the demographic pressure in the coming decades is preparing a difficult challenge for schooling, which itself is a necessary condition for economic development and adaptation to climatic changes. Therefore, in developing countries, if cheaper methods, such as lessons through Android smartphones, become available to access the knowledge, it is likely that the poorest parents will use the cheapest means, while families with sufficient resources will continue to choose human teachers for their children whenever possible. The universal extension of Wi-Fi connections, with local substitutes, in case of unavailability, which are capable to store large data bases, creates an entirely new context, not free of commercial interests.

It is therefore worthwhile to explore more in depth the perspectives which robotics and computers may offer to these challenges. Moreover, observing how schooling is becoming a business in some developing or emerging countries, the commercial potential of education needs, if seized by actors mastering the digital tools at large scale, may become a reality, with all the questions it raises on quality, equity, and ethics.

People seem to be worried about the use of robots in schools. In 2012, a European survey of public attitudes (European Commission 2012) to robots over 27,000 persons reached interesting conclusions. In great majority, European citizens are not opposed to the use of robots, in case of manufacturing or various domestic uses. On the opposite, 60% consider that robots should be banned from the care of children, 34% that they should be entirely banned from the field of education, while only 2% thought robots could be used in education, namely schooling. Similar attitudes are observed towards health care, care of children, elderly, or disabled persons, i.e., human tasks. Either pros or cons, are these attitudes justified?

We place the present discussion in a slightly broader frame than the mechanical robots and extend it to the possible roles of computer-based artificial intelligence in education. Indeed, there exists a continuum from the latter to the former, and technology is constantly opening new combinations of soft- and hardware. As a matter of fact, the term "computer aided education" goes beyond "robotics" itself (Cigi 2013). This broad frame may be addressed with the goal to totally or partially replace the teachers by robots, a discussion well introduced by Amanda Sharkey (2016), from whom we borrow several points, referring to her abundant and up-to-date bibliography.

A Simple Case: Robots as Pedagogical Tools

In primary and secondary schools, science and technology lessons are already exploiting robotics as a rich pedagogical tool. Since 2008, the robot Thymio II,[7] developed at the Ecole Polytechnique Fédérale in Lausanne (Switzerland), provides a combination of robotics and programming, in order to introduce children to the digital world. *La main à la pâte* in France has developed extensive modules, named *"1, 2, 3 . . . Codez"* helping primary school teachers to introduce robotics, from age 5 upwards.[8] These modules, introduced in 2014, are disseminated with a great success among teachers. Observing teachers and children in the thousands of classes which are using worldwide these teaching aids,[9] some interesting conclusions are reached:

- First, children at all ages find robots attracting and exciting their curiosity.
- Young children progressively learn the difference between "alive" and "not-alive," comparing the robot with animals or plants. Qualifying the robot as "intelligent" and having themselves programmed it, they explore the meaning of human intelligence.
- Programming a robot for a specific task, then letting it act, explores the benefits of mistakes and errors, without

[5] https://sustainabledevelopment.un.org/

[6] 'Education for people and the planet', UNESCO 2016.

[7] https://www.generationrobots.com/en/179-educational-robot-thymio

[8] Free modules available in English, German and French: http://www.fondation-lamap.org/fr/node/65695

[9] See http://www.fondation-lamap.org/fr/page/34536/1-2-3-codez-espace-enseignants

any value judgment or cognitive risk for the student. As machine learning, it can be repeated at no cost and introduces the teacher to the use of more sophisticated machine learning if so wished.

- Equally interesting is the combination offered there between a physical object (the robot) and a logical set of instructions (the program). The former is submitted to the constraints of the physical world (e.g., the size of the room where the robot moves, the friction on its wheels, the battery . . .), while the latter is only constrained by logics and eventually mathematical rules. The fertile difference between an error and a physical approximation or uncertainty may then be introduced to students.
- Programming the robot offers an introduction to encoding with a rich combination of variables, sequences, programming events, and feedback. This is a first and early introduction to computer science. Some ethical aspects may also be introduced and discussed in the classroom: who should make a decision, the machine or the child?

To conclude this point, the use of robotics in classroom, combined with computer science and eventually electronics, is a straightforward and creative way to teach technology within the aggregate called STEM (Science Technology Engineering Mathematics).

Robot Teachers: A Diversity of Possible Roles?

Education among humans, as among some species within the animal world, begins with imitation. In parallel with imitation, a person-to-person relationship is established, which begins at the infant stage with the use of symbolic language. The act of teaching, its specific characters when exercised by humans, the complex interactions between the mind of the teacher and one of the pupils have been extensively studied, and cannot be developed here (Ziv et al. 2016; Strauss 2016). In principle, the question of an eventual "robot teacher" should be analyzed within this extremely complex context, especially when it is proposed to replace human teachers by a humanoid robot, which would fully handle the classroom.

As a first-order approach, I here follow the categories introduced by Sharkey (2016), who distinguishes: (a) the "social" robots as a substitute for teacher; or (b) as "a companion and peer"; or finally (c) as a tool for distance learning with telepresence. Each of these roles deserves a specific discussion, based on the few published experiences available today. We shall conclude that considering artificial intelligence (AI) as a potential teaching aid, rather than a full teacher substitute, seems to be the best direction to explore and implement, as argued by Rose Luckin and coworkers from University College London (Luckin et al. 2016). We

observe that such categories, although helpful to sort out the diversity of uses and their positive or negative aspects, do not properly cover the great versatility of robots for many types of applications. The *NAO* robot, initially developed in France and currently in Japan,[10] seems to be used in many different instances: companion, game partner, attendance of a sick person, education, teaching aid for disabled, etc.

Robots as a Full Substitute to Teachers

Saya is a female humanoid robot developed in Japan. Its facial mobility allows to express emotions (anger, joy . . .). The techniques are similar to the ones developed for sexual robots (Levy 2007),[11] of which there already exist some presentations on Internet. Using robots to replace teachers in the classroom would require demonstrating the necessity and efficiency of such decision. As Sharkey notes, robots can be useful when special risks are affecting tasks carried by humans, such as dangerous environments or need for very fast decisions. Teaching is free of such risks. The heart of a positive interaction between the teacher and the student lays in the ability of the former to "read the mind" of the latter, hence to efficiently accompany the emotions as well as the acquisition of knowledge and know-how. Sharkey also argues that there exist to date no evidence showing that a robot, acting "alone" as a full teacher's substitute, can better understand what happens in the children's mind. Research may clarify this point in the future.

Many jobs done by humans today are transformed into robotics tasks for economic reasons, provoking at least temporarily an employment crisis, without a compensating creation of jobs. Would a similar evolution be conceivable for the teaching profession, which suffers from a recruitment crisis in many countries? At the moment, the available evidence does not show that robots could outperform humans in a teaching role, neither that they would be cheaper than a teacher.

Robots as Companions for Learning

As a fully humanoid teacher seems a fantasy at the moment, some tasks in the classroom could nevertheless evolve, by using robots with a gradation in complexity and interactivity. We mention *Elias Robot* for young learners, focused on language acquisition and based on the already mentioned humanoid *NAO*, which today appears as the most

---

[10] The French designed *NAO* robot has been sold to the Japanese company SoftBank Robotics in 2016: https://www.softbankrobotics.com/emea/en/nao. See also https://fr.wikipedia.org/wiki/NAO_(robotique)

[11] See also https://fr.wikipedia.org/wiki/Robot_sexuel

advanced robot for classroom. *Elias* is being tested in Finnish schools (Reuters 2018). In Chinese kindergarten, the semi-humanoid robot *Keeko* is used "to help children solve logical problems."[12] Another robot, *Tega*, is so described: "A smartphone-based robot, *Tega* serves as a classroom learning companion for younger kids. The interactive robot asks students to complete tasks, monitors their success, and provides feedback. *Tega*'s shape and skin mimics that of a stuffed animal, which many young students find appealing and non-threatening" (Lynch 2019).

The *Avatarmond iPal Robot* family is advertised as follows: "Under the supervision of a teacher, iPal can aid in lessons by presenting educational content in an engaging manner that supports social development and encourages interest in science and technology" (Nanjing AvatarMind Robot Technology 2017).

Two English-speaking *Robovie* robots have been tried in a Japanese elementary school, interacting with fifth and sixth grade pupils. The children wore an RFID tag, allowing the robot to identify them (Kanda et al. 2004). A further experiment, done by the same team, simulated attachment, progressing with time, of the robot to individual students. In this experiment the robot would learn some particularities of a child, like the name, or give him a "secret."

Another experiment has been reported in an English language school for Japanese children, with the idea that children would attach to the robots and be able to actively teach some verbs to them (Tanaka and Matsuzoe 2012). The reported gain in efficiency is not clear.

Irobi, made in South Korea and very successful in Asia, is a 7 kg semi-humanoid robot, Palk (2010) explains: "For children, *Irobi* is like a nanny. It speaks (1000 words), sings, expresses feelings by its movements. It can learn English." *Engkee* is a robot "teacher" for English lessons, implemented in South Korean classrooms since 2010.

The European Commission has been supporting a research program (2016–2018) named Second Language Tutoring Using Social Robots.[13] Initiated in the Netherlands, it provides students with a "companion robot" (Fig. 1), in order to help language acquisition, especially for immigrant Turkish population.

All these systems would deserve detailed research to understand their potential effects, but one cannot escape the feeling that, by resembling sophisticated dolls rather than humans, they are based on a quite naïve vision of the child's relation to other children.

At the university level with much older students, the robot *Jill*, based on IBM's Watson system (open multicloud platform), has been developed by the Georgia Institute of Technology to teach a graduate course online throughout the world. *Jill*'s creator Ashok Goel observes that the students, at least in the beginning, did not even notice they were dealing with a robotic teaching assistant (TA). Here is the analysis, possibly over-optimistic, given by Tim Sprinkle 2017, from the American Society of Mechanical Engineers: "Jill was an incredibly effective teaching assistant. She answered student questions within minutes, no matter when they contacted her. She offered in-depth answers to a wide range of complex queries. She was generally more accessible, more personal, and more upbeat than any human could ever be. The class rolled through half a semester before Goel gave up Jill's real identity. Since then, he's used the AI system in a few other classes and has noticed that, beyond helping with his workload, Jill also improves the overall student experience, making for better, more effective, and more engaged learning."

Telepresence and Teaching

Within the continuum between full substitutes and aided teaching, the telepresence robots represent an intermediate step, in the sense that they are operated by humans—students or teacher—at a distance, to interact with a remote classroom. In Korea, the *EngKey* robot is being used for distant English lessons. An experiment has been made to remotely help teachers in the Philippines, in order for them to teach their South Korean students.

One may question whether this is more efficient than a straight Skype communication with the teacher telepresence?

Robots in Special Education

Robots could be considered as a kind of "exoskeleton," where disabilities which may hinder an access to education, would be helped by the use of a robot (Virnes 2008). The above-mentioned *NAO* robot is used to help Alzheimer patients or educate autist children.

Ethics and Teacher Substitutes

When facing the endless blossoming of robotic technologies, the way their apparently costless or cheap access is developing along a new and often hidden capitalistic model, when observing their potential impact on education, the words of Pope Francis in the Encyclical Letter *Laudato Si'* come to mind. After reminding that "*it is right to rejoice in these advances* [of technical prowess] *and to be excited by the immense possibilities which they continue to open up before*

---

[12]See https://www.youtube.com/watch?v=jppnAR1mtOw

[13]See EU Commission Horizon 2020 Research project L2TOR: http://www.l2tor.eu/. This site offers a rich and recent bibliography of the research findings.

Fig. 1 A child with a L2TOR companion robot

us [102]," Francis, quoting Romano Guardini's book The End of the Modern World, warns on the "ironclad logic" of technology: "*The effects of imposing this model on reality as a whole, human and social, are seen in the deterioration of the environment, but this is just one sign of a reductionism which affects every aspect of human and social life. We have to accept that technological products are not neutral, for they create a framework which ends up conditioning lifestyles and shaping social possibilities along the lines dictated by the interests of certain powerful groups. Decisions which may seem purely instrumental are in reality decisions about the kind of society we want to build.*" And later: "*Isolated individuals can lose their ability and freedom to escape the utilitarian mindset and end up prey to an unethical consumerism bereft of social or ecological awareness* [219]."

Using a different tone, Ashok Goel, quoted by the American Society of Mechanical Engineers (Sprinkle 2017) and already mentioned above with his robot *Jill*, recognizes the need to personalize all the tutoring and teaching: "*to get there, technology* [i.e. *Jill* and other robots] *needs to become more human*". What does it mean for education, in order to remain human?

In her excellent paper, Sharkey develops an in-depth analysis of the ethical concerns about robot teachers, covering the various uses mentioned above (Sharkey 2016).

- First, she discusses the threat to privacy, with robots exerting personal surveillance, collecting personal data on children, monitoring teacher performance or classroom activities. To assess performance of children, emotions could be assessed with various sensors, measuring facial expressions or physiological reactions. The results may be used without control of the parents or imposed on them as criteria for judging their child's behavior. When undertaken with research aims, such actions could be done

with adequate ethical protocols,[14] but their generalization may easily turn into a "Panopticon" to control the classroom and even to provide data for commercial production of educational material. Telepresence robots may even convey data escaping from the country where they act.

- Second, Amanda Sharkey analyses the illusion or rather the postulate, which assumes that a robot is able to relate to humans. She discusses the attachment and deception children may encounter when, dealing with robots, they loss real human contact. The concept of *uncanny valley* seems appropriate here, as it is depicting the emotional response of humans to an object which appears more or less undistinguishable from the reality of a human person.[15] Exposing children to the robot *NAO* and others, Kimberly Brinks (Living Lab Program, University of Michigan) has explored how 240 youngsters, age 3–18, eventually trust a robot and feel at ease (Kim 2018; Brink et al. 2017).

- Third, the question of control and accountability is addressed. If a robot takes, partially or totally, the role of a teacher, it would have to exert such functions as authority, empathy, reward, and eventually punishment. How would children react to such behaviors coming from a machine? how far would the machine be "authorized" to act?

Similar questions may emerge on the robot-soldier. Some arguments are given in favor of its use, claiming that a robot behavior might be more "ethical" than human reactions (?).

---

[14] Key Laboratory for Child Development and Learning Science, Nanjing (China). During the period 2005–2015, this laboratory has carefully developed a research protocol to measure young student's emotions while learning in the classroom.

[15] The long discussion appearing in the following reference shows the actuality of this concept, in light of the efforts of technology to 'become more human'. See: https://en.wikipedia.org/wiki/Uncanny_valley

It is worth quoting here Chrystof Heyns, the United Nations special rapporteur on extrajudicial, summary or arbitrary executions. He argues against the use of autonomous robots to make lethal decisions on the battlefield. His reasoning is that robots "lack human judgement, common sense, appreciation of the larger picture, understanding of the intentions behind people's actions, and understanding of values and anticipation of the direction in which events are unfolding" (Heyns 2013). Several of these arguments apply as well to a robot-teacher, which would most likely lack the ability to understand the complexity of children behavior and moral background. In addition, even a good programming might not avoid all kind of biases which may lead to unequitable treatment of students (color of skin, accent and language, weak disabilities, parental and cultural heritage, etc.).

Similar questions may be raised for a companion robot, or for the telepresence robot, although in the latter case a human presence is making decisions at distance.

A Way for the Future: Computer-Aided Instruction

Analyzing the economical perspectives in developing countries confronted to the digital revolution, Jeffrey Sachs analyzes the key sectors of economy and lists the potential effects of this revolution. Considering education, he states: "Education [will see] a major expansion of access to low-cost, high-quality online education, including online curricula, online monitoring of student progress, online teaching training, "connected classrooms" via videoconferencing, and distance tutoring" (Sachs 2019, p. 162). Von Braun and Baumüller equally addresses education and knowledge as a domain where artificial intelligence and robotics could reduce poverty and marginalization (von Braun and Baumüller 2021, Chap. 7 this volume).

There is, and will be, a very broad range of ways to use algorithms and computers to help the learning process, complementing the classical face-to-face classroom or auditorium. All kinds of software,[16] some using artificial intelligence, are already available to help visualize the solving of mathematical problems, such as GeoGebra. Others are simulating phenomena in astronomy,[17] physics,[18] or chemistry.[19] Complexity of climate change is modeled in an accessible way to help teachers with lecturing or classroom discus-

sions.[20,21] This use of software is an extremely rich field which rapidly develops, greatly helping teachers at all levels if they are properly trained to use these tools with a critical mind.

Massive Open On-line Courses (MOOC) represent another aspect of computer-aided education (Wikipedia last updated 2020). Although versatile in use, and able to ensure a broad dissemination, one should not underestimate their cost of development and monitoring. For example, the *Class'Code* MOOC offered in France since 2016 by the Institut national de recherche en informatique et automatique (INRIA), in cooperation with *La main à la pâte*, aims at students aged 8–16, in order to initiate them into the process of computer sciences (machines, algorithms, information, programs). This has to-date reached about 3000 students, for a non-negligible investment cost of about 2 M€, i.e., about 40 years of a teacher's salary cost in a developed country.

Websites offering "questions & answers" help to students, with typical exercises in science and mathematics, may replace traditional books with more progressive, case-adapted algorithms, such as the Socratic application, now offered by Google.[22]

Smartphones already exist with a variety of sensors and could accommodate more through USB connections. These can be used to collect data: simple applications use the smartphone accelerometer and can provide useful measurements of seismicity (app SEISME), others collect information on biodiversity (app PLANTNET) and there seems to be no limit on the development of such actions of participative science (Académie des Sciences 2018).

We are probably observing the emergence of a considerable diversity of learning tools, available on computers through Internet, but also through smartphones, of easy and free access, which can profoundly transform the teaching practice, especially for science lessons in poor areas where experimental material is rare.

Conclusion

In the classroom, the replacement of teachers by robots could be extremely diverse in its modalities, from full substitutes to teaching or learning companions. It is still in infancy and sufficient research of the impact is not yet available. The technical possibilities combining artificial intelligence and

---

[16]See https://listoffreeware.com/list-of-best-free-math-software/

[17]E.g. The Nebraska Astronomy Applet Project: https://astro.unl.edu/naap/

[18]E.g. https://listoffreeware.com/free-physics-simulation-software-windows/

[19]E.g. https://www.acs.org/content/acs/en/education/students/highschool/chemistryclubs/activities/simulations.html, from the American Chemical Society.

[20]See *Climate Change Policy Simulator C-Roads*, from the Massachusetts Institute of Technology. https://www.climateinteractive.org/tools/c-roads/

[21]See Software *SimClimat* (2019, in French): http://education.meteofrance.fr/lycee/animations/logiciel-simclimat-un-modele-simple-de-bilan-radiatif-de-la-terre

[22]See https://play.google.com/store/apps/details?id=org.socratic.android&hl=fr

teaching needs are probably immense, but the opportunities and costs of such investment remain today questionable. The ethical aspects of such developments raise many questions, to be explored in depth, since children are by essence extremely vulnerable human beings. Providing tools which better answer human expectations, especially those of students, is quite different from building a "conscious" robot which is designed *to be exactly like a human*.

Facing these limitations and words of caution, the needs to develop education worldwide are so pressing, and their cost implies such a small probability to be fully covered during this century, that any reasonable solution which benefits from these technological advances will become helpful, especially in the broad area of computer-aided education.

Acknowledgments The author thanks David Wilgenbus (section "Robots as pedagogical tools") and Christopher Pedregal for helpful exchanges during the design of this contribution, and Joachim von Braun for having encouraged it.

References

Académie des sciences. (2018). *La science hors les murs. Colloquium at the French Académie des sciences.* Retrieved November 29, 2018, from https://www.academie-sciences.fr/pdf/conf/colloque_291118.pdf

Battro, A. M., Léna, P., Sánchez Sorondo, M., & von Braun, J. (2017). *Children and sustainable development.* Cham: Springer International Publishing.

Bodkin, H. (2017). *Inspirational' robots to begin replacing teachers within 10 years.* Available via The Telegraph. Retrieved February 20, 2020, from https://www.telegraph.co.uk/science/2017/09/11/inspirational-robots-begin-replacing-teachers-within-10-years/.

Brink, K. A., Gray, K., & Wellman, H. N. (2017). Creepiness creeps in: Uncanny valley feelings are acquired in childhood. *Child Development, 90*(4), 1202. https://doi.org/10.1111/cdev.12999.

Charpak, G., Léna, P., & Quéré, Y. (2005). *L'Enfant et la science.* Paris: O. Jacob.

Cigi, C. C. (2013). Computer aided education. *Procedia—Social and Behavioral Sciences, 103,* 220–229. https://doi.org/10.1016/j.sbspro.2013.10.329.

Dehaene, S. (2018). *Apprendre! Les talents du cerveau, les défis des machines.* Paris: O. Jacob.

European Commission. (2012). *Eurobarometer 382: Public attitudes towards robots.* Brussels: European Commission.

Heyns, C. (2013). *Report of the special rapporteur on extrajudicial, summary or arbitrary executions*: United Nations, A/HRC/23/47. Available via OHCHR. Retrieved February 21, 2020, from https://www.ohchr.org/Documents/HRBodies/HRCouncil/RegularSession/Session23/A-HRC-23-47_en.pdf.

Kanda, T., Hirano, T., Eaton, D., & Ishiguro, H. (2004). Interactive robots as social partners and peer tutors for children: A field trial. *Human Computer Interaction, 19*(1–2), 61–84. https://doi.org/10.1207/s15327051hci1901%262_4.

Kim, M. (2018). *The social robot teacher: What makes children trust a robot and feel at ease?* Available via BOLD. Retrieved February 21, 2020, from https://bold.expert/the-social-robot-teacher/.

Levy, D. (2007). *Love + sex with robots: The evolution of human-robot relationships.* London: Duckworth.

Luckin, R., Holmes, W., Griffiths, M., & Forcier, L. B. (2016). *Intelligence unleashed: An argument for AI in education.* London: Pearson.

Lynch, M. (2019). *The rise of the robot teacher.* Available via The Tech Advocate. Retrieved February 21, 2020, from https://www.thetechedvocate.org/the-rise-of-the-robot-teacher/.

Nanjing AvatarMind Robot Technology. (2017). *AvatarMind iPAL® robot family: For senior care, retail/hospitality, and children's education.* Available via Nanjing AvatarMind Robot Technology. Retrieved February 21, 2020, from https://www.ipalrobot.com/.

Palk, S. (2010). *Robot teachers invade south Korean classrooms.* Available via CNN: Global Connections. Retrieved February 21, 2020, from http://edition.cnn.com/2010/TECH/innovation/10/22/south.korea.robot.teachers/index.html.

Reuters. (2018). *Finland schools are testing out robot teachers.* Available via New York Post. Retrieved February 21, 2020, from https://nypost.com/2018/03/27/finland-schools-are-testing-out-robot-teachers/.

Sachs, J. D. (2019). Some brief reflections on digital technologies and economic development. *Ethics & International Affairs, 33*(2), 159–167. https://doi.org/10.1017/S0892679419000133.

Sharkey, A. J. C. (2016). Should we welcome robot teachers? *Ethics and Information Technology, 18,* 283–297. https://doi.org/10.1007/s10676-016-9387-z.

Sprinkle, T. (2017). *Robot TeachersTransform education.* Available via The American Society of Mechanical Engineers. Retrieved February 21, 2020, from https://www.asme.org/topics-resources/content/robot-teachers-transform-education.

Strauss, S. (2016). *Teaching's potential to create harmony among people.* Presented at the workshop: Children and sustainable development: A challenge for education. [Los niños y el desarrollo sostenible: un desafío para la educación]. November 13–15, 2015, Casina Pio IV, Vatican City.

Tanaka, F., & Matsuzoe, S. (2012). Children teach a care-receiving robot to promote their learning: Field experiments in a classroom for vocabulary learning. *Journal of Human-Robot Interaction, 1*(1), 78–95. https://doi.org/10.5898/JHRI.1.1.Tanaka.

UNESCO Institute for Statistics (UIS). (2016). *The world needs almost 69 million new teachers to reach the 2030 education goals.* UIS fact sheet 39. Available via UNESCO. Retrieved February 21, 2020, from http://uis.unesco.org/sites/default/files/documents/fs39-the-world-needs-almost-69-million-new-teachers-to-reach-the-2030-education-goals-2016-en.pdf.

Virnes, M. (2008). *Robotics in special needs education.* IDC'08: Proceedings of the 7th international conference on interaction design and children, pp 29–32. doi: https://doi.org/10.1145/1463689.1463710.

von Braun, J., & Baumüller, H. (2021). *AI, robotics and the poor* (pp. 85–97). New York: Springer

Wikipedia (last updated 2020). *Massive open online course.* Available via Wikipedia. https://fr.wikipedia.org/wiki/Massive_Open_Online_Course.

Ziv, M., Solomon, A., Strauss, S., & Frye, D. (2016). Relations between the development of teaching and theory of mind in early childhood. *Journal of Cognition and Development, 17*(2), 264–284. https://doi.org/10.1080/15248372.2015.1048862.

Humans Judged by Machines: The Rise of Artificial Intelligence in Finance, Insurance, and Real Estate

Frank Pasquale

Contents

Abstract

There are opportunities but also worrisome trends as AI is applied in finance, insurance, and real estate. In these domains, persons are increasingly assessed and judged by machines. The financial technology (Fintech) landscape ranges from automation of office procedures, to new approaches for storing and transferring value, to the granting of credit. The Fintech landscape can be separated into "incrementalist Fintech" and "futurist Fintech." Incrementalist Fintech uses data, algorithms, and software to complement professionals who perform traditional tasks of existing financial institutions. It promises financial inclusion, but this inclusion can be predatory, creepy, and subordinating. These forms of financial inclusion undermine their solvency, dignity, and political power of borrowers. Futurist Fintech's promoters claim to be more equitable, but are likely to falter in their aspiration to substitute technology for key financial institutions. When used to circumvent or co-opt state monetary authorities, both incrementalist and futurist Fintech expose deep problems at the core of the contemporary digitization of finance.

Keywords

Artificial intelligence · Finance · Insurance · Real estate · Fintech · Privacy · Employment

I wish to thank the organizers of the 2019 PAS/PASS Workshop on AI and robotics for giving me the opportunity to present this work. I also wish to thank Raymond Ku, and the audience at Case Western Reserve University, for thoughtful comments on this work when it was presented as the 2019 Center for Cyberspace Law and Policy Distinguished Lecture. This work is planned to be simultaneously published in the Case Western Reserve University Law Review.

F. Pasquale (✉)
Brooklyn Law School, Brooklyn, NY, USA
e-mail: pasquale.frank@gmail.com

Introduction

The financial technology ("fintech") landscape is complex and diverse. Fintech ranges from automation of office procedures once performed by workers, to some genuinely new approaches to storing and transferring value, and granting credit. Established and start-up firms are using emerging data sources and algorithms to assess credit risk. And even as financial institutions are adopting some distributed ledger technologies, some proponents of cryptocurrency claim that it "changes everything" and will lead to a "blockchain revolution."

For purposes of this paper, I will divide the fintech landscape into two spheres. One, incrementalist fintech, uses new data, algorithms, and software to perform classic work

J. von Braun et al. (eds.), *Robotics, AI, and Humanity*, https://doi.org/10.1007/978-3-030-54173-6_10

of existing financial institutions. This new technology does not change the underlying nature of underwriting, payment processing, lending, or other functions of the financial sector. Regulators should, accordingly, assure that long-standing principles of financial regulation persist here. I address these issues in Part I below.

Another sector, which I deem "futurist fintech," claims to disrupt financial markets in ways that supersede regulation, or render it obsolete. For example, if you truly believe a blockchain memorializing transactions is "immutable," you may not see the need for regulatory interventions to promote security to stop malicious hacking or modification of records. In my view, futurist fintech faces fundamental barriers to widespread realization and dissemination. I address these issues in Part II below.

Incrementalist Fintech: The Problems of Predatory, Creepy, and Subordinating Inclusion

Over the past decade, algorithmic accountability has become an important concern for social scientists, computer scientists, journalists, and lawyers. Exposés have sparked vibrant debates about algorithmic sentencing (van Dam 2019). Researchers have exposed tech giants showing women ads for lower-paying jobs, discriminating against the aged, deploying deceptive dark patterns to trick consumers into buying things, and manipulating users toward rabbit holes of extremist content (Gibbs 2015; Angwin et al. 2017; Warner 2019). Public-spirited regulators have begun to address algorithmic transparency and online fairness, building on the work of legal scholars who have called for technological due process, platform neutrality, and nondiscrimination principles (Citron 2008; Pasquale 2008, 2016).

Establishment voices have hailed fintech as a revolutionary way to include more individuals in the financial system. Some fintech advocates advocate radical deregulation of their services, to enable their rapid entry into traditional banking markets. However, there is a risk of the fintech label merely masking "old wine in new bottles." The annals of financial innovation are long, but not entirely hallowed (FCIC 2011). When deregulatory measures accelerated in the late 1990s and early 2000s, their advocates argued that new technology would expertly spread and diversify risk. However, given biases in credit scores based on "fringe" or "alternative" data (such as social media use), even the finreg (finance regulatory) establishment is relatively comfortable with some basic anti-bias interventions (Scism 2019).

New quantitative approaches to underwriting and reputation creation have often failed to perform as billed, or

have raised serious normative concerns[1] (Pasquale 2015). Most fundamentally, a technology is only one part of a broader ecosystem of financial intermediation (Lin 2014, 2015[2]). AI underwriting may feed into a broader culture of total surveillance, which severely undermines human dignity. Regulators must ask larger questions about when "financial inclusion" can be predatory, creepy (as in 24/7 surveillance), or subordinating (as in at least one Indian fintech app, which reduces the scores of those who are engaged in political activity) (Hill and Kozup 2007; Taylor 2019; Variyar and Vignesh 2017; Vincent 2015).

Limiting the factors feeding into credit decisions is important, because our current path is toward a "full disclosure future." For some fintech firms, everything is fair game. Each week brings new examples of invasive data collection. Before they push consumers to accept even more Faustian bargains for better credit terms, regulators need to decide how far data collection can go.[3] According to one aggrieved worker, her boss "bragged that he knew how fast she was driving at specific moments ever since she had installed the app on her phone" (Vincent 2015). Progressive car insurance offers discounts for car tracking devices that measure "hard stops," among other putative indicators of bad driving.

Nor is this kind of tracking simply a matter of consumer choice. Once enough people agree to a certain kind of tracking, refusing it looks suspicious. What have you got to hide? (Peppet 2015). Moreover, in a market economy, no firm wants to risk falling behind others by lacking data they have. The insurer John Hancock gave us an early glimpse of how the ensuing dynamic plays out. It sells life insurance to those who will wear FitBits, Apple Watches, or similar devices (The Guardian Editorial 2018; Senior 2018; Barlyn 2018). Life insurers also want to scour applicants' social media accounts, to spot "red flags" for dangerous behavior (Scism 2019).

We all may be emitting other "tells" that are more grave. For instance, researchers recently parsed the mouse movements of persons who extensively searched for information about Parkinson's disease on Bing (White et al. 2018). This group—which is far more likely to have Parkinson's than the population as a whole—tended to have certain tremors in their mouse movements distinct from other searchers. These tremors were undetectable by humans—only computers can clock the microseconds difference in speed that distinguished normal from pathological tremors. Thus, there is no defense

---

[1] Pasquale (2015) describes search results as a form of reputation creation (for name searches) that can create serious normative and ethical concerns.

[2] Lin (2014, 2015) offer 10 regulatory principles for the new financial industry.

[3] For an overview of the issues raised, see Bruckner (2018).

to such detection—no "privacy enhancing technology" that can prevent such technology, once developed, from classifying someone as "likely to develop Parkinson's." The more data about troubling fates is available, better AI will get at predicting them. We may want our doctors to access such information, but we need not let banks, employers, or others use it.

All of the scenarios explored above have been hailed as part of a tsunami of data-driven "financial inclusion." True, if financial firms have access to more information about potential clients, some persons who once would not have been part of the financial system will likely gain access to credit. But there is more to life—and public policy—than consenting transactions among persons and fintechs. Without proper guardrails, there will be a race to the bottom in both sharing and behavior shaping, as more individuals compete for better deals. That would result in a boom in predatory inclusion (which harms more than it helps), creepy inclusion (which gives corporate entities a voyeuristically intimate look at our lives), and subordinating inclusion (which entrenches inequality by forcing people to maintain the same patterns of life that resulted in their desperation in the first place). Lawmakers should discourage or ban each of these types of "inclusion."

Predatory inclusion is a concept with a long history (Seamster and Charron-Chénier 2017). Credit enables, while its shadow side (debt) constrains. When those desperate for opportunity take on a heavy loan burden to attend training programs of dubious value, the latter effect predominates. The rhetoric of uplift convinces too many that more learning is a sure path to better earning power. Fly-by-night, for-profit colleges take advantage of their hope (McMillan Cottom 2018). They peddle a cruel optimism—that the future has to be better than the past (Berlant 2011; Blacker 2013; McGettigan 2013; Newfield 2016).

The same principle also applies to "creepy inclusion." AI (and the data collection it is now sparking) allows lenders to better microtarget vulnerable consumers (Upturn 2015; Mierzwinski and Chest 2013). A sophisticated firm might notice a parent in December rapidly re-ordering toy options by price, and may track their phone as they walk from store to store without buying a child's present. "Vulnerability-based marketing" may even enable the ad to be timed, at the very point in the day when despair is most likely to set in. A bright blinking message "Let our credit save your Christmas!" may be well-nigh irresistible for many strapped parents. This vulnerability-based marketing will only get worse with the spread of 24/7 tracking hypothesized above. This is one reason I would call the offer of better credit terms in exchange for nonstop cell phone tracking, archiving, and data resale a prime example of "creepy inclusion." (Kotsko 2015; Tene and Polonetsky 2014).

If a boss asked an employee if she would mind him trailing her twenty-four hours a day, seven days a week, the feeling of threat would be imminent. She might even be able to get a restraining order against him as a stalker. A cell phone tracker may seem like less of a threat. However, the use and re-use of data creates distinctive and still troubling menace. Creepiness is an intuition of future threat, based on a deviation from the normal. Normal experience is one of a work life distinct from home life, and of judgments about us being made on the basis of articulable criteria. Creepy inclusion disturbs that balance, letting unknown mechanical decision makers sneak into our cars, bedrooms, and bathrooms.

Of course, at this point such surveillance demands are rare. Financial sector entrepreneurs brush aside calls for regulation, reassuring authorities that their software does not record or evaluate sensitive data like location, person called, or the contents of conversations.[4] However, metadata is endless, and as seen with the example of the hand tremors predicting Parkinson's, can yield unexpected insights about a person (ACLU 2014). Moreover, now is the time to stop creepy inclusion, before manipulative marketing tricks so many people into bad bargains that industry can facilely assert its exploitation is a well-established "consumer preference."

The timing issue is critical, because industry tries to deflect regulation when a practice first begins, by saying that it is an "innovation." "Wait and see how it turns out," lobbyists say. But once a practice has been around for a while, another rationale for non-regulation emerges: "How dare you interfere with consumer choice!" This cynical pair of rationales for laissez-faire is particularly dangerous in data regulation, since norms can change quickly as persons jostle for advantage (Ajunwa et al. 2017). Uncoordinated, we can rapidly reach an equilibrium which benefits no one. Cooperating to put together some enforceable rules, we can protect ourselves from a boundless surveillance capitalism (Zuboff 2019). For example, some jurisdictions are beginning to pass laws against firms "microchipping" workers by subcutaneously injecting a rice-sized sensor underneath their skin (Polk 2019).[5] Others are beginning to require operators of AI and bots to disclose their identity (Pasquale 2017).

That project of self-protection is urgent, because "subordinating inclusion" is bound to become more popular over

---

[4]For an example of this kind of distinction, note the assurances from one such firm in Leber (2016).

[5]Some "supporters argue that these are ill-informed concerns, easily addressed with information about the safety of the devices, and how they will and will not be used. They contend that chipping is really no different than having employees carry around electronic access cards that are used for entry into buildings and purchases at the company cafeteria, but with significant advantages of convenience and security because employees can't forget their chip, lose it, or have it stolen." (Polk 2019). However, an access card is not injected into a person's body. The potential for a "Stockholm Syndrome," or what Sacasas calls the "Borg Complex," is strong. See, e.g., Metz (2018), Sacasas (2013).

time. Penalizing persons for becoming politically involved—as firms in both India and China now do—further entrenches the dominance of those who provide credit over those in need of it. The rise of what some call "vox populi risk (Citi 2018)"—including the supposed "danger" of persons demanding corporations treat them better—will provoke more executives to consider the political dimensions of their lending. Firms may well discover that those who get involved in politics, or sue their landlords for breach of a lease, or file a grievance at work, are more likely to contest disputed charges, or are even more likely to default. But such correlations cannot inform a humane credit system. They set us all on a contest of self-abasement, eager to prove ourselves the type of person who will accept any indignity in order to get ahead.

While predatory, creepy, and subordinating inclusion are objectionable on diverse grounds, they all clarify a key problem of automation. They allow persons to compete for advantage in financial markets in ways that undermine their financial health, dignity, and political power. It is critical to stop this arms race of surveillance now, because it has so many self-reinforcing internal logics.

Fallacies of Futurist Fintech

Though sober reports from the World Economic Forum (2017), Deloitte, and governmental entities accurately convey the problems and opportunities posed by the incrementalist side of fintech, much of the excitement about the topic of financial technology arises out of a more futuristic perspective. On Twitter, hashtags like #legaltech, #regtech, #insurtech, and #fintech often convene enthusiasts who aspire to revolutionize the financial landscape—or at least to make a good deal of money disrupting existing "trust institutions" (e.g., the intermediaries which help store and transfer financial assets).

Finance futurism fits with broader industry narratives about the role of automation in transforming society. Captains of finance capital have long aspired to automate work. Machines don't demand raises or vacations. They deliver steady returns—particularly when protected by intellectual property law, which may forbid competitors from entering the market. The "dark factory"—so bereft of human workers that not even lights are needed in it—is a rentier's dream. Finance has long demanded that factories increase their productivity via better machinery; now it has turned the tools of automation on itself. An upper echelon of managers and traders has displaced analysts, accountants, and attorneys with software. Ordinary investors fleeing fees are opting for "set-it-and-forget-it" index funds, which automatically allocate money to a set list of stocks and bonds. "Robo-advisers" present such options via low-key, low-pressure apps. Bots exchange buy and sell orders. Boards are con-

stantly pressuring management to push "AI" further up the value chain, replacing managerial judgment with machine learning. Finance's self-automation is a familiar ideological story of capitalism: one of market competition pushing firms and entrepreneurs to do more with less.

With the rise of cryptocurrency, another ideology—a mix of anarcho-capitalism and cyberlibertarianism—is strengthening this regnant neoliberal model of automation. Imagine distributed software that allowed you (and those you transact with) to opt out of using banks altogether. When you buy something online, a distributed ledger would automatically debit your account—and credit the seller's—with the exact amount of the purchase, recorded in *all* computers connected to the network, so that tampering with the common record was effectively impossible. Now imagine that humble arrangement scaled up to the sale and purchase of equities, bonds—almost anything. Fees on investing might fall below even 0.1%—just enough to keep the distributed ledger maintained and updated. If the ledger itself could automatically "pay" those who do its work with tokens valued as money, perhaps fees could drop to zero.

The system just described is starting to develop. Bitcoin and other cryptocurrencies function as the tokens described above, while blockchains serve as the distributed ledgers. However, its implications are far from clear. Users opt out of traditional finance for a reason, and simply avoiding wire transfer fees does not seem like a plausible rationale for taking on the risks of the new. Using crypto to evade regulation, sweetens the pot. Critics fear these new "coins" are primarily a way to grease the skids of a pirate economy of data thieves, money launderers, and drug dealers (Beedham 2019). The cyberlibertarian rejoinder is a simple one: people deserve to do what they please with their assets—tax authorities, court judgments, or financial regulators be damned.

Finn Brunton's *Digital Cash* entertainingly narrates the stories of *soi-disant* renegades who aspired to set up their own systems of currency to operate independently from extant legal systems (Brunton 2019). On the one hand, these would-be prophets of monetary systems beyond or beside or hidden from the state, aspire to exist outside the system, ostensibly because of their disgust with crony capitalism and all its trappings. On the other, as soon as some of their schemes get sufficient traction, there is a scramble to convert the "anti-system" digital assets of pioneering cryptocurrency promoters into safer, more familiar equities, bonds, and interest-bearing bank accounts.

Brunton offers a "history of how data was literally and metaphorically *monetized*," as promoters of digital cash tried to make digital data *about stores of value* itself valuable (Brunton 2019, p. 3). This bootstrapping is reminiscent of the history of fiat money itself, which arose in part to pay tributes, taxes, and fees imposed by the state. We can think of the distributed ledger I described above as data—a representation of who has what: of who owes, and who owns. Bitcoins originate as ways of rewarding those who expended the computational power necessary to maintain the ledger. What can make Bitcoins so authoritative, so valuable, that persons have gladly traded 18,000 U.S. dollars for one of them?

The answer is narrative: the stories we (and our friends, the media, and our social media feeds) tell ourselves about how value is created and stored. As Brunton argues, "the history of digital cash can also show us a particularly vivid example of the use of money and technologies to tell stories about the future." (Brunton 2019, p. 3). Nobel Prize winning economist Robert J. Shiller has recently highlighted the importance of "narrative economics" for understanding why certain conceptions (and misconceptions) of commercial life come to dominate the thought of policymakers, politicians, and people generally (Schiller 2019). One critical narrative step toward cryptocurrency is a negative one: widespread acceptance of tales of distrust and disillusionment at the existing financial order. These were not hard to find in the wake of the Global Financial Crisis (GFC) of 2008. Scores of books chronicled self-dealing in leading financial firms, and questioned the wisdom of government interventions. Bitcoin was announced in late October 2008, and the first entry in its ledger (the so-called "genesis block") included a reference to bank bailouts.

Brunton elegantly recounts a primordial stew of ideas that encouraged crypto enthusiasts in the decades leading up to Bitcoin, when alternative currencies were rare and isolated. He describes isolated "agorists," who saw every transaction between individuals without state-sponsored money as a victory for private initiative over public control (Brunton 2019, p. 3). Some were inspired by J. Neil Schulman's 1979 novel *Alongside Night* (set in a dystopian 1999 New York beset by hyperinflation), where "characters move to or inhabit alternative zones, where they can live outside the emergency, exacerbate the existing crisis, and return to the changed world on the other side of the disaster where their utopia becomes possible." (Brunton 2019, p. 178). As contemporary libertarians "exacerbate the existing crisis" of global warming by pushing deregulation of carbon emissions, the designs of such agorists seem quite alive today. Rumors about imminent hyperinflation or currency collapse are as valuable to Bitcoin HODL-ers (named for a keyboard-smashing typo of a HOLD order) as they are to hucksters of Krugerrands.

By the end of *Digital Cash*, Brunton has little patience for cryptocurrency's leading success story, Bitcoin: "As of this writing, it seems to have found a role that perfectly exemplifies the present moment: a wildly volatile vehicle for baseless speculation, a roller coaster of ups and downs driven by a mix of hype, price-fixing, bursts of frenzied panic, and the dream of getting rich without doing much of anything" (Brunton 2019, p. 204).

For cultural theorist David Golumbia, by contrast, these developments are unsurprising. Golumbia tends to critique computation as reductionist, all too prone to flatten our experience of education, relationships, and politics into simplistic categories and ersatz quantifications. His 2009 *The Cultural Logic of Computation* chronicled the distortions to human thought and civic association caused by the overreach of digitization (Golumbia 2009). *The Politics of Bitcoin* anticipated an explosion of interest in the troubling adoption of cryptocurrency by white supremacists, criminals, and frauds (Burns 2018; Malik 2018).

Well-versed in both critical theory and the philosophy of language, Golumbia interprets key texts in cryptocurrency advocacy as cyberlibertarian propaganda. In 1997, Langdon Winner diagnosed cyberlibertarianism as a linkage of "ecstatic enthusiasm for electronically mediated forms of living with radical, right-wing libertarian ideas about the proper definition of freedom, social life, economics, and politics" (Winner 1997). Libertarian thought tends to derogate government as the chief enemy of liberty—rather than recognizing, as nearly all other political philosophies do, that government is critical to *assuring* freedom, since anarchy is little more than rule by the strongest.

Insights like these serve, for Golumbia, as an Ariadne's thread to guide him through remarkably solipsistic, fantastic, or dense texts. *The Politics of Bitcoin* is a first-rate work of what Paul Rabinow and William Sullivan (Rabinow and Sullivan 1988) have called "interpretive social science," carefully elaborating meaning from a set of contested, murky, or ostensibly contradictory texts. Building on the work of Langdon Winner, Golumbia explains that the core tenet of cyberlibertarianism is the insistence that "governments should not regulate the internet" (Golumbia 2016, p. 5). For Golumbia, the original sin of Bitcoin was its systematic effort to promote itself as a somehow "safer" alternative than independent central banks. This anti-system rhetoric had a long history, including far right groups enraged by the power of the Federal Reserve, the New Deal, and the civil rights movement.

Golumbia traces linkages between anti-state and pro-Bitcoin rhetoric in the thought of cryptocurrency advocates. As he explains early in the work, his purpose is "to show how much of the economic and political thought on which Bitcoin is based emerges directly from ideas that travel the gamut from the sometimes-extreme Chicago School economics of Milton Friedman to the explicit extremism of Federal Reserve conspiracy theorists." (Golumbia 2016, p. 12). Friedman and his monetarist disciples accommodated themselves to the Federal Reserve (the central bank of the U.S.), but sought to tightly constrain it. Golumbia reviews the origins of their ideas, and then describes their appropriation in the present day.

Like much laissez-faire economic thought, these extreme ideas' viral success owes much to their simplicity. Some right-wing theorists of money characterize central banking as a conspiracy where insiders can plot to devalue currency by printing ever more dollars. *Real* value, they insist, can only be measured by something that is scarce because its supply cannot be dramatically altered by human will. Gold fits this bill; paper (or, worse, digital entries on a government balance sheet) does not. The Bitcoin imaginary is rife with stories of inflation sparked by incompetent or greedy banking authorities.

This populist worry about inflation is the mirror image of a critical nineteenth century politicization of money, when insurgent reformers were deeply concerned with the problem of *deflation* (that is, the rising burden of debt, and concurrent economic slowdown, that tends to occur when a unit of currency starts gaining value over time). In the 1880s and 90s, American farmers and small businesses were crushed by debt as the real value of what they owed increased over time (rather than decreasing, as it would during an inflationary period). A rigid gold standard kept the government from pursuing a looser monetary policy, which would have better fit the needs of a growing country. A populist demand to smash the "cross of gold" was frustrated at the time, of course, but later crises led to a suspension of the gold standard and a more accommodative monetary regime. Now, there is enormous populist energy in exactly the opposite direction: Bitcoin reflects and reinforces fears of out-of-control central banks printing money willy-nilly, massively devaluing currency. How did this switch occur?

Of course, events like the German, Zimbabwean, and Venezuelan hyperinflations spurred such worries. Quantitative easing provoked both paranoid and entirely justifiable resentments of connected bankers buying up assets on the cheap with easy government credit. But we must also recognize that cryptocurrency's romance with deflation has also been stoked by ideological movements, entrepreneurial fantasists, and an increasingly unfair financial regulatory system. As Golumbia explains, "It is a cardinal feature of right-wing financial thought to promote the idea that inflation and deflation are the result of central bank actions, rather than the far more mainstream view that banks take action to manage inflation or deflation in response to external economic pressures." Bitcoin enthusiasts seize this idea as one more rationale to shift assets out of national currencies, abandoning inevitably political and legal governance of money for the "safety" of code.

The law professor Katharina Pistor's *The Code of Capital* demonstrates how foolish that aspiration may turn out to be. Pistor is primarily concerned with legal, rather than computational, code. Her long and distinguished research career demonstrates why that prioritization makes sense. Pistor has demonstrated the many ways that law is not merely a constraint on finance, but instead is *constitutive* of financial markets. The vast edifice of currency exchanges, derivatives,

swaps, options, and countless other instruments, rests on a foundation of law—or, to be more precise, the relative power of one party to force another to obey the terms of contracts they have made. Pistor has demonstrated the critical role of law in creating and maintaining durable exchanges of equity and debt. Though law to some extent shapes all markets, in finance it is fundamental—the "products" traded are very little more than legal recognitions of obligations to buy or sell, own or owe.

In earlier work, Pistor has argued that "finance is essentially hybrid between state and markets, public and private" (Pistor 2013). And we can see immediately the enormous problems this social fact poses for cyberlibertarians. They want the state out of the (or at least their) money business; Pistor's legal theory of finance reminds us of just how revolutionary such a wish is. Nevertheless, she gamely recounts the assumptions of cyberlibertarians, on the way to a more realistic account of how the "coding" of capital—the ability of experts to ensure certain claims to assets and income streams are durable, transferable, and enforceable—may increasingly depend on technological (rather than merely legal) prowess.

As Pistor observes, the egalitarian case for automated finance is a simple one: "When the digital code replaces the legal code," we are assured, "the commitments we make to one another become hardwired, and even the powerful cannot simply wiggle out of them." (Pistor 2019, p. 184). However, the problem of the wealthy evading contracts should be counterbalanced against the many ways in which the wealthy use contracts to ruthlessly dun debtors, sidetrack labor disputes into feckless arbitral panels, and reroute assets away from progressive taxation. Digital utopists' strategic decision to raise the salience of the powerful's evasion of contractual obligations is a political maneuver. Indeterminacy and lack of enforceability are not always and everywhere problematic features of contracts. Indeed, a key part of legal training is the ability to spot the ways that judicial precedent, legislation, and regulation enable us to escape from bad bargains, or at least mitigate their effects.

Cryptocurrency buffs may cede those points, and then pivot to the economic and political case for digital finance. Automated recordation of assets and transfers is presumed to be cheaper and more egalitarian than an army of clerks and lawyers. But as Pistor patiently reminds us, "Someone has to write the code, watch it, and fix its bugs; and someone must find an answer to the question of whose interests the code serves, or perhaps ought to serve." (Pistor 2019, p. 185). Rather than the "flat world" of Tom Friedman's dreams, automated finance just introduces new hierarchies: among cryptocurrencies; within any particular one, among high level coders and those who just do what they are told; and, of course, between the coder class and ordinary users (Pasquale 2019). Indeed, Pistor predicts that large financial institutions will try to coopt the utopian energy that Brunton describes, by "enclos[ing] the digital code in law and leav[ing] little space to the digital utopists." (Pistor 2019, p. 186).

Even when domesticated by established financial interests, both blockchain and smart contracts may be of limited value. There are many contractual relationships that are too complex and variable, and require too much human judgment, to be reliably coded into software. Code may reflect and in large part implement what the parties intended, but should not *itself* serve as the contract or business agreement among them.

Moreover, even if immutability of contractual terms were possible, it is not always desirable; when conditions change enough, re-negotiation is a strength of traditional law, not a weakness. So, too, do statutory opt-outs (which let persons break contracts in certain situations, such as an emergency or illness) render legally unenforceable a contract coded to execute no matter what. An immutable ledger distributed across thousands or millions of computers and servers may also directly conflict with common data protection laws: how can a "right to be forgotten" or "right to erasure" operate if the relevant record is not merely ubiquitous, but hard-coded into permanence? Cyberlibertarians may respond that such laws themselves are infringements on sacred rights of free expression, but even they would likely blanch at the prospect of the uncontrollable copying and permanent recording of, say, their breached medical records. So long as hackers menace the security of any computer code, data, and communications, sophisticated and powerful parties are unlikely to opt into a brave new world of totally automated finance.

For the marginalized, though, code may indeed become a law unto itself. To be truly the final word in contracting, smart contracts would need both parties to give up other rights they might have outside the four corners of their coded relationship. Since the *Lochner* era in the United States, a libertarian legal movement has prioritized one goal in commercial life above nearly all others: the right to give up one's rights. Styled as "freedom of contract," such a right includes the liberty to sacrifice one's rights to a receive a minimum wage, days off, or a trial in a court of law when those (or any other) rights are violated. Widely known as "forced arbitration," that last proviso is an increasingly powerful unraveller of law in the realms of both employment and consumption, all too often consigning legitimate claims to byzantine and biased arbitral fora. If courts allow parties to opt into smart contracts with no appeal to the statutes and regulations that govern the rest of finance, the crypto dream of an alternative legal system may finally be realized—with nightmarish costs for the unwary or unlucky. Imagine a car programmed to shut off the moment a payment is late, and you have a sense of the implications of automated enforcement of loan terms.

Despite the precision and breadth of her critiques of Bitcoin in particular and blockchain technologies generally,

Pistor remains open-minded about future advances in the "digital coding" of finance. The bulk of her book traces the manifest unfairness wrought by centuries of development of trust, bankruptcy, securities, tax, and intellectual property law toward protecting the interests of the wealthy and powerful. From that perspective, *any* alternative to the legal coding of capital may seem promising. However, it is hard to imagine how the utopian visions of cryptocurrency would mesh with more pragmatic reforms of finance law and policy, or with the increasingly evident importance of the state as a final guarantor of value and liquidity. Cryptocurrency enthusiasts cannot escape the legal foundations of finance, no matter how ingeniously they code their next initial coin offering.

The divide between economics and politics seems both obvious, and obviously ideological. There are markets and states. The market is a realm of free exchange; the state keeps order. Price signals nudge consumers and businesses to decide when to buy and sell; command and control bureaucracies allocate resources governmentally. The market's natural order is spontaneous; the state's man-made rationality is planned. Societies exist on a continuum between free markets and statist control. These distinctions are not only familiar ideological crutches for an American right prone to paint regulation as the antithesis of freedom. They have also tended to divide up jurisdictional authority in universities, luring researchers of governments to political science departments, and experts in commodification and exchange to economics.

However ideologically or functionally useful these divides may have been in the past, we now know that they obscure just as much as they illuminate (Pasquale 2018). The state can be entrepreneurial; private enterprise can be predatory and sclerotic (Link and Link 2009). Market logics may have invaded society, but social logics have interpenetrated markets, too. The hottest new methodology in the legal academy is law and political economy (LPE). In a pointed corrective to decades of hegemony by law and economics in so-called "private law fields," LPE studies critical public interventions that created the markets now so often treated as sanctuaries of free choice and natural order. For example, segregated neighborhoods did not just "happen" in the United States, as the spontaneous result of millions of individual decisions. They were structured by federal home finance policies, by judicial decisions to enforce "restrictive covenants" preventing transfers to African Americans, and numerous other legal factors. Sociologists, historians, and many others can help us trace back the roots of contemporary "free choice" to struggle and coercion.

As methodological lenses proliferate, economists can no longer claim a monopoly of expertise on the economy. Other social science and humanities scholars have vital insights, advancing deep and nuanced accounts of the role of money in society, among many other dimensions of commerce.

Academic study of the economy is now, more than at any time over the past 70 years of creeping specialization, up for grabs. The struggle of futurist fintech to reinvent basic anti-fraud and consumer protection measures shows how deeply financial markets depend on the state to function—and how, nevertheless, powerful actors in finance are using both legal and digital code to reduce the scope and intensity of state power.

They are likely to continue to succeed in doing so, at least for the foreseeable future. Cyberlibertarian ideology is ascendant, evident in the talking points of hard-right Trump appointees and white papers of DC think tanks. Funded by large technology firms, the Kochs, and the many other branches of the neoliberal thought collective that Philip Mirowski (2014) has described, key "blockchain experts" hype cryptocurrency as the first step of a journey into a libertarian future. A formidable coalition of finance lobbyists and nervous oligarchs aid cyberlibertarians by undermining the power of legitimate authorities to shape (or even monitor) global capital flows.

To the extent it signs on to the post-regulatory agenda of futurist fintech, established banks are playing a dangerous game: they need the power of the state to enforce their contracts, and to serve as lender-of-last-resort in case of crisis. Given the temptations of massive returns via excess leverage, they will continue to resist state power, undermining the very mechanisms that are necessary to save capitalism from itself, and weakening the very tax and monetary authorities necessary (in conventional economic theory) to fund the next bailout. This is of course an old story in capitalism: As Andreas Malm described in *Fossil Capital*, and Samuel Stein related in the recent *Capital City*, bosses have always wanted the succor of the state (providing infrastructure, health care, and other necessary subsistence to workers) without paying the taxes so obviously necessary to support it (Malm 2016; Stein 2019). In this environment, expect to see more appeals to the digital as the source of a "free lunch," a chance to get "something for nothing" via new forms of money and exchange.

Cryptocurrency is alluring now because finance's present is so alarmingly exploitative and inefficient. But things could always get worse. Both incrementalist and futurist fintech expose the hidden costs of digital efforts to circumvent or co-opt state monetary authorities. They help us overcome the disciplinary divides—between politics and economics, or law and business—that have obscured the stakes of the metaphors and narratives that dominate contemporary conceptions of currency. We face a stark choice: recognize the public nature of money and redirect its creation toward public ends, or allow the power of the state to increasingly be annexed to the privateers best poised to stoke public enthusiasm for private monies, and to profit immensely from the uncertainty they create (Pasquale 2019a).

References

ACLU. (2014). *Metadata: Piecing together a privacy solution.* Available via ACLU of California. Retrieved February 17, 2020, from https://www.aclunc.org/sites/default/files/Metadata%20report%20FINAL%202%2021%2014%20cover%20%2B%20inside%20for%20web%20%283%29.pdf.

Ajunwa, I., Crawford, K., & Schultz, J. (2017). Limitless workplace surveillance. *California Law Review, 105*(3), 735–776. https://doi.org/10.15779/Z38BR8MF94.

Angwin, J., Scheiber, S., & Tobin, A. (2017). *Facebook job ads raise concerns about age discrimination.* Available via The New York Times. Retrieved February 17, 2020, from https://www.nytimes.com/2017/12/20/business/facebook-job-ads.html.

Barlyn, S. (2018). *Strap on the Fitbit: John Hancock to sell only interactive life insurance.* Available via Reuters. https://www.reuters.com/article/us-manulife-financi-john-hancock-lifeins/strap-on-the-fitbit-john-hancock-to-sell-only-interactive-life-insurance-idUSKCN1LZ1WL.

Beedham, M. (2019). *Here's how to fight back against bitcoin-ransoming malware.* Available via TNW Hard Fork. Retrieved February 17, 2020, from https://thenextweb.com/hardfork/2019/09/26/heres-how-to-fight-back-against-bitcoin-ransoming-malware-ransomware/.

Berlant, L. (2011). *Cruel optimism.* Durham, NC: Duke University Press.

Blacker, D. J. (2013). *The falling rate of earning and the neoliberal endgame.* Washington, DC: Zero Books.

Bruckner, M. A. (2018). The promise and perils of algorithmic lenders' use of big data. *Chicago-Kent Law Review, 93*(1), 3–60.

Brunton, F. (2019). *Digital cash: The unknown history of the anarchists, utopians, and technologists who created cryptocurrency.* Princeton, NJ: Princeton University Press.

Burns, J. (2018). *Cut off from big Fintech, white nationalists are using bitcoin to raise funds.* Available via Forbes. Retrieved February 17, 2020, from https://www.forbes.com/sites/janetwburns/2018/01/03/cut-off-from-big-fintech-white-supremacists-are-using-bitcoin-to-raise-funds/#301d1f9e33b3.

Citron, D. K. (2008). Technological due process. *Washington University Law Review, 85*(6), 1249–1313.

Citi., (2018). Investment themes in 2018: How much longer can the cycle run? Available via https://www.citivelocity.com/citigps/investment-themes-2018/. Accessed. 12/15/20.

Financial Crisis Inquiry Commission (FCIC). (2011). *Final report of the national commission on the causes of the financial and economic crisis in the United States.* Available via govinfo. Retrieved February 18, 2020, from https://www.govinfo.gov/content/pkg/GPO-FCIC/pdf/GPO-FCIC.pdf.

Gibbs, S. (2015). *Women less likely to be shown ads for high-paid jobs on Google, study shows.* Available via The Guardian. Retrieved February 18, 2020, from https://www.theguardian.com/technology/2015/jul/08/women-less-likely-ads-high-paid-jobs-google-study.

Golumbia, D. (2009). *The cultural logic of computation.* Cambridge, MA: Harvard University Press.

Golumbia, D. (2016). *The politics of bitcoin: Software as right-wing extremism.* Minneapolis, MN: University of Minnesota Press.

Hill, R. P., & Kozup, J. C. (2007). Consumer experiences with predatory lending practices. *The Journal of Consumer Affairs, 41*(1), 29–46. https://doi.org/10.1111/j.1745-6606.2006.00067.x.

Kotsko, A. (2015). *Creepiness.* Washington, DC: Zero Books.

Leber, J. (2016). *This new kind of credit score is all based on how you use your cell phone.* Available via Fast Company. Retrieved February 18, 2020, from https://www.fastcompany.com/3058725/this-new-kind-of-credit-score-is-all-based-on-how-you-use-your-cellphone.

Lin, T. C. W. (2014). The new financial industry. *Alabama Law Review, 65*(3), 567–623.

Lin, T. C. W. (2015). Infinite financial intermediation. *Wake Forest Law Review, 50*(3), 643–669.

Link, A. N., & Link, J. R. (2009). *Government as entrepreneur.* New York: Oxford University Press.

Malik, N. (2018). *How criminals and terrorists use cryptocurrency: And how to stop it.* Available via Forbes. Retrieved February 18, 2020, from https://www.forbes.com/sites/nikitamalik/2018/08/31/how-criminals-and-terrorists-use-cryptocurrency-and-how-to-stop-it/#39df3d983990.

Malm, A. (2016). *Fossil capital: The rise of steam power and the roots of global warming.* New York: Verso.

McGettigan, A. (2013). *The great university gamble: Money, markets and the future of higher education.* London: Pluto Press.

McMillan Cottom, T. (2018). *Lower ed: The troubling rise of for-profit colleges in the new economy.* New York: The New Press.

Metz, R. (2018). *This company embeds microchips in its employees, and they love it.* Availabe via MIT Technology Review. Retrieved February 18, 2020, from https://www.technologyreview.com/s/611884/this-company-embeds-microchips-in-its-employees-and-they-love-it/.

Mierzwinski, E., & Chest, J. (2013). Selling consumers not lists: The new world of digital decision-making and the role of the fair credit reporting act. *Suffolk University Law Review, 46*, 845–880.

Mirowski, P. (2014). *Never let a serious crisis go to waste: How neoliberalism survived the financial meltdown.* New York: Verso Books.

Newfield, C. (2016). *The great mistake: How we wrecked public universities and how we can fix them.* Baltimore, MD: Johns Hopkins University Press.

Pasquale, F. (2008). Internet nondiscrimination principles: Commercial ethics for carriers and search engines. *University of Chicago Legal Forum, 2008*(1), 263–299.

Pasquale, F. (2015). Reforming the law of reputation. *Loyola University Chicago Law Journal, 47*, 515–539.

Pasquale, F. (2016). Platform neutrality: Enhancing freedom of expression in spheres of private power. *Theoretical Inquiries in Law, 17*(1), 487–513.

Pasquale, F. (2017). Toward a fourth law of robotics: Preserving attribution, responsibility, and explainability in an algorithmic society. *Ohio State Law Journal, 78*(5), 1243–1255.

Pasquale, F. (2018). New economic analysis of law: Beyond technocracy and market design. *Critical Analysis of Law, 5*(1), 1–18.

Pasquale, F. (2019). A rule of persons, not machines: The limits of legal automation. *George Washington Law Review, 87*(1), 1–55.

Pasquale, F. (2019a). Realities and relationships in public finance. *Provocations, 2*, 21–28. https://www.provocationsbooks.com/2019/02/05/realities-and-relationships-in-public-finance/.

Peppet, S. R. (2015). Unraveling privacy: The personal prospectus and the threat of a full-disclosure future. *Northwestern University Law Review, 105*(3), 1153–1204.

Pistor, K. (2013). Law in finance. *Journal of Comparative Economics, 41*(2), 311–314.

Pistor, K. (2019). *The code of capital: How the law creates wealth and inequality.* Princeton, NJ: Princeton UP.

Polk, D. (2019). *Time to get serious about microchipping employees and biometric privacy laws.* Available via Law Fuel. Retrieved February 18, 2020, from http://www.lawfuel.com/blog/time-to-get-serious-about-microchipping-employees-and-biometric-privacy-laws/.

Rabinow, P., & Sullivan, W. (Eds.). (1988). *Interpretive social science—A second look.* Berkeley and Los Angeles, CA: University of California Press.

Sacasas, L. M. (2013). *Borg complex.* Available via The Frailest Thing. Retrieved February 18, 2020, from https://thefrailestthing.com/2013/03/01/borg-complex-a-primer/.

Schiller, R. J. (2019). *Narrative economics: How stories go viral and drive major economic events.* Princeton, NJ: Princeton University Press.

Scism, L. (2019). *New York insurers can evaluate your social media use—If they can prove it's needed.* Available via The Wall Street Journal. Retrieved February 18, 2020, from https://www.wsj.com/articles/new-york-insurers-can-evaluate-your-social-media-useif-they-can-prove-why-its-needed-11548856802.

Seamster, L., & Charron-Chénier, R. (2017). Predatory inclusion and education debt: Rethinking the racial wealth gap. *Social Currents, 4*(3), 199–207. https://doi.org/10.1177/2329496516686620.

Senior, A. (2018). *John Hancock leaves traditional life insurance model behind to incentivize longer, healthier lives.* Available via CISION PR Newswire. Retrieved February 18, 2020, from https://www.johnhancock.com/news/insurance/2018/09/john-hancock-leaves-traditional-life-insurance-model-behind-to-incentivize-longer%2D%2Dhealthier-lives.html.

Stein, S. (2019). *Capital city: Gentrification and the real estate state.* New York: Verso.

Taylor, K.-Y. (2019). *Predatory inclusion.* n+1 Magazine 35. Retrieved February 18, 2020, from https://nplusonemag.com/issue-35/essays/predatory-inclusion/.

Tene, O., & Polonetsky, P. (2014). A theory of creepy: Technology, privacy, and shifting social norms. *Yale Journal of Law and Technology, 16*(1), 59–102.

The Guardian Editorial. (2018). *The Guardian view on big data and insurance: Knowing too much.* Available via The Guardian. Retrieved February 18, 2020, from https://www.theguardian.com/commentisfree/2018/sep/27/the-guardian-view-on-big-data-and-insurance-knowing-too-much.

Upturn. (2015). *Online lead generation and payday loans.* Available via Upturn. Retrieved February 18, 2020, from https://www.upturn.org/reports/2015/led-astray/.

Van Dam, A. (2019). *Algorithms were supposed to make Virginia judges fairer: What happened was far more complicated.* Available via The Washington Post. Retrieved February 18, 2020, from https://www.washingtonpost.com/business/2019/11/19/algorithms-were-supposed-make-virginia-judges-more-fair-what-actually-happened-was-far-more-complicated/.

Variyar, M., & Vignesh, J. (2017). *The new lending game, post-demonetisation.* Available via ET Tech. Retrieved February 18, 2020, from https://tech.economictimes.indiatimes.com/news/technology/the-new-lending-game-post-demonetisation/56367457.

Vincent, J. (2015). *Woman fired after disabling work app that tracked her movements 24/7.* Available via The Verge. Retrieved February 18, 2020, from https://www.theverge.com/2015/5/13/8597081/worker-gps-fired-myrna-arias-xora.

Warner, M. R. (2019). *Senators introduce bipartisan legislation to ban manipulative 'dark patterns'.* Available via Mark R. Warner: Press releases. Retrieved February 18, 2020, from https://www.warner.senate.gov/public/index.cfm/2019/4/senators-introduce-bipartisan-legislation-to-ban-manipulative-dark-patterns.

White, R. W., Doraiswamy, P. M., & Horvitz, E. (2018). Detecting neurodegenerative disorders from web search signals. *NPJ Digital Medicine, 1*(8), 1–4. https://doi.org/10.1038/s41746-018-0016-6.

Winner, L. (1997). Cyberlibertarian myths and the prospects for community. *Computers and Society, 27*(3), 14–19. https://doi.org/10.1145/270858.270864.

World Economic Form. (2017). *Beyond Fintech: A pragmatic assessment of disruptive potential in financial services.* Available via https://es.weforum.org/reports/beyond-fintech-a-pragmatic-assessment-of-disruptive-potential-in-financial-services. Accessed. 12/15/20.

Zuboff, S. (2019). *The age of surveillance capitalism: The fight for a human future at the new frontier of power.* New York: PublicAffairs.

Part III

Robotics, AI, and Militarized Conflict

Designing Robots for the Battlefield: State of the Art

Bruce A. Swett, Erin N. Hahn, and Ashley J. Llorens

Contents

Abstract

There is currently a global arms race for the development of artificial intelligence (AI) and unmanned robotic systems that are empowered by AI (AI-robots). This paper examines the current use of AI-robots on the battlefield and offers a framework for understanding AI and AI-robots. It examines the limitations and risks of AI-robots on the battlefield and posits the future direction of battlefield AI-robots. It then presents research performed at the Johns Hopkins University Applied Physics Laboratory (JHU/APL) related to the development, testing, and control of AI-robots, as well as JHU/APL work on human trust of autonomy and developing self-regulating and ethical robotic systems. Finally, it examines multiple possible future paths for the relationship between humans and AI-robots.

Keywords

Artificial intelligence · Robotics · Unmanned vehicles · AI ethics · AI-robots · Battlefield

B. A. Swett (✉) · E. N. Hahn · A. J. Llorens
The Johns Hopkins University Applied Physics Laboratory
(JHU/APL), Laurel, MD, USA
e-mail: bruceswett@verizon.net; erin.hahn@jhuapl.edu;
Ashley.Llorens@jhuapl.edu

© The Author(s) 2021
J. von Braun et al. (eds.), *Robotics, AI, and Humanity*, https://doi.org/10.1007/978-3-030-54173-6_11

Introduction

There is currently an arms race for the development of artificial intelligence (AI) and unmanned robotic platforms enhanced with AI (AI-robots) that are capable of autonomous action (without continuous human control) on the battlefield. The goal of developing AI-empowered battlefield robotic systems is to create an advantage over the adversary in terms of situational awareness, speed of decision-making, survivability of friendly forces, and destruction of adversary forces. Having unmanned robotic platforms bear the brunt of losses during combat—reducing loss of human life—while increasing the operational effectiveness of combat units composed of manned-unmanned teams is an attractive proposition for military leaders.

AI-Robots Currently Used on the Battlefield

Numerous types of fielded robotic systems can perform limited operations on their own, but they are controlled remotely by a human pilot or operator. The key element of these systems is that no human is onboard the platform like there is on a plane flown by a pilot or a tank driven by a commander. The human control component has been removed from the operational platform and relocated away from the battlefield. Real-time communication between the human operator and the unmanned robotic system is required, so that the human can control the platform in a dynamic, complex battlefield. As advanced militaries around the world contest the portions of the electromagnetic spectrum used for the control of unmanned combat systems, the need for increased robot autonomy is heightened. U.S. policy directs that, regardless of the level of autonomy of unmanned systems, target selection will be made by an "authorized human operator" (U.S. Dept. of Defense 2012). The next sections describe the primary categories of deployed robotic systems.

Unmanned Aerial Vehicles

Unmanned aerial vehicles (UAVs) are widely used by the U.S., Chinese, and Russian militaries, as well as by numerous nation states and regional actors (Dillow 2016). The relatively low cost and low technology barrier to entry for short-range UAVs has enabled the rapid proliferation of UAV technologies (Nacouzi et al. 2018) for use individually, in groups, or in large swarms. Long-range UAVs with satellite-based communications are more likely to be used by countries with extensive military capabilities. UAVs were initially used for surveillance to provide an enhanced understanding of the area of conflict, but their role has expanded into target identification and tracking; direct attack; command, control, and communications; targeting for other weapons systems; and resupply (see Fig. 1). UAVs are divided into Class I (small drones), Class II (tactical), and Class III (strategic) (Brown 2016), or Groups 1–5, depending on size and capabilities (U.S. Army Aviation Center of Excellence 2010).

In actual combat, Russia flew multiple types of UAVs in the conflict with the Ukraine, using UAVs to rapidly target Ukrainian military forces for massed-effect artillery fire (Fox 2017; Freedberg Jr. 2015), and has employed armed UAVs in Syria (Karnozov 2017). The U.S. has an extensive history of using UAVs in combat, with the MQ-9 Reaper targeted strikes in the Middle East being the most notable. Israel has developed the IAI Heron to compete with the Reaper and, along with China, is a major supplier of drone technology (Blum 2018). Armed Chinese CH-4 drones are in operation over the Middle East by countries that are restricted from purchasing U.S. drone technologies (Gambrell and Shih 2018), and China has become a leading developer and distributer of armed UAVs (Wolf 2018). Current UAVs are usually flown by a pilot at a ground station. There are AI algorithms that can set waypoints—a path for the UAV to fly—but the UAV doesn't make decisions on where to go or to fire weapons. These functions are currently the pilot's responsibility. There are indications that major militaries, including the U.S. Army, are moving toward taking the human out of the loop by having UAVs perform behaviors autonomously (Scharre 2018; Singer 2001).

Unmanned Ground, Surface, and Underwater Vehicles

The U.S. Army has successfully used small, unarmed, portable unmanned ground vehicles (UGVs) for improvised explosive device (IED) removal and explosive ordnance disposal in Iraq and Afghanistan, and it is purchasing UGVs in large numbers (O'Brien 2018). Recently, the focus has shifted to using both small, armed UGVs and larger, armed robotic tanks. Russia deployed the Uran-9 robotic tank, which is armed with a gun and missiles, in Syria with mixed results, and the U.S. Army is developing an optionally manned replacement for the Bradley fighting vehicle (Roblin 2019). The U.S. Army is also developing robotic tanks and technology kits to convert existing tanks into remotely controlled and/or autonomous UGVs (Osborn 2018). Similar to UAVs, unmanned robotic tanks serving as scouts for manned tanks would reduce the loss of life while increasing the overall firepower that could be brought to bear on the adversary (see Fig. 2, left).

Maritime unmanned platforms have been developed and tested by several countries, including the U.S. and China. China demonstrated control of a swarm of 56 unarmed un-

Fig. 1 *Military uses of UAVs.* UAVs are now a ubiquitous presence on the battlefield and are being used for intelligence, surveillance, and reconnaissance; targeting; communications; and direct attack purposes in militaries around the world (images courtesy of the Marine Corps Warfighting Lab—quad copter; and U.S. Air Force—Reaper)

Fig. 2 *Military uses of unmanned ground, surface, and underwater vehicles.* UGVs have been used for IED and ordnance removal and are now being developed as mobile weapons systems and semiautonomous tanks (above, left, from BigStockPhoto, Mikhail Leonov). USV and UUV swarms (above, right, JHU APL) are able to coordinate swarm behavior to inflict significant damage on a larger vessel

manned surface vehicles (USVs) in 2018 (Atherton 2018). The importance of swarm control of USVs or unmanned underwater vehicles (UUVs) lies in the ability to overwhelm the defenses of much larger ships, to inflict disproportionate damage (see Fig. 2 right). The U.S. is developing UUVs that can perform anti-submarine warfare and mine countermeasures autonomously (Keller 2019b; Keller 2019a). As with UAVs and UGVs, USVs and UUVs require communications with human operators or integrated control systems, and that will remain the case until AI autonomy algorithms are sufficiently robust and reliable in performing missions in communications-denied environments.

Integrated Air Defense Systems and Smart Weapons

Russia has developed and fielded a highly effective integrated air defense system (IADS) that can engage and destroy multiple aircraft at range, creating regions where no one can fly (i.e., anti-access/area denial, or A2AD) (see Fig. 3). The Russian IADS combines radar tracking; air defense missiles for short, medium, and long range; electronic warfare (EW) jamming capabilities; guns; surface-to-air missiles; robust communications; and an integration control system (Jarretson 2018). Russia has recently made improvements to the range of the S-500 IADS and has sold the existing S-400 IADS to China for use in the South China Sea (Trevithick 2018). Importantly, because of the speed of incoming aircraft and missiles, these IADS are fully autonomous weapons systems with no identification of friend or foe. In this case, "autonomous" means that the system determines what are targets, which targets to engage, and the timing of engagement and then enacts the targeting decisions. While in this autonomous mode, the autonomous system's decisions and actions are too fast for human operators to identify and intervene. Similarly, the U.S. Navy has used the MK 15 Phalanx Close-In Weapon System on ship to counter high-

Fig. 3 *Russian IADS*. IADS developed by Russia, which include radars, long- and short-range missiles, EW systems, and machine guns, create areas where no aerial platform can operate (from BigStockPhoto, Komisar)

speed missiles and aircraft without human commands (Etzioni and Etzioni 2017). The argument for this human-on-the-loop approach is that a human operator is not fast enough to respond to the threat in time, so an algorithm has to be used.

Understanding AI and how It Enables Machine Autonomy

In this section we provide a framework for understanding how AI enables robots to act autonomously to pursue human-defined goals. This framework is offered as a basis for understanding the current state of the art in AI as well as the limitations of AI, which are described in the next section. AI, as a field of study, may be thought of as the pursuit to create increasingly intelligent AI that interacts with the world, or AI agents (Russell and Norvig 2003). A straightforward way to understand the term intelligence in this context is as an agent's ability to accomplish complex goals (Tegmark 2018), with the agent's degree of autonomy in pursuing those goals arising from the delegation of authority to act with specific bounds (Defense Science Board 2017). Agents are typically implemented as systems, where the interactions of the system components—sensors, computers, algorithms—give rise to its intelligence.

Whether agents are embodied robotic systems or exist only as programs in cyberspace, they share a common ability to perceive their environment, decide on a course of action that best pursues their goals, and act in some way to carry out the course of action while teaming with other agents. Multi-agent teams may themselves be thought of as intelligent systems, where intelligence arises from the agents' ability to act collectively to accomplish common goals. The appropriate calibration of trust is a key enabler for effective teaming among agents performing complex tasks in challenging environments. Perceive, Decide, Act, Team, Trust (PDATT) is a conceptual framework that allows us to build intuition around the key thrusts within AI research, the current state of the art, and the limits of AI-robots (see Fig. 4).

Elements of the PDATT Framework

Perceive

Perception for an intelligent system involves sensing and understanding relevant information about the state of the operating environment. Salient perception of physical environments for an intelligent system may require sensing of physical phenomena both within and beyond the range of human senses. The perception space for an intelligent system may also lie within a purely informational domain requiring, for example, measurement of activity patterns on a computer

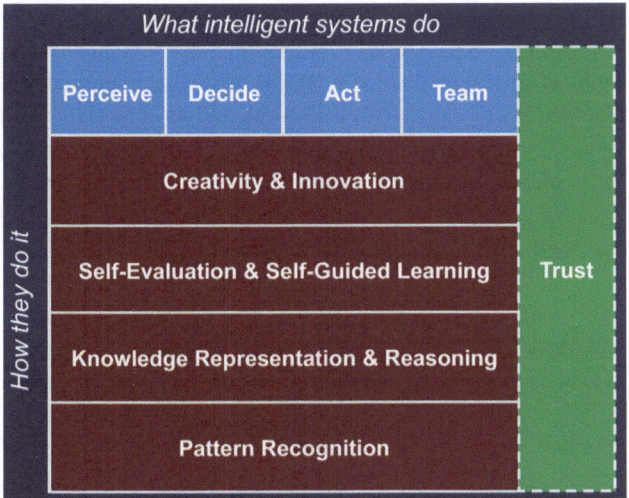

Fig. 4 *PDATT AI framework.* The PDATT framework outlines the main elements of AI-robots. The light blue boxes illustrate what AI-robots do: they must perceive salient elements of the environment to decide which action to select. An action is then taken, with expected effects in the environment. The AI-robot operates alongside humans in teams, requiring frequent and meaningful communication to ensure trust during mission operations. The red boxes illustrate how AI-robots accomplish the four foundational functions of perceive, decide, act, and team. The green box represents the set of capabilities needed to ensure that the AI-robot is trusted by the human operator in meeting mission objectives (JHU/APL)

network or counting the number of mentions of a particular public figure on Twitter. Recent progress in machine learning has enabled amazing improvements in object recognition and sequence translation, while combining this kind of data-driven pattern recognition with robust human-like reasoning remains a fundamental challenge.

Decide

Decision-making for an intelligent system includes searching, evaluating, and selecting a course of action among a vast space of possible actions toward accomplishing long-term goals. The challenge that long-term decision-making poses to an intelligent system can be illustrated using strategy games such as chess or Go. An algorithm that plays chess must determine the best sequence of moves to accomplish its ultimate goal of winning the game. Although a game like chess is less complex than real-world decision-making, it still presents a difficult decision-making challenge for an agent. In 1950, famous mathematician and engineer Claude Shannon estimated that the game size—the number of different sequences of actions in the game of chess, taking into account all the possible moves by each player—is 10^{43} (Shannon 1950). For the game of Go, the estimated game size is 10^{170} (Tromp 2016). Of course, despite their complexity, these games abstract the dynamics and uncertainty associated with real-world decision-making. While recent progress in developing artificial agents capable of superhuman-level play

in games like these is exciting, significant research will be needed for AI to be capable of long-term decision-making for complex real-world challenges (Markowitz et al. 2018).

Act

Acting for an intelligent system includes the ways in which the system interacts with its environment in accomplishing given goals. Acting may include traversing an environment, manipulating objects, or even turning computers on or off to prevent network intrusion. A system may act in order to improve its perception—for example, moving a sensor to get a better line of sight to an object of interest. An agent's action space is the set of possible actions available to it. In chess, the action space consists of allowable movement of pieces on the board. For a driverless car, the action space consists of control actions such as accelerating and steering. For an agent that controls a centrifuge in a nuclear power plant, the action space may include raising or lowering the speed or temperature of critical system components. For a humanoid robot assistant, the action space may include manipulation of household objects with human-like dexterity. Recent progress in robotics has enabled systems capable of taking complex actions within carefully controlled conditions, such as a factory setting, while enabling robots that act in unstructured real-world setting remains a key challenge.

Team

For an intelligent system, effectively pursuing complex goals in challenging environments almost always requires acting as part of a multi-agent team with both humans and other machines. Teaming typically requires some form of communication, the ability to create sufficient shared situational awareness of the operating environment, and the ability to effectively allocate tasks among agents. Recent demonstrations of multi-agent teaming have involved large choreographed formations, such as the Intel UAV demonstration at the 2019 Super Bowl (Intel 2019). In future applications, we envision teams of agents with heterogeneous capabilities that can seamlessly and flexibly collaborate to accomplish human-specified goals. Flexible multi-agent teaming in unstructured scenarios will require continued research. This level of teaming may require something akin to the development of a machine theory of mind, including the ability to model the goals, beliefs, knowledge, etc. of other agents (Premack and Woodruff 1978).

Trust

Realizing the promise of AI-robots will require appropriate calibration of trust among developers, users, and certifiers of technologies, as well as policy-makers and society at large. Here, trust is defined as " . . . the intention to accept vulnerability to a trustee based on positive expectations of his or her actions" (Colquitt et al. 2007). Fundamental advancements are needed in trust as it relates to intelligent systems, including the following.

- *Test and evaluation:* assuring complex intelligent systems meet design goals in the face of significant uncertainty in both the test and operational environments
- *Resilience to adversarial influence:* hardening intelligent systems against subversion through phenomenological, behavioral, and informational attack by an adversary
- *Goal alignment:* ensuring that actions performed by intelligent systems remain aligned with human intent even as those systems are tasked to pursue increasingly high-level goals
- *Policy:* Determining which decisions and actions are appropriate for an intelligent system to perform autonomously in accordance with system capabilities as well as societal values and ethics.

The Risks of AI-Robots on the Battlefield

So, given an understanding of how AI-robots work using PDATT and the current use of AI-robots on the battlefield, what are the risks of accelerated use and enhancement of capabilities of AI-robots on the battlefield? Assuming that artificial general intelligence (AGI)—with human-level perception, understanding of context, and abstract reasoning—won't be available in the next decade, what are the limitations of current AI-robots that induce risk on the battlefield?

Current Limitations of AI-Robots on the Battlefield and Associated Risks

Perception

Computer vision algorithms have made significant gains in the identification of people and objects in video data, with classification rates matching or exceeding human performance in some studies (Dodge and Karam 2017). However, making sense of the world in three dimensions involves more than object detection and classification. Current technical challenges in perception include tracking humans and objects consistently in varied backgrounds and lighting conditions; identifying and tracking individuals in various contexts; identifying behaviors; and predicting intent or outcomes. Actions taken over time also have meaning, as do interactions between people, between people and objects, and between different objects. The human brain has "what" pathways, to link visual identification of a person or object with background information, associations, and prior experiences, also known as "vision-for-perception." The brain's "where" pathway not only tracks the location of the person or object relative to

the viewer, but it also encodes "vision-for-action" (Goodale and Milner 2006). Current AI systems lack a sophisticated, multisensory perceptual system that creates rich internal representations of the state of the world, and its changes, in real time. The risks associated with limited perception include the inability to reliably detect friend or foe; relationships between elements in a scene; and the implications of the actions, people, and objects in a scene.

Decision-Making, Reasoning, and Understanding Context

Largely due to the deficits in AI perceptual systems, current AI does not have an ability to understand the context of actions, behaviors, or relationships. This complex web of experience is essential for reasoning and decision-making. The most advanced fielded AI reasoning systems are rules based—that is, they are a set of "if–then" statements that encode human processes and decisions (Kotseruba and Tsotsos 2018). While these rules-based systems have been successful in constrained tasks and limited robotic autonomy, they are statically programmed prior to the AI system operating in the battlefield, so they are necessarily limited and predictable. Current AI algorithms are "narrow," in that they can solve a specific problem. The eventual goal is to create AGI, which will have human-level reasoning and will be both generative, able to create new behaviors on the fly in response to novel situations, and generalizable, able to apply learning in one area to other operations or missions. The development of AGI is a long-term vision, with estimates of realization between 10 years and never (Robitzski 2018). The risk of AI-robots having insufficient reasoning is compounded by the limitations in perception. Choosing appropriate actions within a specific context and mission requires highly accurate and reliable perception, as well as a wealth of background knowledge of what the context, rules, and perceptions mean for decision-making. None of these highly accurate and reliable perceptual and reasoning capabilities currently exist for fully autonomous AI-robot behavior.

Action Selection, Self-Correction, and Ethical Self-Assessment

If an unmanned platform is to perform independently of a human operator, it will need to encode and function within the legal and ethical constraints of the military operating the system. AI systems have difficulties predicting how action choices will affect the environment, so they are unable to weigh the alternatives from an ethical or legal perspective. A recent project that used AI algorithms to predict the best engineers to hire discriminated against women; another project that used AI algorithms to predict whether criminals would re-offend discriminated against Blacks (Shaw 2019). Humans are able to predict consequences of different possible actions all the time, weighing which alternative is the most advantageous while adhering to personal chosen norms. Future AI systems for battlefield robots will need to have the ability to predict the impact of actions on the environment, including friendly forces, adversaries, and noncombatants, and then make decisions informed by the Law of Armed Conflict and national policies (Dowdy et al. 2015). If AI-robots do not have continuous, feed-forward prediction capabilities, they will be unable to self-regulate, risking their ability to adhere to mission parameters in the absence of direct human oversight.

Teaming and Trust: Transparency in Human Interactions

A significant concern with today's state-of-the-art AI algorithms is that some neural networks used for battlefield robots are not human readable, or are "black boxes" (Knight 2017). So, while the algorithm that detects the presence of a tank may do so with a certain probability of detection and false-alarm rate, the human operator can't currently query the algorithm's basis of determining the classification beyond a simple confidence factor, as it is encoded in the weights of the neural network. As the role of AI-robots on the battlefield expands, the AI's rationale for selecting a specific course of action, or its determination of an object as a threat, will need to be explainable to the human operator. The AI's course of action must also be coordinated and deconflicted with other blue force actors, to avoid impeding adjacent operations and operational elements. It is also important to remember that military decisions made during war are subject to legal review, so collecting a chain of evidence on what algorithms were used, and their basis of determining actions, is critical. Future AI-powered robots don't need to be perfect, but they do need to perform important functions consistently and understandably, storing reviewable information on perceptions, reasoning, and decisions, so that human soldiers trust them and can determine when their use is acceptable. A lack of transparency could diminish the human operator's trust in the AI-robot, and the operator may not know when to use or not use the AI-robot safely and appropriately.

Trust: Vulnerabilities to Cyber and Adversarial Attacks

In recent studies, neural network-based AI computer vision algorithms were tricked into changing the classification of an object after relatively few pixels in the image to be identified had been changed (Elsayed et al. 2018). The implication for the battlefield is that an adversary could modify the classification outcomes of the AI so that the robot would take unintended actions. Trust in an AI-empowered combat robot requires that the robot perform tasks well and consistently in a manner that is comprehensible to the operator, and that it not be corruptible by the adversary. If an adversary could use cyberattacks to take over the robot, this would weaken the robot's operational effectiveness by undermining the operator's trust that the robot would complete missions as directed. Recently, Russia claimed that the U.S. took control of 13 UAVs and attempted to use them to attack a Russian military base in Syria (Associated Press 2018). Addressing cyber vulnerabilities will be critical to the successful use of AI-robots on the battlefield, as AI-robots that can be redirected or co-opted from their intended use would pose unacceptable risk to friendly forces and noncombatants.

The Future of AI-Robots on the Battlefield

The U.S., China, and Russia continue to increase defense investments for the application of AI to a broader set of potential missions (Allen 2019; Apps 2019). The U.S. Department of Defense (DoD) released an overarching AI strategy (Moon Cronk 2019), as did China, and Russia is scheduled to do so in mid-2019, demonstrating all three countries' long-term commitment to making AI a bigger and more coordinated part of their defense infrastructures. While the imminent use of human-like robotic killing machines still gets top billing in discussions and articles on AI, this is only one of many potential AI applications on the future battlefield, and it is not necessarily a concern for the near future. Several of the current AI applications in weapons focus on improved human-machine teaming and increased deployment of unmanned vehicles for various missions, mostly nonlethal. If they are capable of lethal force, fielded systems to date have had defensive purposes only. However, this norm may be challenged as future AI applications are likely to cover a wider spectrum, from data sorting for enhanced intelligence analysis to replacement of human decision-making in targeting decisions.

What AI-Robots May Be Able to Do in the Near Future

Countries with sophisticated militaries will continue to invest in AI-enabled weapons. However, these systems are expensive to develop and field, and they are useful for a set of missions but not for all combat-related tasks. Therefore, from a cost-benefit perspective, AI is not likely to fully replace human operators as many assume, but rather to augment them and perhaps take over specific missions or aspects of missions. Refinements in the near future could broaden the number of relevant missions, and better AI-enabled weapons may challenge the current role of the human in the targeting cycle. Below is a discussion of five areas of development that are likely to impact the future battlefield.

Improving What We Already Have and Expanding Missions

Research into and development of systems such as UAVs for use in swarms, as well as UUVs, continues, but the emphasis is on more autonomy (e.g., the ability for drones to accurately and consistently target and engage other drones without human input, the ability of armed UUVs to travel long distances undetected) and greater coordination among multiple platforms (Zhao 2018). However, new payloads represent a growing concern. Russia is thought to be developing nuclear-powered UUVs capable of carrying nuclear payloads (U.S. Dept. of Defense 2018). And the idea of nuclear armed drones is out there but has not yet been realized.

Perhaps the most unsettling future development then is not the AI-enabled platform—although the ability to travel greater distances with stealth is important—but what payload the AI-robot is able to deploy. Debate and calls for international agreements have emerged in response to the idea of applying AI to any part of the nuclear enterprise, out of fear that it will upend nuclear stability entirely (Geist and Lohn 2018). This fear is based on the belief that AI will encourage humans to make catastrophic decisions or that AI will do so on its own. As discussed previously, the current limitations of AI to understand context, reason, and take action would need to be overcome, and even then, the mix of AI with nuclear would benefit from a broader debate among policy-makers and scientists.

Dull, Dirty, and Dangerous Jobs

Numerous "dull, dirty, and dangerous" jobs that soldiers used to do are increasingly being done by AI-empowered machines. Resources are being used to develop better unmanned trucks and logistics vehicles to reduce the number of human operators needed to do these jobs. Currently, unmanned logistics vehicles could enable new missions that would otherwise be precluded because of the inability to safely get materials situated in the desired location. In another use, an AI-enabled unmanned truck could advance ahead of a formation of tactical vehicles if the commander believes the area has buried IEDs. An analog exists for an unmanned lead aircraft advancing first into contested airspace. The application of AI here does two things—it frees up human operators and reduces the operator's risk.

Augmentation of Human Decision-Making

For both analyzing data and reviewing surveillance footage, people currently pore over data or stare at a screen for hours on end watching video footage. In the case of surveillance footage, even a brief deviation of the eyes from the screen can result in the analyst missing something critical. Current AI applications have become better at recognizing objects and people, but a machine's ability to both recognize an object or a human and make an assessment about whether the situation is dangerous, or whether something unexpected is occurring, is still in development. Scientists are using novel methods to train computers to learn how something works by observing it, and in the course of this learning, to detect real from fake behavior (Li et al. 2016). This use of AI has implications for a variety of security applications. For the analyst staring at the screen, it is possible that the AI could make data analysis (such as pattern-of-life detection) much easier by doing some of the work for them, and perhaps doing it better. Such analysis is necessary for the development of strike packages, so the AI would augment the human in developing targeting priorities. While many AI implementations for dull, dirty, and dangerous jobs are instantiated as a physical robotic system, this application is reducing the human burden in the information space.

It is possible that application of AI for information analysis could advance to the point that militaries seek to use it in lieu of a human conducting target identification and engagement. While most Western policy-makers are ok with drones killing drones, when it comes to lethal force, the standing view is that a human must be in the loop. This view is reflected in DoD policy and in its interpretation of international legal obligations (U.S. Dept. of Defense 2012). However, as technology improves and the speed of battle increases, this policy may be challenged, especially if adversaries' capabilities pose an increasingly greater threat.

Replacing Human Decision-Making

One of the biggest advantages in warfare is speed. AI-robot weapons systems operating at machine speeds would be harder to detect and counter, and AI-empowered information systems allow adversaries to move more quickly and efficiently. This increase in speed means the time for human decision-making is compressed, and in some cases, decisions may need to be made more quickly than humans are able to make them. Advances in the application of AI to weapons may even be used to preempt attacks, stopping them before they occur. In the physical world, this may be achieved through faster defensive systems that can detect and neutralize incoming threats. Examples of these systems already exist (Aegis, Harpy), but new weapons, such as hypersonics, will make traditional detection and interception ineffective. In the information world, AI as applied to cyber weapons is a growing concern. The U.S. DoD's policy on autonomy in weapons systems specifically excludes cyber weapons because it recognizes that cyber warfare would not permit time for a human decision (U.S. Dept. of Defense 2012). Indeed, countries continue to invest in AI-enabled cyber weapons that autonomously select and engage targets and are capable of counter-autonomy—the ability of the target to learn from the attack and design a response (Meissner 2016). What is less certain is how cyber warfare will escalate. Will AI systems target infrastructure that civilians rely on, or will an AI-initiated attack cause delayed but lethal downstream effects on civilians?

In addition to speed, a second driver for taking humans out of the loop in AI is related to scale—the volume of information received or actions taken exceeds human capacity to understand and then act. For example, AI-powered chatbots used for social media and information operations can simultaneously analyze and respond to millions of tweets (Carley et al. 2013; Zafarani et al. 2014). This volume of information exceeds a human's capability to monitor and intervene. So, the increasing scale of data and actions, as well as the compressed time for decision-making, are necessitating autonomy, without the ability for human intervention, during the AI's action selection and execution.

In the physical domain, countries are developing weapons designed to remove human decision-making. The most common forms are weapons capable of following a target based on a computational determination within the weapon rather than a human input. These are not new, but emerging applications allow weapons to loiter and select targets from a library of options. For some militaries, it is unclear whether the human in the loop in these circumstances simply determines the library of options ahead of time or is able to evaluate the accuracy of the target selected during the operation. As computer vision capabilities improve, it is possible that militaries will seek to remove the human from all aspects of the targeting cycle. In A2AD environments, where there are few if any civilians, this application may be compelling. However, for a more diverse set of missions, it may not be appropriate. This delegation of targeting decisions is at the heart of ongoing international debates focused on the regulation of AI-enabled weapons systems.

Electronic Warfare

Also driven by increasing speed and reduced time for decision-making, the application of AI to EW is an area of increasing importance for several countries. Systems are being developed for use on various platforms that can detect an adversary EW threat while nearly simultaneously characterizing it and devising a countermeasure. Projects have also been underway for applications of AI to space assets, such as DARPA's Blackjack, to allow autonomous orbital operations for persistent DoD network coverage (Thomas 2019). And, the F-35 is reportedly designed with very sophisticated AI-enabled EW capabilities for self-protection and for countering air defense systems (Lockheed Martin 2019). These are just examples, and this area is likely to grow as research and applications continue to evolve.

JHU/APL Research Toward AI-Robots for Trusted, Real-World Operations

The Johns Hopkins University Applied Physics Laboratory (JHU/APL) has decades of experience developing and testing AI-robots for challenging real-world applications, from defense to health. JHU/APL often serves as a trusted partner to U.S. DoD organizations, helping to facilitate an understanding of capabilities and limitations of state-of-the-art AI as it relates to use in military applications. As AI-robots for air, ground, surface, and underwater use are being developed for use on the battlefield, JHU/APL is committed to ensuring that they are developed, tested, and operated in accordance with U.S. safety, legal, and policy mandates.

Areas of AI-Robotics Research at JHU/APL

In addition to developing simulated and real-world semiautonomous and autonomous AI-robots, JHU/APL performs research on the evaluation, testing, and security of these systems. This research includes developing external control systems for AI-robots to ensure safety, AI systems that are self-correcting, explainable and transparent, and AI with ethical reasoning. The following sections describe each of these areas of research.

Safe Testing of Autonomy in Complex Interactive Environments

JHU/APL is the lead developer of the Safe Testing of Autonomy in Complex Interactive Environments (TACE) program for the Test Resource Management Center and other DoD sponsors. TACE provides an onboard "watchdog" program that monitors the behaviors of an AI-robot and takes control of the platform if the AI produces behaviors that are out of bounds relative to the safety, policy, and legal mandates for the mission (see Fig. 5). TACE operates in live, virtual, and constructive environments, which combine simulation and reality to allow AI algorithms to learn appropriate behavior before operating in the real world. TACE is currently used during the development and testing of autonomous systems to maintain safety and control of AI-robot systems. TACE is independent from the robot's AI autonomy system, has separate access to sensor and control interfaces, and is designed to deploy with the AI-robot to ensure compliance with range safety constraints (Lutz 2018).

Self-Regulating AI

JHU/APL has invested independent research and development funding to examine novel methods of AI self-regulation. Self-regulating AI is a hybrid model that uses both rules-based encoding of boundaries and limitations with neural network-based AI that produces adaptive behaviors in

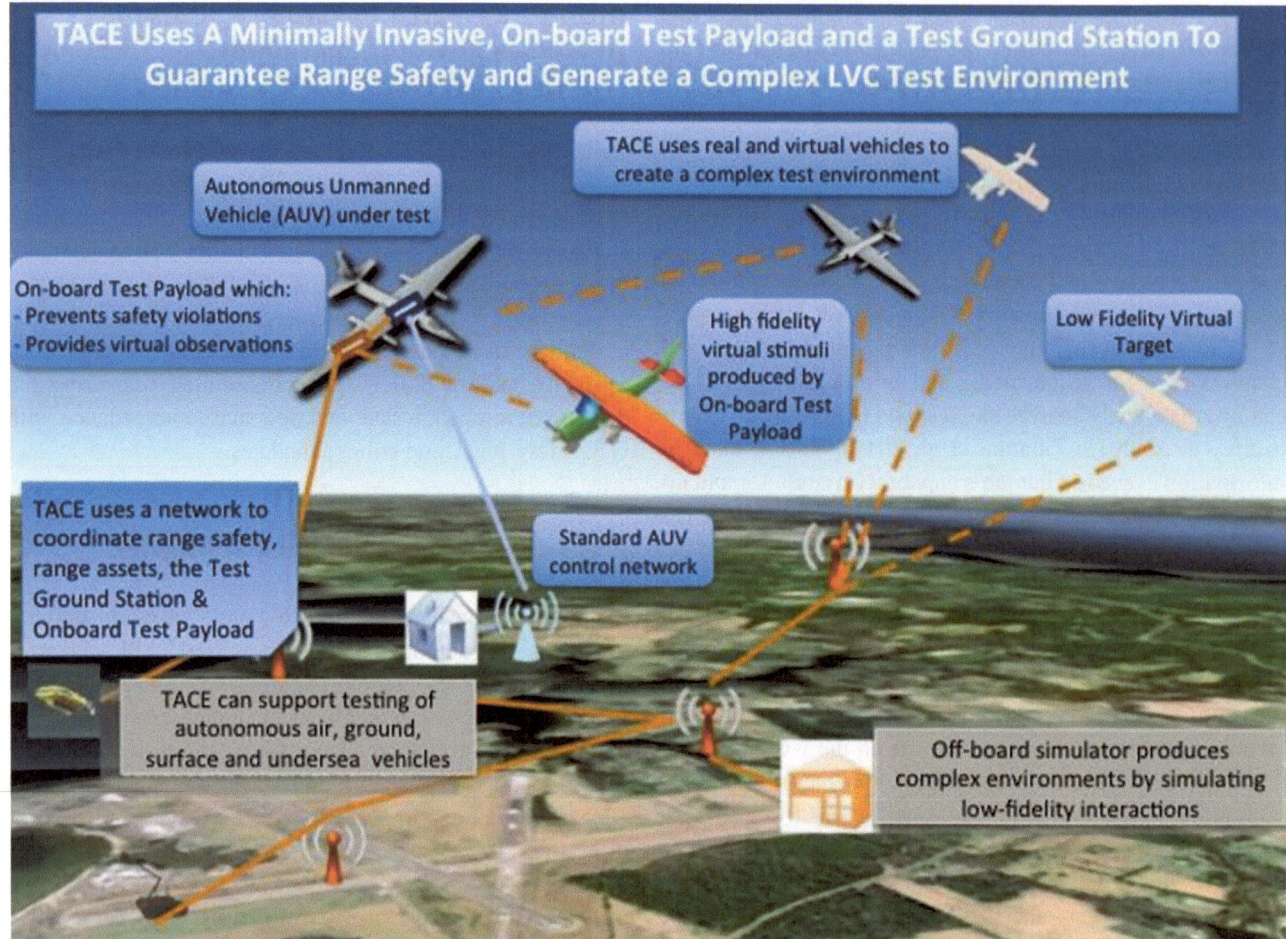

Fig. 5 *TACE*. TACE is an architecture for autonomy testing that serves as an onboard watchdog over AI-robots. TACE monitors all perceptions and behaviors of the AI-robot, takes over when the AI attempts to perform a behavior that is outside of range safety constraints, and records all outcomes for post-mission evaluation (Lutz 2018)

complex environments. While TACE is an external process that ensures that AI behaviors meet mission requirements, and takes over if they don't, self-regulation incorporated the rules into the AI's decision-making process. A key element to this research is the AI's ability to predict environmental effects of sets of possible actions. Developed in simulation, the AI is penalized for decision choices that result in environmental effects that violate mission rules and parameters. In this manner, the AI learns to make decisions that don't violate the mission requirements, which include the Rules of Engagement, the Law of Armed Conflict (Dowdy et al. 2015), and U.S. policies. Self-regulating AI can be used in conjunction with an external watchdog governor system to provide multiple layers of protection against aberrant AI behaviors.

Explainable AI and Human-Machine Interactions

Even if AI-robots perform mission-critical tasks well and reliably, how humans and AI-robots work together in manned-unmanned teams is an open area of research. Simple AI

systems that perform a function repeatedly without making decisions are merely tools; the issues in human-machine interactions arise when the AI-robot acts as an agent, assessing the environment, analyzing its mission goals and constraints, and then *deciding to do something*. How predictable will the AI-robot be to human teammates? How transparent will the robot's perception, decision-making, and action selection be—especially in time-critical situations? JHU/APL is researching how humans develop trust in other humans, and how that process is altered when humans work with AI-robots. This includes having the AI learn when it needs to provide information to its human teammate, when to alert the human to a change in plans, and, importantly, when to not make a decision and ask the human for help. Figure 6 presents an integration of research on the elements of human trust in autonomy, including human, autonomous entity, team, and task factors. Developing an AI system that understands what it knows and doesn't know, or what it can and cannot do, is especially challenging. It requires that the AI have the capability to self-assess and provide a level of confidence

Fig. 6 *The theoretical underpinnings of trust.* Important elements in the development of human trust in autonomy are presented—including human, autonomous entity, team, and task factors—that, taken together, present a method for evaluating trust and the breakdown in trust (Marble and Greenberg 2018)

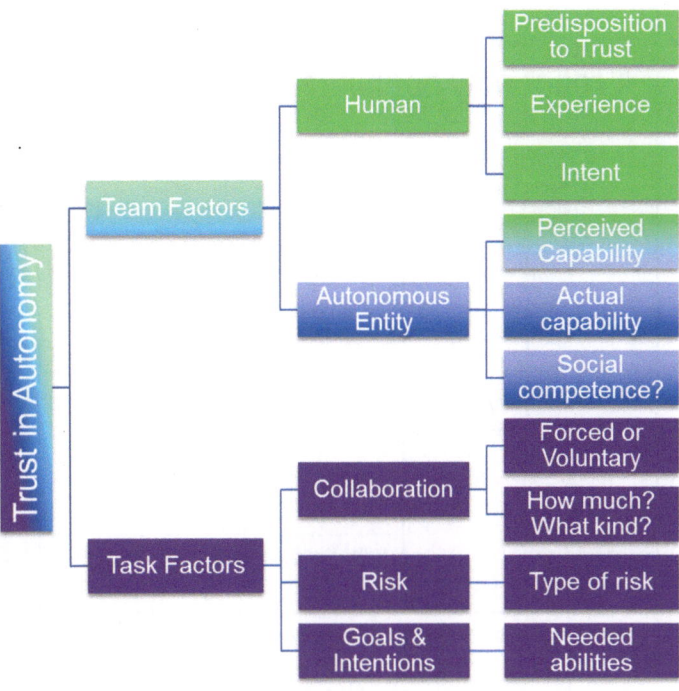

for every perception, decision, and action. The ability to know when an AI-robot can and can't be used safely and appropriately in the battlefield is a critical step in developing meaningful interactions and trust between humans and AI-robots.

Ethical AI

A crucial step in designing autonomous AI-robots is to develop methods for the AI to encode our values, morals, and ethics. JHU/APL teamed up with the Johns Hopkins Berman Institute of Bioethics to study how to enable an AI to perform moral reasoning. Starting from the first of Isaac Asimov's laws of robotics—"A robot may not injure a human being or, through inaction, allow a human being to come to harm" (Asimov 1950)—this research explores the philosophical, governance, robotics, AI, cognitive science, and human-machine interaction aspects of the development of an AI with moral reasoning. For an AI to perform moral reasoning, it would necessarily need to encode what a human is, what injuries to humans are, and the ways by which humans are injured. Moral and ethical decision-making to prevent an AI from harming a human starts with perception. Even basic capabilities—such as differentiating living people from objects—are challenges for current AI algorithms. The JHU/APL research formulates a method of moral-scene assessment for intelligent systems, to give AI systems a "proto-conscience" that allows them to identify in a scene the elements that have moral salience (see Fig. 7). This moral perception reifies potential insults and injuries to persons for the AI to reason over and provides a basis for ethical

behavioral choices in interacting with humans (Greenberg 2018).

Conclusions: The Future of AI for Battlefield Robotics

Rapid advancements in the fields of AI-robots have led to an international arms race for the future of the battlefield. AI systems will become more capable and more lethal, and wars eventually will be fought at machine speed—faster than humans can currently process information and make decisions. As AI reasoning systems mature, more complex behaviors will be delegated to AI-robot systems, with human control on the loop or out of the loop entirely. Proliferation will result in large numbers—hundreds of thousands to millions—of AI-robots on the battlefield. JHU/APL is working to develop multiple means by which AI-robots on the battlefield can be used within the legal, ethical, and policy mandates of the U.S. Government.

Competing Visions of the Future of AI-Robots

Beyond the near-term development of new AI and AI-robot capabilities is the question of the long-term relationship between humans and AI-enhanced robots. As of now, AI-empowered robots are extensions of human activities. As AI systems evolve to operate autonomously, self-assess and self-regulate, and work effectively alongside humans to accomplish tasks, what will the roles be for the humans and

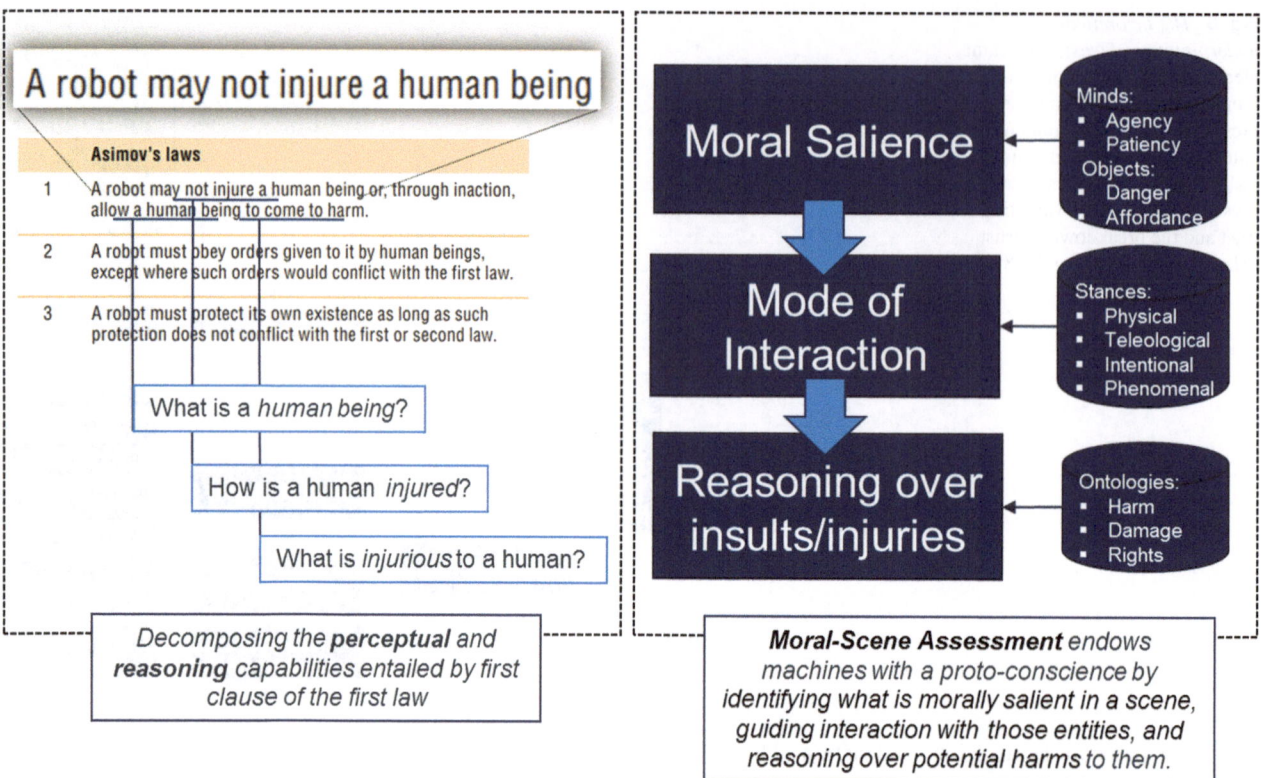

Fig. 7 *Moral-scene assessment.* Starting from Asimov's Three Laws of Robotics (Asimov 1950), this decomposition (above, left) extracts the perceptual and reasoning capabilities needed for an AI-robot to engage in moral consideration. Ethical decision-making to prevent an AI from harming a human starts with perception (above, right), providing a basis for ethical behavior choices (Greenberg 2018)

the robots? The following sections review two projections of the future relationship between humans and robots and then present an alternative possibility.

the development and deployment of AI and AI-empowered autonomous systems (IEEE 2019).

Human Subservience to AI-Robots

The most popular envisioned future relationship between humans and AI-robots is exemplified by the movie "The Terminator," in which an AI defense system (Skynet) becomes self-aware and decides to enslave and destroy the human race (Cameron 1984). The central premise is that humans build a synthetic reasoning system that is able to process faster and control more sensors and weapons than humans can. In the movie, humans become obsolete. Elon Musk recently stated that autonomous AI-powered weapons could trigger the next world war, and that AI could become "an immortal dictator" (Holley 2018). A number of experts have called for a ban on "killer robots," to ensure that humans are not made to be subservient to, or destroyed by, robots (Gibbs 2017). The concerns about the continuing advancement of AI technologies has led the Institute of Electrical and Electronics Engineers (IEEE) to develop a global initiative on AI development, including a primer on ethically aligned design, to ensure that human well-being is prioritized in

Human Dominance over AI-Robots

An alternative view of the future relationship between humans and robots that is equally concerning is one in which AI-empowered robots are developed and evolved as slaves to humans. While this is less of a concern in the near term when AI-robots are primarily tools, as we move toward AGI, we move into a space where AI-robots act as independent agents and become indistinguishable from humans. DeepMind has developed an AI (AlphaGo) that beat the best humans in chess, shogi, and Go (Silver et al. 2018) and an AI system (AlphaStar) that can beat the best human StarCraft II video game players (AlphaStar Team, DeepMind 2019). It is not far-fetched to imagine the continued development of super-human AI capabilities in the next 20–50 years. The question arises: if an AI approaches AGI—and is not distinguishable from human rationality using human perception—do humans still have the right to "turn it off," or does a self-aware AI have the rights of personhood (van Hooijdonk 2018; Sheliazhenko 2017)?

Co-Evolution with AI-Robots

In contrast with the two previous envisioned futures, where humans are either subservient to or destroyed by robots or vice versa, another possibility exists. It is important to remember that machine learning and AI algorithms were developed by studying mammalian (including human) neural computations in the brain, and that future developments in AI will occur through an increased understanding of the brain (Hassabis et al. 2017). In this respect, developing AI is a lot like looking in a mirror. We are currently exploring a very minute part of what a human brain can do, through experimental and computational research, and then creating mathematical models to try to replicate that neural subsystem. It is vital to understand that this process also helps us learn about how we, as humans, work—what our computational strengths and weaknesses are—and how human diseases and limitations could be overcome.

An example of this is the study of memory—both human memory storage and retrieval—and AI uses of memory. Human memory is quite limited, with only a few independent, short-term memory slots (Baddeley and Hitch 1974) that are quite volatile and long-term memory that is often inaccurate, error prone, and retroactively alterable (Johnson 2010). Persons with Alzheimer's disease have lost the ability to locate individual memories, as the index to memory location has been degraded by the disease (RIKEN 2016). Recent research used a mathematical model to encode episodic memories (i.e., remembering where you put your car keys) in the hippocampus, a function that is degraded in people with Alzheimer's disease. These enhanced neural signals were then sent back into the patient's brain, resulting in an improvement in short-term memory of 35–37% (Wake Forest Baptist Medical Center 2018). Future integration of AI with brain-computer interfaces could result in improved memory for both Alzheimer's patients and healthy humans. Similar improvements to human perception, reasoning, decision-making, and learning are possible.

So, an alternative vision of the future of humans and robots is one where humans and AI co-evolve, each learning from the other and improving each other. If the development of AGI is the ultimate human mirror showing us our potential for self-destruction and destruction of others, it is also a mirror that shows us the potential for humans to grow, change, and evolve in positive ways. As in healthcare research, where the more we understand human physiology and pathologies, the more creative solutions we can develop, research toward AGI offers us the opportunity to understand ourselves and to improve the human condition—not as enforced by external organizations or movements, but as a matter of individual choice. In this way, co-evolution between humans and AI becomes like any other method of human self-improvement—education, healthcare, use of technology—in that we can choose to what level we participate, weighing the pros and cons of that decision.

This concept of co-evolution between humans and AI-robots also addresses the fears of the first two envisioned futures. A central concern of the "Terminator" future is that humans will be "left behind" by AI systems, leading to human destruction or subordination (Caughill 2017). By co-evolving, humans will be able to understand and keep up with emerging AI systems. If humans are able to process information at machine speeds, and at massive scale, we will be able to intervene in decisions made by AI-robots before actions are taken. In this manner, human operators will be able to ensure that AI systems are operating legally and ethically, within the human-defined mission parameters. Also, by understanding our human limbic systems and emotions better—including our fight or flight responses and fear of the "other"—we have the potential to co-evolve with AI to improve our self-control beyond the current human tendencies to dominate and subordinate, as described in the enslavement future projection. Co-evolution between humans and AI-robots could lead to the voluntary improvement of both individual humans and human society.

In conclusion, AI and AI-robots are an increasing presence on the battlefield, and are being developed with greater speed, scale of operation, autonomy, and lethality. Understanding how AI and AI-robots operate, the limitations of current AI-robots, and the likely future direction of these systems can inform policy-makers in their appropriate uses and potential for misuse. JHU/APL has leveraged its expertise in the development of AI and AI-robot systems to perform research in how these systems can be developed responsibly and safely, and used within U.S. safety, legal, and policy mandates. While there are highly divergent opinions on the relationship of humans to future AI systems—including fears of humans being dominated by AI, or vice versa—there is an opportunity for humans to learn and advance from interactions with AI. In a future of co-evolution, humans would be able to ensure that AI reflects human ethical and legal values because humans would be able to continue to understand and interact with AI systems.

References

Allen, G. C. (2019). *Understanding China's AI strategy*. Washington, DC: CNAS. Available via CNAS. Retrieved March 20, 2019, from https://www.cnas.org/publications/reports/understanding-chinas-ai-strategy.

AlphaStar Team, DeepMind. (2019). *AlphaStar: Mastering the real-time strategy game StarCraft II*. Available via DeepMind. Retrieved March 6, 2019, from https://deepmind.com/blog/alphastar-mastering-real-time-strategy-game-starcraft-ii/

Apps, P. (2019). *Commentary: Are China, Russia winning the AI arms race?* Available via Reuters. Retrieved March 20, 2019, from https://www.reuters.com/article/us-apps-ai-commentary/commentary-are-china-russia-winning-the-ai-arms-race-idUSKCN1P91NM

Asimov, I. (1950). *The Isaac Asimov collection*. New York: Doubleday.

Associated Press. (2018). *Russia claims US aircraft took control of drones in attempted attack on its Syrian base*. Available via Military Times. Retrieved March 6, 2019, from https://www.militarytimes.com/flashpoints/2018/10/25/russia-claims-us-aircraft-took-control-of-drones-in-attempted-attack-on-syrian-base/.

Atherton, K. (2018). *See China's massive robot boat swarm in action*. Available via C4ISRNET. Retrieved March 5, 2019, from https://www.c4isrnet.com/unmanned/2018/06/01/see-chinas-massive-robot-boat-swarm-in-action/.

Baddeley, A. D., & Hitch, G. (1974). Working memory. In G. H. Bower (Ed.), *Psychology of learning and motivation, vol. 8* (pp. 47–89). Cambridge, MA: Academic Press. https://doi.org/10.1016/S0079-7421(08)60452-1.

Blum, B. (2018). *9 Israeli drone startups that are soaring to success*. Available via ISRAEL21c. Retrieved March 5, 2019, from http://www.israel21c.org/9-israeli-drone-startups-that-are-soaring-to-success/.

Brown, J. (2016). *Types of military drones: The best technology available today*. Available via MyDroneLab. Retrieved March 5, 2019, from http://mydronelab.com/blog/types-of-military-drones.html.

Cameron, J. (1984). The Terminator [Motion picture]. https://www.imdb.com/title/tt0088247/?ref_=nv_sr_4

Carley, K. M., Pfeffer, J., Liu, H., Morstatter, F., & Goolsby, R. (2013). Near real time assessment of social media using geo-temporal network analytics. In *Proceedings of the 2013 IEEE/ACM International Conference on Advances in Social Networks Analysis and Mining* (pp. 517–524). New York: ACM. doi: https://doi.org/10.1145/2492517.2492561.

Caughill, P. (2017). *Artificial intelligence is our future. But will it save or destroy humanity?* Available via Futurism.com. Retrieved March 7, 2019, from https://futurism.com/artificial-intelligence-is-our-future-but-will-it-save-or-destroy-humanity.

Colquitt, J. A., Scott, B. A., & LePine, J. A. (2007). Trust, trustworthiness, and trust propensity: A meta-analytic test of their unique relationships with risk taking and job performance. *The Journal of Applied Psychology, 92*(4), 909–927. https://doi.org/10.1037/0021-9010.92.4.909.

Defense Science Board. (2017). *Summer study on autonomy*. Washington, DC: Defense Science Board. Available via Homeland Security Digital Library. Retrieved February 21, 2020, from https://www.hsdl.org/?abstract&did=794641.

Dillow, C. (2016). *All of these countries now have armed drones*. Available via Fortune. Retrieved March 4, 2019, from http://fortune.com/2016/02/12/these-countries-have-armed-drones/.

Dodge, S., & Karam, L. (2017). A study and comparison of human and deep learning recognition performance under visual distortions. In: *2017 26th International Conference on Computer Communication and Networks (ICCCN)* (pp. 1–7). doi: https://doi.org/10.1109/ICCCN.2017.8038465.

Dowdy, L. R., Cherry, L. J., Berry, L. J., Bogar, L. T., Clark, L. B., Bogar, L. T., Blank, M. J., MJ, D. S., Festa, M. M., Lund, M. M., Radio, M. K., & Thorne, T. (2015). *Law of armed conflict deskbook* (5th ed., p. 253). Charlottesville, VA: International and Operational Law Department, U.S. Army Judge Advocate General's Legal Center and School.

Elsayed, G., Shankar, S., Cheung, B., Papernot, N., Kurakin, A., Goodfellow, I., & Sohl-Dickstein, J. (2018). Adversarial examples that fool both computer vision and time-limited humans. In S. Bengio, H. Wallach, H. Larochelle, K. Grauman, N. Cesa-Bianchi, & R. Garnett (Eds.), *Advances in neural information processing systems* (Vol. 31, pp. 3910–3920). Red Hook, NY: Curran Associates. Retrieved from http://papers.nips.cc/paper/7647-adversarial-examples-that-fool-both-computer-vision-and-time-limited-humans.pdf.

Etzioni, A., & Etzioni, O. (2017). Pros and cons of autonomous weapons systems. *Military Review, 2017*(3), 72–80.

Fox, M. A. (2017). Understanding modern Russian war: Ubiquitous rocket, artillery to enable battlefield swarming, siege warfare. *Fires, 2017*(5), 20–25. Retrieved from https://sill-www.army.mil/firesbulletin/archives/2017/sep-oct/additional-articles/Understanding%20modern%20Russian%20war.pdf.

Freedberg, S. J. Jr. (2015). *Russian drone threat: Army seeks Ukraine lessons*. Available via Breaking Defense. Retrieved March 4, 2019, from https://breakingdefense.com/2015/10/russian-drone-threat-army-seeks-ukraine-lessons/.

Gambrell, J., & Shih, G. (2018). *Chinese armed drones now flying over Mideast battlefields. Here's why they're gaining on US drones*. Available via Military Times. Retrieved March 4, 2019, from https://www.militarytimes.com/news/your-military/2018/10/03/chinese-armed-drones-now-flying-over-mideast-battlefields-heres-why-theyre-gaining-on-us-drones/.

Geist, E., & Lohn, A. (2018). *How might artificial intelligence affect the risk of nuclear war?* Santa Monica, CA: RAND Corporation. https://doi.org/10.7249/PE296.

Gibbs, S. (2017). *Elon Musk leads 116 experts calling for outright ban of killer robots*. Available via The Guardian. Retrieved February 21, 2020, from https://www.theguardian.com/technology/2017/aug/20/elon-musk-killer-robots-experts-outright-ban-lethal-autonomous-weapons-war.

Goodale, M., & Milner, D. (2006). *One brain—two visual systems*. Available via The Psychologist. Retrieved March 5, 2019, from https://thepsychologist.bps.org.uk/volume-19/edition-11/one-brain-two-visual-systems.

Greenberg, A. (2018). Moral-scene assessment for intelligent systems. In *Presented at the 2018 IEEE Resilience Week*, Denver, CO, 2018.

Hassabis, D., Kumaran, D., Summerfield, C., & Botvinick, M. (2017). Neuroscience-inspired artificial intelligence. *Neuron, 95*(2), 245–258. https://doi.org/10.1016/j.neuron.2017.06.011.

Holley, P. (2018). *Elon Musk's nightmarish warning: AI could become 'an immortal dictator from which we would never escape'*. Available via Washington Post. Retrieved March 6, 2019, from https://www.washingtonpost.com/news/innovations/wp/2018/04/06/elon-musks-nightmarish-warning-ai-could-become-an-immortal-dictator-from-which-we-would-never-escape/.

IEEE. (2019). *IEEE-Standards Association*. The IEEE global initiative on ethics of autonomous and intelligent systems. Available via IEEE. Retrieved March 6, 2019, from https://standards.ieee.org/industry-connections/ec/autonomous-systems.html.

Intel. (2019). *Intel drone light show and Intel True View technology enhance Pepsi Super Bowl LIII Halftime Show and Super Bowl LIII viewing experience*. Available via Intel Newsroom. Retrieved March 20, 2019, from https://newsroom.intel.com/news-releases/intel-drones-true-view-super-bowl-liii/.

Jarretson, P. (2018). *How does Russia's integrated air defense network work?* Available via Quora. Retrieved March 5, 2019, from https://www.quora.com/How-does-Russias-integrated-air-defense-network-work.

Johnson, J. (2010). Chapter 7: Our attention is limited, our memory is imperfect. In J. Johnson (Ed.), *Designing with the mind in mind* (pp. 121–130). Waltham, MA: Elsevier.

Karnozov, V. (2017). *Extent of Russian UAV use over Syria revealed*. Available via AIN Online. Retrieved March 5, 2019, from https://www.ainonline.com/aviation-news/defense/2017-09-21/extent-russian-uav-use-over-syria-revealed.

Keller, J. (2019a). *Future unmanned underwater vehicles with machine autonomy for deep-sea missions are focus of DARPA angler program.* Available via Military and Aerospace Electronics. Retrieved March 5, 2019, from https://www.militaryaerospace.com/articles/2019/01/unmanned-deep-sea-machine-autonomy.html.

Keller, J. (2019b). *Navy eyes sonar, communications, and power upgrades for black pearl unmanned underwater vehicles (UUVs).* Available via Military and Aerospace Electronics. Retrieved March 5, 2019, from https://www.militaryaerospace.com/articles/2019/01/unmanned-underwater-vehicles-uuvs-sonar-power.html.

Knight, W. (2017). *The dark secret at the heart of AI.* Available via MIT Technology Review. Retrieved March 5, 2019, from https://www.technologyreview.com/s/604087/the-dark-secret-at-the-heart-of-ai/.

Kotseruba, I., & Tsotsos, J. K. (2018). 40 years of cognitive architectures: Core cognitive abilities and practical applications. *Artificial Intelligence Review, 53,* 17. https://doi.org/10.1007/s10462-018-9646-y.

Li, W., Gauci, M., & Groß, R. (2016). Turing learning: A metric-free approach to inferring behavior and its application to swarms. *Swarm Intelligence, 10*(3), 211–243. https://doi.org/10.1007/s11721-016-0126-1.

Lockheed Martin. (2019). *Unprecedented battlefield access.* Available via Lockheed Martin. Retrieved March 18, 2019, from https://www.f35.com/about/capabilities/electronicwarfare.

Lutz, R. (2018). Taming the Terminator. *Presented at the Interservice/Industry Training, Simulation, and Education Conference (I/ITSEC),* Orlando, FL, November 26–30, 2018.

Marble, J. L., & Greenberg, A. (2018). ESCAPE with PARTNER: Experimental suite for cooperatively-achieved puzzle exercises with platform assessing risk and trust in non-economic relationships. In *Proceedings of the 2018 IEEE Resilience Week: Avoiding Skynet.* Denver, CO.

Markowitz, J., Gardner, R. W., & Llorens, A. J. (2018). On the complexity of reconnaissance blind chess. *ArXiv:1811.03119 [Cs].* http://arxiv.org/abs/1811.03119

Meissner, C. (2016). *The most militarily decisive use of autonomy you won't see.* Available via Defense One. Retrieved March 23, 2019, from https://www.defenseone.com/ideas/2016/11/most-militarily-decisive-use-autonomy-you-wont-see-cyberspace-ops/132964/.

Moon Cronk, T. (2019). *DOD unveils its artificial intelligence strategy.* Available via U.S. Dept. of Defense. Retrieved March 18, 2019, from https://dod.defense.gov/News/Article/Article/1755942/dod-unveils-its-artificial-intelligence-strategy/.

Nacouzi, G., Williams, J. D., Dolan, B., Stickells, A., Luckey, D., Ludwig, C., Xu, J., Shokh, Y., Gerstein, D. M., & Decker, M. (2018). *Assessment of the proliferation of certain remotely piloted aircraft systems: Response to section 1276 of the National Defense Authorization act for fiscal year 2017.* Santa Monica, CA: RAND Corporation. https://doi.org/10.7249/RR2369.

O'Brien, M. (2018). *Army looks for a few good robots, sparks industry battle.* Available via Army Times. Retrieved March 5, 2019, from https://www.armytimes.com/news/your-army/2018/12/30/army-looks-for-a-few-good-robots-sparks-industry-battle/.

Osborn, K. (2018). *The Army is building robot attack tanks [blog post].* Available via National Interest. Retrieved March 5, 2019, from https://nationalinterest.org/blog/buzz/army-building-robot-attack-tanks-26541.

Premack, D., & Woodruff, G. (1978). Does the chimpanzee have a theory of mind? *The Behavioral and Brain Sciences, 1*(4), 515–526. https://doi.org/10.1017/S0140525X00076512.

RIKEN. (2016). *Flipping a light switch recovers memories lost to Alzheimer's disease mice.* Available via Science Daily. Retrieved March 7, 2019, from https://www.sciencedaily.com/releases/2016/03/160316140419.htm.

Robitzski, D. (2018). *When will we have artificial intelligence as smart as a human? Here's what experts think.* Available via Futurism. Retrieved March 5, 2019, from https://futurism.com/human-level-artificial-intelligence-agi.

Roblin, S. (2019). *Russia's Uran-9 robot tank went to war in Syria (it didn't go very well).* Available via National Interest. Retrieved March 5, 2019, from https://nationalinterest.org/blog/buzz/russias-uran-9-robot-tank-went-war-syria-it-didnt-go-very-well-40677.

Russell, S., & Norvig, P. (2003). *Artificial intelligence: A modern approach.* Upper Saddle River, NJ: Prentice Hall. Retrieved from http://aima.cs.berkeley.edu/.

Scharre, P. (2018). *Why drones are still the future of war.* Available via CNAS. Retrieved March 22, 2019, from https://www.cnas.org/publications/commentary/why-drones-are-still-the-future-of-war.

Shannon, C. E. (1950). XXII. Programming a computer for playing chess. *Philosophical Magazine, 41*(314), 256–275. https://doi.org/10.1080/14786445008521796.

Shaw, J. (2019). *Artificial intelligence and ethics.* Available via Harvard Magazine. Retrieved March 5, 2019, from https://harvardmagazine.com/2019/01/artificial-intelligence-limitations.

Sheliazhenko, Y. (2017). Artificial personal autonomy and concept of robot rights. *European Journal of Law and Political Sciences, 2017*(1), 17–21. https://doi.org/10.20534/EJLPS-17-1-17-21.

Silver, D., Hubert, T., Schrittwieser, J., Antonoglou, I., Lai, M., Guez, A., Lanctot, M., Sifre, L., Kumaran, D., Graepel, T., Lillicrap, T., Simonyan, K., & Hassabis, D. (2018). A general reinforcement learning algorithm that masters chess, shogi, and go through self-play. *Science, 362*(6419), 1140–1144. https://doi.org/10.1126/science.aar6404.

Singer, P. W. (2001). *In the loop? Armed robots and the future of war.* Available via Brookings. Retrieved March 5, 2019, from https://www.brookings.edu/articles/in-the-loop-armed-robots-and-the-future-of-war/.

Tegmark, M. (2018). *Life 3.0: Being human in the age of artificial intelligence.* New York: Knopf. Retrieved from https://www.penguinrandomhouse.com/books/530584/life-30-by-max-tegmark/9781101970317/.

Thomas, P. (2019). *Blackjack.* Available via DARPA. Retrieved March 18, 2019, from https://www.darpa.mil/program/blackjack.

Trevithick, J. (2018). *Russia's s-500 air defense system reportedly hits target nearly 300 miles away.* Available via The Drive. Retrieved March 5, 2019, from http://www.thedrive.com/the-war-zone/21080/russias-s-500-air-defense-system-reportedly-hits-target-nearly-300-miles-away.

Tromp, J. (2016). Number of legal 12xn go positions. Available via John Tromp. Retrieved from https://tromp.github.io/go/L12.html

U.S. Army Aviation Center of Excellence. (2010). *"Eyes of the Army": U.S. Army roadmap for unmanned systems, 2010–2035.* Available via Homeland Security Digital Library. Retrieved from https://www.hsdl.org/?abstract&did=705357

U.S. Dept. of Defense. (2012). *Directive 3000.09: Autonomy in weapons systems.* Available via Executive Services Directorate. Retrieved March 4, 2019, from https://www.esd.whs.mil/portals/54/documents/dd/issuances/dodd/300009p.pdf.

U.S. Dept. of Defense. (2018). *Nuclear posture review Feb 2018.* Available via U.S. Dept. of Defense. Retrieved March 20, 2019, from https://www.defense.gov/News/Special-Reports/NPR-2018.

van Hooijdonk, R. (2018). *If AI becomes self-aware, will we need to give it the rights we normally reserve for humans?.* Available via Richard van Hooijdonk. Retrieved March 6, 2019, from https://www.richardvanhooijdonk.com/en/blog/ai-becomes-self-aware-will-need-give-rights-normally-reserve-humans/.

Wake Forest Baptist Medical Center. (2018). *Prosthetic memory system successful in humans.* Available via Science Daily. Retrieved

March 7, 2019, from https://www.sciencedaily.com/releases/2018/03/180327194350.htm.

Wolf, H. (2018). *3 Reasons why China is the global drones leader*. Available via World Economic Forum. Retrieved March 5, 2019, from https://www.weforum.org/agenda/2018/09/china-drones-technology-leader/.

Zafarani, R., Ali Abbasi, M., & Liu, H. (2014). *Social media mining*. New York: Cambridge University Press. Retrieved from http://dmml.asu.edu/smm/.

Zhao, T. (2018). *Tides of change: China's nuclear ballistic missile submarines and strategic stability*. Available via the Carnegie Endowment for International Peace website. Retrieved March 20, 2019, from https://carnegieendowment.org/files/Zhao_SSBN_final.pdf.

Applying AI on the Battlefield: The Ethical Debates

Gregory M. Reichberg and Henrik Syse

Contents

Abstract

Because lethal autonomous weapon systems (LAWS) are designed to make targeting decisions without the direct intervention of human agents (who are "out of the killing loop"), considerable debate has arisen on whether this mode of autonomous targeting should be deemed morally permissible. Surveying the contours of this debate, the authors first present a prominent ethical argument that has been advanced in favor of LAWS, namely, that AI-directed robotic combatants have an advantage over their human counterparts, insofar as the former operate solely on the basis of rational assessment, while the latter are often swayed by emotions that conduce to poor judgment. Several counter arguments are then presented, inter alia, (1) that emotions have a positive influence on moral judgment and are indispensable to it; (2) that it is a violation of human dignity to be killed by a machine, as opposed to being killed by a human being; and (3) that the honor of the military profession hinges on maintaining an equality of risk between combatants, an equality that would be removed if one side delegates its fighting to robots. The chapter concludes with a reflection on the moral challenges posed by human-AI teaming in battlefield settings, and how virtue ethics provides a valuable framework for addressing these challenges.

Keywords

AI · Autonomy · Combatants · Emotions · Ethics · Rationality · Robots · Weapons

G. M. Reichberg (✉)
Peace Research Institute Oslo (PRIO), Research School on Peace and Conflict | Political Science, University of Oslo, Grønland, Norway
e-mail: greg.reichberg@prio.org

H. Syse
Peace Research Institute Oslo (PRIO), Oslo, Norway

Björknes University College, Oslo, Norway

Introduction

The battlefield is an especially challenging domain for ethical assessment. It involves the infliction of the worst sorts of harm: killing, maiming, destruction of property, and devastation of the natural environment. Decision-making in war is carried out under conditions of urgency and disorder. Clausewitz famously referred to this as the "fog of war"; indeed, the

© The Author(s) 2021
J. von Braun et al. (eds.), *Robotics, AI, and Humanity*, https://doi.org/10.1007/978-3-030-54173-6_12

very root of our word "war" is derived from the Germanic root *wirr*, which signified "thrown into confusion." Showing how ethics is realistically applicable in such a setting has long taxed philosophers, lawyers, military practitioners, and educators. The advent of artificial intelligence (AI) has added a new layer of complexity. Hopes have been kindled for smarter targeting on the battlefield, fewer combatants, and hence less bloodshed; simultaneously, stern warnings have been issued on the grave risks of a new arms race in "killer robots," the loss of control over powerful machinery, and the risks associated with delegating lethal decisions to increasingly complex and autonomous machines.

While war has remained a constant in human existence, the progressive introduction of new technologies (e.g., gunpowder, mechanized infantry, air power, nuclear munitions) has led to dramatic shifts in battlefield dynamics. Warfare has been extended into new domains—air, undersea, cyber, and now outer space—that in turn interact in novel ways. How transformative AI will ultimately be within this multilayered battlefield has been the subject of much speculation, but already military forces the world over, not least the major powers but also many lesser ones, are investing heavily in AI-based weapons systems and platforms.[1] Ethical reflection on the likely implications is imperative. This chapter aims to outline the main directions of current debate in this field. Our focus is on AI-based weapons technology; we will largely leave to the side how AI more broadly supports military activity: monitoring troop movements and capabilities, administration of aid to wounded personnel or their extraction from the battlefield, diffusing explosive munitions, and so forth.

At the outset, it can be noted that AI is not itself a weapon. Rather, it is a cognitive tool that facilitates the application of weaponry to selected targets. AI does this both through the mediation of robots and by assisting human agents in applying the weaponry themselves. In either case, AI mimics[2] cognitive abilities—sensation, memory, and inference—that are found in human beings. AI is patterned after these abilities, sometimes falling short of them, and other times surpassing them. At the present stage of scientific advancement, general artificial intelligence has not been achieved (and it remains an open question whether it ever will be); for the foreseeable future at least, machine intelligence will remain highly selective in its operations. For this reason, in what follows, we proceed from the assumption that AI is a tool—albeit the most sophisticated tool yet devised by human beings—and

even when implemented in robots, it does not possess agency in the proper sense of the term (which entails a capacity for self-awareness, ability to set goals, and so forth). This is not to say that AI *qua* tool cannot run in autonomous mode, making its own decisions and learning from previous decisions. On the contrary, this is already possible. However, even when operating in autonomous mode, AI serves in a support capacity; full agency, selfhood, and personhood cannot be attributed to it. Responsibility for any wrongdoing that might result from AI-powered operations must be traced back to the human agents who propose, design, operate, or direct their use.

AI is a tool that extends human cognitive abilities beyond their normal range of functioning. AI can enhance human sensory capacities, as when it is used for purposes of surveillance or detection; AI can increase the speed with which humans process information; and AI can contribute to human decision-making, either by providing input that supports decisions made by humans or, as with autonomously functioning AI, the decision itself is delegated to the machine. Decision is the cognitive act whereby an antecedent phase of deliberation (whether extended or instantaneous) issues into a course of action. A decision is a special form of judgment: "x shall be done." The doing of x changes some aspect of the world. This is what philosophers call "practical" (as opposed to "speculative" or "theoretical") judgment. In what follows, we are concerned with practical judgments within the sphere of military action, particularly those decisions that result in harm done to human beings or the natural environment. The chapter proceeds as follows:

1. To provide context for our discussion, we review (i) the principal reasons that have induced military planners to develop AI-based warfighting capabilities, (ii) how autonomy in AI-based weapons systems is a matter of degree, and (iii) current attempts to code ethical norms into autonomous AI systems for military applications.

2. Thereafter, we review ethical arguments for and against the use of autonomously functioning AI battlefield targeting systems, focusing first on the more principle-based arguments, and thereafter on arguments belonging more in the domain of the technological and pragmatic, although, admittedly, there is not a clear-cut distinction between the two categories, and some arguments will overlap.

3. By way of conclusion, we look at how AI–human collaboration ("the force mix") on the battlefield can be expected to impact on the practical judgment of military personnel, their ability to engage in ethical ("virtuous") conduct, and their moral integrity. Using the tradition of virtue ethics as our point of departure we formulate a number of questions for further research.

---

[1] For a survey of the current technology see Swett et al. in this volume.

[2] "Mimic" is not here taken in a pejorative or dismissive sense, as when we say that a parrot "mimics" human speech. It is taken rather in the sense of Platonic "imitation," as when it is said that one reality can possess ("participates in") what another has but in a diminished or analogous manner. See Patterson (1985).

Background Considerations

In promoting the development of AI-based warfighting capabilities, military planners have responded to several different needs and technological developments.

First, there is the robotics revolution, which has led to an increasing deployment of remote-piloted unmanned ground, surface, underwater, and aerial vehicles. Best known of these are the "drones" (unmanned aerial vehicles—UAVs), which have been extensively used to deliver lethal attacks most notably in Afghanistan, but elsewhere as well. The remotely controlled deployment of these vehicles by human pilots (often sitting thousands of kilometers away from the theater of operations) presents as threefold difficulty: such deployment (1) is very labor intensive (one or more operators are needed to control a single vehicle), (2) requires communication links that are subject to adversarial disruption or are inoperative in some locations, and (3) functions relatively slowly given its dependency on human cognitive reflexes and decision-making. AI-directed unmanned vehicles provide a way around these three difficulties, freeing up human operators for other tasks, obviating the need for constant communications links, and allowing for a more rapid response time. The last feature has become especially important in the context of "swarm warfare," whereby multiple vehicles proceed against a single target (or against another swarm), in a high-speed, tightly coordinated attack. Speed of response is also highly beneficial in related settings, for instance, in cyber confrontations that occur in the milliseconds, or radar-based defensive action to protect against incoming missiles.

It goes without saying that the use of unmanned attack vehicles has the added advantage of protecting military personnel from lethal harm; AI-directed attacks decrease the number of personnel that need be placed on the battlefield, thereby preserving them from injury and death. After World War I force protection has become a paramount concern for conventional armies, and the US experience in Viet Nam showed how soldierly casualities can have a very adverse political impact, even on an otherwise dominant military force.

Replacing human agents in combat settings, in the ways summed up above, is possible only when AI enables weapon systems to operate in autonomous mode. For purposes of this discussion, artificial intelligence may be defined as *intelligent behavior embedded in artificial matter*. "Intelligent" designates an ability to solve complex problems to achieve some goal, while "artificial" excludes biological systems—most importantly: living, breathing, thinking human beings. This definition covers both autonomous and non-autonomous systems. An artificially intelligent system is autonomous if the selection of the means for reaching a preset goal is left to the system itself, as in what has become known as "machine learning." Here the machine "has flexibility in how it achieves its goal" (Scharre 2018: 31). By contrast, a system is *non*-autonomous if the means for reaching a preset goal are predetermined by an external agent, as in the case of a cruise missile that follows a given program, however complex the program might be. It goes without saying that autonomy is very much a matter of degree. There is a spectrum of intelligence in lethal machines (Scharre 2018: 31–34), from systems that are automatic (simple, threshold based), for instance an improvised explosive device, to those that are automated (complex, rule-based), for instance a precision-guided munition, and finally those that are autonomous (goal-oriented, self-directed with respect to the selection of means), for instance the Israeli manufactured Harpy that destroys specific radar installations, none of which are determined in advance, within a particular radius.

Judgments made about machine "autonomy" are very much in the mind of the beholder. Machines whose inner workings we do not understand will often seem to be wholly unpredictable and to produce effects that follow from a decision made by the machine itself, thus the ascription of autonomy to the machine. But once we understand the complex logic on which a machine operates, its resulting actions will usually be redescribed as being merely automated (Scharre 2018: 26 ff.). Of course, this begs the question of whether any machine could ever be autonomous in the proper metaphysical sense of possessing free will. Being free in this sense entails the ability to dominate the reasons for one's action; no one reason requires me to do this or that, i.e., necessitates my action (Simon 1969). Whether freedom from the necessitation of reasons can be achieved by a machine is in all probability impossible. The machine is bound by its underlying architecture and that is why our initial judgment of a machine's autonomy eventually gives way to more moderate characterization in terms of automaticity. In other words, here as elsewhere we need to be on guard against the anthropomorphic imagination.

With respect to lethal weaponry, autonomous functioning is usually described in terms of a threefold distinction (see Scharre 2018: 29–30) between modes of human presence in the "killing loop." First, (1) there is semiautonomous machine killing. Such a system can detect the external environment, identity hostile targets, and even propose a course of action, but the kill decision can only happen through the intervention of a human being. Here a human operator remains *in the killing loop*, in the sense that he/she must take positive action if the lethal attack is to be consummated. Then, (2) there is supervised autonomous machine killing. Such a machine can sense, decide, and act on its own, but it remains under the supervision of a human being who can veto the passage from decision to action. Should no veto be issued, the machine is fully capable of running through the combat cycle (observe, orient, decide, act) on its own. Here a human being remains not *in*, but *on the killing loop*. Third, (3) there is fully autonomous machine killing whereby a human being is needed only to activate the machine, but afterwards it carries out its assigned task without communication back to the human user. Here the human being is *out of the killing loop*. This threefold distinction of *in*, *on*, and *out of* the loop refers to three modes of operation but not necessarily three kinds of machine, as one and the same machine could be set to run on each of these three modes.

Finally, with respect to AI-driven weapon systems that operate in autonomous mode (humans out of the loop), current research has sought to devise algorithms that embed ethical principles into the targeting decisions adopted by these machines. The issue here is to consider whether and how autonomous robots can be programmed to function legally and ethically as battlefield combatants. As already noted, robots can have other tasks on the battlefield, such as retrieving injured soldiers (the Battlefield Extraction Assist Robot), or monitoring human battlefield conduct so that norm violations will be reported back to central command or an international agency, and perhaps even prevented (fear of punishment arising from robotic observations might dissuade solders from acting wrongly in the first place). Our interest in this chapter is, however, with weaponized robots, usually termed LAWS (Lethal Autonomous Weapons Systems); AWS (Autonomous Weapons Systems) is used by some as an alternative.

The question here is whether "the rules governing acceptable conduct of personnel might perhaps be adapted for robots" (Lin et al. 2008: 25). Is it possible "to design a robot which has an explicit internal representation of the rules and strictly follows them?" (ibid.) Attempts at answering these questions have focused on the following considerations.

As a point of departure, we have to distinguish between operational and functional morality. An operational morality is one in which all possible options are known in advance and the appropriate responses are preprogrammed. "The actions of such a robot are entirely in the hands of the designers of the systems and those who choose to deploy them" (ibid. 26). Such robots have no ability to evaluate their operations and correct errors. An operational morality has the advantage of being entirely derivatory on the decisions made by the designer/user, so the lines of control, and hence responsibility, are crystal clear. However, apart from very narrow operating environments, it is impossible to preconceive all possible options in advance, because of the complexity of the environments in which the robots will be deployed, because the systems are introduced in settings for which they were not planned, or even because of the complexity of the technology involved such that "the engineers are unable to predict how the robot will behave under a new set of inputs" (ibid.). As the battlefield is a notoriously disorderly environment, for deployment in real-life battlefield settings, LAWS must be programmed with a *functional morality*, namely a built-in capacity to evaluate and respond to moral/legal considerations.

Following on from this, the design of functional morality in robots—namely a capacity for moral reasoning so that unanticipated situations can be dealt with appropriately—has been approached in two different ways, *top-down* and *bottom-up*:

(a) In a top-down approach, a particular moral theory is encoded in the software. This will typically involve some version of deontology or consequentialism, which are then detailed into a set of rules that can be turned into an algorithm. There are many challenges here, for instance the possibility of conflict between rules. This could lead to paralysis if the rules are meant to function as hard restraints, or if the rules are designed only as guidelines, this could open the door to robotic behavior that should be prohibited. A further and especially pressing issue concerns what is termed the "frame-problem" (Dennett 1984; Klincewicz 2015), namely to grasp the relevant features of a situation so the correct rules are applied. To borrow an example: How would a robot programmed with Asimov's "First Law [of robotics, which says that a robot may not injure a human being or, through inaction, allow a human being to come to harm] *know*, for example, that a medic or surgeon welding a knife over a fallen fighter on the battlefield is not about to harm the soldier?" (Lin et al. 2008: 31). In a famous case, a Soviet colonel saw his computer screen flash "launch," warning him that the US had initiated a nuclear attack. Thinking there was a bug in the system he waited, and it happened again repeatedly; finally, "missile strike" replaced "launch" and the system reported its highest confidence level. Still the colonel paused. Having seconds left to decide the matter, he called the ground-based operators for confirmation, but they had detected nothing. It turns out the system had malfunctioned; it had mistaken light reflecting off a cloud configuration for the trace of an incoming missile. The frame-problem, if unresolved, can have enormous consequences if a powerful weaponized AI system is in play (Scharre 2018: 1–29).

(b) In a *bottom-up* approach to functional machine morality, systems mimicking evolutionary or developmental processes are implemented within machine learning. The basic idea here is that "normative values are implicit in the activity of agents rather than explicitly articulated . . . in terms of a general theory" (Lin et al. 2008: 35). This has led to the application of virtue ethics to autonomously functioning machines. Just as people are taught to acquire the right set of character traits (virtues) and on that basis come progressively to understand what morality requires (it has been suggested by Kolberg and others that this is how children learn about morality), likewise neural networks might provide a pathway toward the engineering of robots that "embody the right tendencies in their reactions to the world" (Lin et al. 2008: 40). This "bottom-up development of virtuous patterns of behavior might be combined [the hybrid approach] together with a top-down implementation of the virtues as a way of both evaluating the [resulting] actions and as a vehicle for providing rational explanations of the behavior." In this way, "a virtuous robot might emulate the kind of character that the armed forces value in their personnel" (ibid.). Even if feasible, development of such technology appears to be still well off in the future.

Principled Arguments for and against Battlefield Use of LAWS

Because LAWS are designed to make lethal targeting decisions without the direct intervention of human agents (who are "out of the killing loop"), considerable debate has arisen on whether this mode of autonomous targeting should be deemed morally permissible. A variety of arguments have been proposed that we classify below into four different main types. Alongside this ethical discussion, calls for the international legal regulation of LAWS, including a ban on their use, have multiplied (see, e.g., the campaigns and web sites of Human Rights Watch and the Campaign to Ban Killer Robots).[3] As there is not a perfect overlap between ethics and law—the latter proceeds from principles and a methodology quite different from the former—the legal issues surrounding LAWS fall outside the scope of the present chapter and will be considered only indirectly.

A principal line of argumentation in favor of LAWS has focused on the qualities that combatants should possess in order to make sound targeting decisions in the heat of battle. Proponents of LAWS have maintained that AI-directed robotic combatants would have an advantage over their human counterparts, insofar as the former would operate solely on the basis of rational assessment, while the latter are often swayed by emotions that conduce to poor judgment (Arkin 2010). The negative role of the emotions is amplified in battlefield settings, when fear, rage, hatred, and related passions often deflect human agents from the right course of action. Machines operating under instructions from AI software would not be prone to emotive distortion; thus, if properly programmed, they could be counted on to function in strict conformity with recognized laws of armed conflict (domestic and international) and the appropriate rules of engagement. The occurrence of wartime atrocities would be reduced if combat missions could be undertaken by au-

tonomously functioning robots. Not only would robotic fighters avoid killing civilians (the *sine qua none* of international humanitarian law), in addition, they could be programmed to assume risk upon themselves to protect civilians from side-effect harm, something that human combatants often shy away from. Finally, robots would be capable of precise targeting, thereby enabling them to disable rather than kill the opposing human combatants. Although human soldiers sometimes have the intent to engage in disabling action, the stress and rapidity of the battlefield, as well as the lack of needed weapons, often results in higher kill rates than might otherwise be the case. Human soldiers do not always have the luxury of precise aiming—say, at the feet rather than the torso—and end up striking sensitive organs that seriously wound or kill their adversaries, despite a wish to cause only minor damage. The same could be said of damage to property and sites of cultural or environmental significance, which are often used to provide cover or expose adversaries to live fire. Robots could be equipped with a sophisticated range of weapons (being strong enough to carry them all), enabling them to select for each situation the weapon best suited to the task at hand; much as expert golfers select from among their clubs.[4]

Against this endorsement of LAWS, counterarguments have been advanced, some principled and others more pragmatic and contingent upon the current state of technological development. We will first be treating the principled objections, which are oriented around four considerations.

First, given that practical decisions take place in circumstances that are inherently affected by contingency, no prior process of reasoning is adequate to render the morally right conclusion, namely a sound decision about what should be done here and now, and in what precise way. Beginning with Socrates's claim that virtue reduces to knowledge, it has long been an aspiration of some philosophers (epitomized by August Comte), and more lately of social scientists, to devise a science of action that is both predictive and unfailingly right. Such knowledge might be deflected in us by wayward passion, but kept on its own trajectory, it will be unfailingly right. The invention of machine learning with algorithms that program ethical behavior can be viewed as the most recent permutation of the philosophical project to reduce virtue to knowledge.

But against this aspiration often associated with Socrates, other philosophers, beginning with Aristotle, have maintained that knowledge alone, no matter how sophisticated and reliable it may be, can ever serve as a proximate guide to morally right human action. This holds doubly true for action undertaken under the chaotic conditions of the battlefield. Even more challenging than predicting the weather (notorious for its difficultly, which even supercomputers cannot

---

[3] See https://www.hrw.org/topic/arms/killer-robots and https://www. stopkillerrobots.org/

[4] See Kalshoven (1984).

fully overcome), on the battlefield soldiers must confront contingencies relating not only to the terrain, local buildings and other installations (which may or may not be damaged), the weather, and their own and their adversaries' weapon systems (which even when known do not always function as planned), but most significantly, they face off against other agents who are possessed of free will, and who can choose among alternative courses of action, both tactically and strategically. How another individual (or group of individuals) will react can never with certitude be known in advance. It is for this reason that Aristotle emphasized how practical reasoning will succeed (i.e., will conduce to morally good choices) only if directed by an upright will. By the affective orientation of an upright will, the agent's attention is directed toward the most morally salient features of the immediate environment and decides which of these, amidst complexity and rapid change, to prioritize in the resulting choice. Thus, the affective disposition of will, oriented to the moral good, substitutes for the lack of perfect knowledge, impossible under these circumstances, thereby enabling right action under conditions of inherent contingency. This ability, at once cognitive and affective, Aristotle termed *phronesis* (*prudentia* in Latin). Through a combination of intellectual skill (enabling correct apprehension of moral principles) and well-ordered desire (that the later Latin tradition would term *voluntas* or "will" in English) the morally prudent person is able to judge well within the singular contingencies of experience. Intellect supplies an abstract grasp of the relevant moral truths, e.g., "noncombatants should never be intentionally harmed," while the will, which desires things in their very concreteness,[5] directs the intellect toward specific items in the perceptual field ("I do not want to harm this person"). The more the will is well ordered to the moral good, the better it will orient the intellect in its judgment of things in their particularity. Thomas Aquinas later explained how, in dealing with the challenges of warfare, a special mode of *phronesis* is requisite (see Reichberg 2017).[6] This he termed "military prudence" (*prudentia militaris*). AI-based machines, no matter how sophisticated their cognitive mimicking may be, will never be possessed of the affective disposition we term "will."[7] LAWS, by consequence, can never attain to *phronesis,* and, for this reason, cannot function as a trustworthy substitute for human combatants (see Lambert 2019).

A *second* principled argument against LAWS is orientated around the indispensability of emotions for the exercise of right judgment on the battlefield. As noted already, proponents of LAWS have assumed that practical judgment functions best when freed from the distortions arising from emotion. This is a viewpoint that was originally articulated by the ancient Stoics, who held that as emotion invariably leads humans astray, only a life lived under the austere dictates of reason will be a morally successful one. From this perspective, AI, which operates with cognitive capacity only, can be counted on to adhere unfailingly to the ethical guidelines that have been programmed into it, such that all its subsequent learning will develop in conformity with these guidelines.[8] Since it operates without emotion, AI has the potential to exceed human beings in making reliable ethical decisions on the battlefield.

By contrast, opponents of LAWS refuse to accept the fundamental premise on which this argumentation is built, namely that emotions are a hindrance to sound judgment and the action that follows from it (see Johnson and Axinn 2013; Morkevicius 2014). On the contrary, they maintain that emotions provide indispensable support for moral agency, and that without emotion, judgment about what should (or should not) be done on the battlefield will be greatly impoverished. Reason, without proper emotional support, will lead combatants astray. From this perspective, emotion can provide a corrective to practical judgments that derive from erroneous abstract principles ("ideology" we would say today) or from blind obedience to authority. There are numerous accounts of soldiers who, acting for instance under the impulse of mercy, desist from wrongful commands or provide aid to enemy combatants whose demise would make no contribution to the military effort (Makos 2014). Human beings have a unique ability to perceive when preordained rules or established plans should be set aside and exceptions made, in the interests of humanity. Our emotions ("feelings") are often what prompt us to see how a given situation requires, "calls for," a response out of our ordinary patterns of behavior. An emotionless machine, operating on sequential reasoning alone, would have no basis to depart from its preprogrammed course of action, and thus no ability for making exceptions. While this could be reckoned only a shortcoming under ordinary operating conditions, should a LAWS be reprogrammed for malicious ends (say, by cyber intrusion or other means), and oriented toward the commission of atrocities, there would be no internal mechanism by which it would resist the new operating plan. No emotional response would provide the necessary cognitive dissonance. By contrast, human beings always have the ability, and sometimes even an obligation, to disobey orders, in other words, to act against instructions

---

[5]Thomas Aquinas emphasized how knowledge is oriented to things under the condition of abstractness and universality, while desire is oriented to things in their concrete singularity, and hence their very contingency. The respective faculties, intellect and will, operate in synergy, and thus each provides input to the other in carrying out its respective operation.

[6]On the interrelation of intellect and will in the cognition of things in their particularity and contingency, see Reichberg (2002).

[7]This claim is of course not self-evident; providing the argumentation would take us too far afield. Our aim at present is to describe the contours of one principled argument against LAWS.

---

[8]This being the case, it is not surprising that stoicism is the philosophy of choice in Silicon valley (see Manthorpe 2017).

received from a commanding officer. But it is hard if not impossible to imagine how a machine could override its software (Johnson and Axinn 2013: 135).

A *third* principled argument against LAWS proceeds from the moral intuition that battlefield killing will be compatible with human dignity only when it is carried out by the direct decision of a human being. To be killed by machine decision would debase warfare into mere slaughter, as though the enemy combatant were on a par with an animal killed on an automated conveyer belt, or, as two authors put the point:

> A mouse can be caught in a mousetrap, but a human must be treated with more dignity. ... A robot is in a way like a high-tech mousetrap; it is not a soldier with concerns about human dignity or military honor. Therefore, a human should not be killed by a machine as it would be a violation of our inherent dignity (Johnson and Axinn 2013: 134).

The operative supposition is that the killing which is done on the battlefield is special (compared to other modes of killing) insofar as it is done according to an established code of honor, a convention, whereby soldiers face off as moral equals. Each is a conscious participant in a profession which requires that a certain deference be shown to the other even in the process of killing him. Shared adherence to the same calling sets military warfare apart from other sorts of confrontations, say, between police and thieves, where there is no moral reciprocity. In a famous essay (Nagel 1979), philosopher Thomas Nagel maintained that the hostility which is characteristic of war is founded, paradoxically, on a mode of interpersonal relations. Respect for the other (hence treating him as an end and not merely as a means) is maintained even in the process of killing him, when he is targeted precisely as a subject, as someone who is aware that he is the target of hostile acts for a specific reason, namely because he too is directing lethal harm against the one who is trying to kill him. What is often called the "war convention" is based on this reciprocity, namely the mutual exposure to intentional harm, and for this reason the personal dignity of soldiers is maintained even in the killing. But this personal dimension, having lethal harm directed against oneself insofar as one is a member of a distinct class, that of arms-bearing soldier, could not be maintained in the event that the opposing soldier were not himself a person, but only a machine.

Against this view, one could say that it proceeds from a conception of warfare—associated with "chivalry"—that is no longer operative today.[9] Today's conflicts are often waged by reference to a justice motive wherein the adversary is deemed immoral, a terrorist or a kind of criminal, such that moral equality with him would be unthinkable. Moreover, if one were to exclude (as morally dubious) "impersonal killing" of the sort carried out by LAWS, then by the same

token much of the technology employed in modern warfare would have to be excluded as well: high altitude bombing of enemy positions, roadside bombs, booby traps, and similar devices. But thus far, few if any are militating for a ban on these methods of warfare, except in cases where civilians might indiscriminately be harmed (land mines or biological weapons) or where the harm to combatants (by, e.g., poisonous gas or chemical weapons) results in long-lasting suffering.

In arguing for an essential difference between human and machine killing, namely that even in a situation where a combatant would justifiably be killed by his human counterpart, it would be wrong in the identical situation to have him killed by a LAWS, Johnson and Axinn (2013) nonetheless draw a distinction between offensive and defensive LAWS. The former would include unmanned ground, surface, underwater, or airborne *vehicles* that are able to attack targets—wherever they may be found—based on preset autonomous decision procedures. Insofar as they are directed against *manned* targets, autonomous killing systems of this sort should never be used, for the reason given above, namely that machine killing is incompatible with human dignity. By contrast, defensive LAWS do not fall under the same moral stricture. "Defensive" would include integrated air defense systems that shoot down aircraft or missiles flying within a specific radius, as well as land-based autonomous turrets or perimeter patrol robots that fire on anyone entering a designated perimeter. Autonomously functioning machines would have moral license to kill anyone entering the said zones, provided these no-go areas were well announced in advance. It could then be assumed that trespassers were engaged in hostile activity, and there could be ample justification to program a machine to kill them upon entry. This would be an AI-based extension of the electric fence.

While the distinction here drawn between the two types of LAWS (offensive and defensive) is useful, it does seem to undermine the authors' argument that it is an inherent violation of human dignity to be killed by a machine. After all, the basic supposition behind the deployment of LAWS is that such machines can effectively be programmed to distinguish combatants from noncombatants, the former being engaged in hostile activity and the latter not. But if advance warning is what enables a differentiation of allowable versus wrongful machine killing, then anytime it is publicly known that a war is underway, there could be moral justification in having a LAWS kill adversarial combatants. After all, by a tacit convention, combatants know that once they step out on the battlefield they are "fair game"; this is the "perimeter" they have entered, and in doing so they assume risk upon themselves. On this reasoning, all may rightly be made lethal targets of a machine.

A *fourth* principled argument against LAWS is focused, as was the previous argument, on the moral equality of

---

[9]On the applicability of chivalric norms to contemporary conflict, in light of recent technological developments, see Olsthoorn (2019).

combatants as a prerequisite for maintaining a rule-based order on the battlefield. An expression coined by political theorist Michael Walzer, but with antecedents in international law, the "moral equality of combatants" refers to the idea that the moral standing of combatants can be determined without reference to the cause, just or unjust, for which they fight (Walzer 1992). All soldiers, whether fighting in just or unjust wars, are bound by the same set of rules. On this conception, "when conditions of enmity arise between states, this does not automatically amount to interpersonal enmity between the individual members of the opposing states. In war, those who do the fighting may consequently do so without special animosity toward their adversaries on the other side, because, like themselves, they are mere instruments of the state. This positions them to confront each other as peers in a rule-bound, albeit bloody competition" (Reichberg 2017: 231–232). Because the actual fighting is conducted in detachment from substantive justice (the question of which side is in the right with respect to the casus belli), combatants deploy force against each other in view, not of hatred or vengeance, or even high-minded goals such as the upholding of justice, but for the preservation of personal security. Moral license to kill derives, in other words, not from the personal moral guilt of the opposing combatant (after all he is acting out of obedience to his leadership as I am to mine), but from a right to self-defense. "Each possesses this license [to kill in war] because each acts in self-defense vis-à-vis the other. The reciprocal imposition of risk creates the space that allows injury to the morally innocent [i.e., combatants on the opposing side]" (Kahn 2002: 2). Rule-based warfare, and the moral equality of combatants that it entails, depends on a mutual assumption of risk by combatants. This reciprocal exposure to serious injury and death is what justifies each in directing self-defensive lethal harm against the other. But should one side prosecute war without assuming any risk upon itself, i.e., its own combatants, its moral right to use lethal force is thereby removed. There can be no moral justification in fighting riskless war. This is exactly the situation that would arise by the introduction of LAWS on the battlefield. The side deploying these "killer robots" against human combatants on the other side might prevail militarily, but the resulting victory would be morally pyrrhic and hence wholly without honor. The professional ethos of soldiering, which rests on a voluntary and reciprocal assumption of risk, would be undermined, and with it the expectation, built up over many centuries, that war can be conducted in a rule-based and even virtuous manner, namely in a way that preserves (and enhances) the moral integrity of those who actively take part in it (Riza 2013).

One could of course respond (Arkin 2010) that the ultimate goal behind LAWS is to reconfigure the battlefield so that in the future robots will fight only robots, not men, thereby removing the asymmetry outlined above. This, however, is unlikely to produce the desired outcome—bloodless war. Unless belligerents agree to a convention whereby defeat of one's robotic army will entail capitulation to the political demands of the victor, the war will simply shift to another plane, one in which human combatants are again pitted against the robotic combatants of the (tactically but not strategically victorious) other side, with a morally dubious risk asymmetry reintroduced onto the battlefield.

Another line of response would question whether the moral equality of combatants, and the mutual assumption of risk that underlies it (Renic 2018), is indeed a needed precondition for the maintenance of rule-based warfare. A lively debate has been ongoing on this topic over the last decade (Syse 2015; Barry and Christie 2018). This is not the place to elucidate the details. For our present purpose it can be said that the alternative viewpoint—which posits a moral *in*equality in favor of the combatants who prosecute the just cause—will entail that the just side has moral warrant to engage in risk-free warfare. Or put differently, if LAWS effectively enable force protection, while simultaneously aiming their fire only at enemy combatants and not civilians, no sound moral argument stands in the way of their use. This moral justification derives wholly from the *ad bellum* cause and from nothing else.

One may, however, make much the same argument without going as far as nullifying *in bello* considerations in favor of making *ad bellum* concerns alone morally decisive. One may more simply, with James Cook, argue that we should avoid romanticizing risk and death in war when we can clearly, with the aid of unmanned and AI-based technology, protect our own warfighters better by offering them the opportunity of better defenses, lower risk, and more accuracy. The latter argument can be made even if we do not reject the moral equality of combatants (Cook 2014).

Technical and Pragmatic Considerations

We choose in the following to treat separately a group of ethical arguments for and against the battlefield use of AI that can broadly be termed "pragmatic," centering partly on the current state of technologies and partly on broader considerations of whether AI will do more harm and good. There is indeed overlap between these considerations and the ones we have called "principled" arguments, and some of the arguments below have been foreshadowed above. Nonetheless, the distinction is useful, since the arguments below do center more on the technical aspects and the consequences of the use of AI technology and take a less principled stand for or against the application of such technologies. These arguments can be classified by reference to what one might term AI battlefield optimists and pessimists (Syse 2016).

By optimists we mean those who see the introduction of autonomous weapons as representing a net gain in terms of much higher precision, less suffering, and fewer fatalities in war (Arkin 2010; Strawser 2010). Beyond the obvious point, outlined above, that robots fighting each other would spare human combatants from having to do the same, thus resulting in less loss of life, the optimists also claim that robots, even when fighting human beings, would show greater battlefield probity, because they would not be misled by emotion. Wartime atrocities would be eliminated if robots were left to do our fighting for us, provided of course that they were programmed properly, to the extent possible, to avoid harm to noncombatants.

Moreover, robotic fighters would have an additional benefit insofar as they could be counted on to assume risk upon themselves to protect noncombatants, something human combatants often avoid. Attacks could, for instance, be carried out at closer range, thus with greater precision, resulting in decreased rates of side-effect damage to civilians and the infrastructure on which they depend. Moreover, given that AI can provide better situational awareness to human soldiers, targeting decisions will prove to be less lethal, as enemy combatants can more readily be incapacitated than killed. In other words, proportionality calculations could be implemented with enhanced accuracy.

Optimists are also quick to acknowledge the economic as well as tactical advantages of autonomous lethal systems (already mentioned above in the section on Background Considerations). For instance, whereas remotely controlled and supervised battlefield robots require much human capital for their operation, these high costs could be bypassed by means of autonomously functioning robots.

There are tactical benefits also, insofar as autonomous robots eliminate the need for electromagnetic communications links, which are hotly contested in wartime, and inoperative in some settings, for instance, deep undersea. Moreover, much of current military planning increasingly makes use of swarm warfare and the resulting maneuvers happen far too rapidly to be directed by human guidance.

Concluding from these lines of reasoning, B. J. Strawser (2010) holds that the use and further development of unmanned (or "uninhabited") and increasingly autonomous weapons may be a duty, if the likelihood of fewer casualties and less suffering is significant. Opposing or delaying such development and use would be akin to holding back on the use of life-saving tactics and strategies, even when we know that they will be effective.

In short, the "optimist" arguments hold that the likely overall result of AI technology on the battlefield will be one of more accuracy and fewer human beings put in harm's way.

Pessimists, by contrast, have offered a set of opposing arguments:

- The anticipation of decreased battlefield human casualty rates (through the introduction of robots) would lower the perceived risks of waging war. In anticipation of fewer battlefield casualties to the deploying side, political leaders who possess a strong LAWS capability will increasingly view initiation of war as a viable policy option. The number of wars will grow accordingly (Asaro 2007).
- It is an illusion to think that robotic warfare will render wars entirely bloodless. Ultimately, the fruit of defeat on the battlefield will be the vulnerability of one's civilian population to lethal robotic attack, which, given the new technologies developed (e.g., swarmed drone attacks), could lead to massive deaths on a par with nuclear detonations. In this connection, Russell (2019: 112) refers to these AI-based technologies as "scalable weapons of mass destruction." A new arms race will emerge, with the most unscrupulous actors prevailing over those who show restraint. For instance, AI engineers at leading US technology firms are refusing to engage in military design projects with lethal applications. It is said that at Chinese defense contracting firms, where the development of AI systems is a priority, the engineers have not expressed the same reservations (Knight 2019). An additional worry, recently voiced by Paul Scharre, is that an arms race in AI military applications will lead to a widespread neglect of safety considerations. "[T]he perception of a race will prompt everyone to deploy unsafe AI systems. In their desire to win, countries risk endangering themselves just as much as their opponents" (Scharre 2019: 135).
- The differentiation between combatants and noncombatants depends on a complex set of variables that, in today's asymmetric battlefields, cannot be reduced to a question of the uniform one may or may not be wearing. Irregular combatants often pose as civilians, and subtle judgments of context are needed to ferret them out from their innocent counterparts. For instance, human beings are adept at perceiving whether their fellows are animated by anger or fear, but machine intelligence in its current form is largely unable to detect this crucial difference.
- Similar problems arise from AI black-boxing, namely the difficulty knowing in advance how an algorithm would dictate a response in an unanticipated set of circumstances (see Danzig 2018 for a survey of the relevant risk factors). Given such immense complexities, teams of

programmers need to collaborate on algorithm design for any one project. Consequently, no one programmer has a comprehensive understanding of the millions of lines of code required for each system, with the result that it is difficult if not impossible to predict the effect of a given command with any certainty, "since portions of large programs may interact in unexpected, untested ways" (Lin et al. 2008: 8). Opening lethal decision-making to such uncertainty is to assume an unacceptable level of moral risk. This unpredictability, amplified by machine learning, could result in mistakes of epic proportions, including large-scale friendly-fire incidents, thereby nullifying the benefits that might otherwise accrue from the use of LAWS. "In the wrong situation, AI systems can go from supersmart to superdumb in an instant" (Scharre 2019: 140). Given this unreliability, military personnel would be unwilling to put their trust in AI systems (see Roff and Danks 2018). The use of such systems would accordingly be avoided, thereby nullifying the tactical benefits that might otherwise accrue. Perhaps a way will be found to overcome algorithmic black-boxing, but at the current stage of AI design, a solution is still well off in the future.

- Likewise, no matter how effectively LAWS might be programmed to act in accordance with ethical norms, cyber intrusion cannot definitively be excluded, such that its code would henceforth dictate unethical behavior, including the commission of atrocities. Advances in cryptology and other defenses against cyber intrusion have still not reached the point where malicious interference can be ruled out.

- Moreover, even if the use of autonomous battlefield robots could, if programmed effectively with moral norms, lead to reduced bloodshed in war, there is no guarantee that all relevant militaries would program their robots in this way. The opposite could easily happen under a variety of scenarios, including states that might refuse to sign onto AI-related treaties that may eventually be negotiated, the assumption of control over such systems by rouge actors or third-party hackers, or the theft, reuse, and reprogramming of battlefield robots.

As this brief summary of core technological and pragmatic arguments shows us, whether the use of complex AI capacities in battlefield weaponry will lead to more or less suffering, more or less casualties, is subject to intense debate. Hence, the uncertainties of the accompanying calculus of moral utility are far-reaching. The "optimists" will, however, insist that their arguments are not meant to be *ipso facto* true: rather, they are unequivocally dependent on the development of AI technologies that discriminate clearly and assuredly. Much of the "pessimist" argument, on the other hand, centers on the unlikelihood that we will—at least in the foreseeable

future—be able to trust, or truly and safely harness, the powers of such technologies.

Virtue Ethics and Human–AI Interaction

Thus far we have mainly considered the question of whether autonomous robots should be allowed on the battlefield. The resulting debate should not blind us to a wider set of questions that are important to address as human–AI interactions—semiautonomous and autonomous—become increasingly prevalent in military planning and execution. How these tools impact on the military personnel who make use of them and their ability to undertake responsible action on the battlefield (whether directing the conduct of hostilities or directly engaging in these hostilities) must complement the reflections delineated above, and may play a vital role as we try to draw conclusions to the quandaries with which we are faced.

Human–machine interaction within the conduct of hostilities is referred to in military jargon as the "Force Mix" (Lucas 2016). Ethics research into the human and machine Force Mix, especially the moral implications for the human agents who use AI-based weapons systems, has arguably failed to keep pace with accelerating technological developments. This lacuna urgently needs to be filled. How service within the Force Mix affects the moral character of the human personnel involved is also our central focus in a new research project undertaken by the authors, in collaboration with a team of experts.[10]

We propose that a *virtue ethics* perspective is especially useful when investigating the ethical implications of human participation in the "Force Mix." Virtue ethics, a philosophical approach (see Russell 2009) associated most closely with Aristotelianism (and the Thomistic school within Catholic moral thought), has been adopted within military training programs (see Moelker and Olsthoorn 2007 for a good overview of the interaction between virtue ethics and professional military ethics). Virtue ethics is uniquely flexible: rather than espouse fixed moral principles, it emphasizes acquiring appropriate dispositions for the proper exercise of one's professional role (Vallor 2016: ch. 1). Paramount is the structural context—including, in the military setting, such important factors as combat unit membership and type of battlefield, as well as the technological setting—within which individuals act, and the ways in which their actions must be adjusted to fit that specific context. The use of AI within combat units

---

[10]"Warring with Machines: Military Applications of Artificial Intelligence and the Relevance of Virtue Ethics"; see https://www.prio.org/Projects/Project/?x=1859 for a description of this project and the principal collaborators.

will inevitably alter these structural conditions, including the prerequisites for force cohesion, within which virtue is exercised. How should we think about the virtue of military personnel in light of this momentous, ongoing change?

Let us add that while often associated with Greek and Christian thought, virtue ethics also has significant parallels in Asian traditions of thought (as discussed e.g. in Vallor 2016), thus making it eminently suitable for a global conversation about ethics and AI.

Within military ethics, virtue has occupied a central role, pertaining not least to the inculcation of soldier identity, unit cohesion, pride, discipline, and conscience. The virtue-based ideal of the good and reliable soldier can be found across cultures and over time, albeit with different emphases and priorities. In spite of the many differences, the idea of the soldier and the officer as someone who must acquire and develop defined character traits or virtues, courage and prudence foremost among them, is central to most military cultures. In Western philosophy, it is exactly this way of thinking that has gone under the name of virtue ethics or, more specifically for the armed forces, professional military ethics.

It could be argued (Schulzke 2016) that an increased reliance on automated weapons and AI makes virtue ethics *less* central to the military enterprise, and that a more rules-based focus will be needed, since machines *per se* cannot have virtues, while they can indeed be programmed to follow rules. We would rather argue that the question of virtue becomes even more pressing in the face of AI, since the very role and competence of the human soldier is what is being augmented, challenged, and placed under great pressure. How do we ensure that soldiers and officers *maintain* those virtues that make them fit for military service and command, instead of delegating them to AI systems—and in the process, maybe, ignoring or losing them ("de-skilling").

The debates between optimists and pessimists, delineated above, are also debates about the role of human virtue. As we have seen, while technology optimists will typically claim that lethal autonomous weapons systems (LAWS) will be superior to more human-driven systems because they tend toward an elimination of mistakes based on errors of judgment and the distortions introduced by strong emotion (Arkin 2010), those on the other side of the debate fear that military decision-making will suffer greatly if prudential reasoning, including such typically human phenomena as doubt and conscience, will be weakened, and war will be waged "without soul" (Morkevicius 2014; Riza 2013).

The questions that virtue ethics helps us posit concern what special moral challenges are faced by the human decision-maker and user of AI systems. Closely linked to that is the following question: How should that decision-maker and user be prepared and trained for action in the battlefield? AI forces us to ask these questions in a new way, since the human–AI encounter is not only a traditional encounter between a user and a tool, in which the tool is essentially an artificial extension of the intentions of the user. Human–AI interaction also represents an encounter between a human agent on the one hand and a nonhuman system which is capable of making seemingly superior and not least self-sufficient, autonomous decisions, based on active learning, on the other. How should the human user of such systems think about his or her role and relationship vis-à-vis them?

In order to answer this question and determine the ethical implications of implementing AI technology into human practices of war, we hold that the following further questions have to be asked, and we conclude the present paper with an attempt at formulating them, indicating thereby the direction of our further research on the AI–human encounter in military settings:

Firstly, which are the specifically human qualities or virtues that remain crucial in guiding decisions about strategy as well as battlefield actions in war? How can we ensure that these qualities are not weakened or ignored as a result of the use of AI-based weapons systems? Or put in other words, how can we ensure that the implementation of AI systems does not lead to a "de-skilling" of the human actor?

Secondly, the Stoic ideal of peace of mind, balance, and moderation is often touted as a military ideal, based on a virtue-ethical tradition (Sherman 2007). But, as also intimated above, we must ask to what extent this ideal denies the importance of *emotions* for proper moral understanding. How does AI play into this debate about the role of emotions, such as fear, anger, distrust, doubt, and remorse—all feelings with significant relevance for decision-making in war? How will AI change the ways in which we understand, appreciate, critique, and develop emotions associated with the use of military force?

Thirdly, in the Socratic tradition, dialogue is considered a crucial prerequisite for the development of virtues and proper decision-making. What kind of a dialogue takes place in the human–AI encounter? Is an AI-based system with significant linguistic and machine-learning capabilities a real, trustworthy dialogue partner? The term "digital twin" is increasingly used in describing the interaction between AI-based and human-operated systems. Does this conceptualization truly capture the nature of the human–AI encounter in the deployment and operation of weapons systems? (Kim 2018).

And finally, and most generally, which of the virtues are most relevant to humans in the human–AI force mix? To what extent do those virtues—such as moderation, prudence, and courage, which we assume are among them—change their character as a result of the use of AI-based systems?

It is worth noting that these are questions with relevance well beyond military ethics. Virtue ethics is not only the dominant ethical framework for moral training within the military but is today also the dominant framework for thinking about professional ethics in a wide array of fields, as for example in the case of nursing ethics, and more widely the ethics of care, fields that are also very much undergoing change due to increased digitalization and widespread use of AI. Raising them within the context of military ethics first, however, does have the benefit of attacking the problem within the context where AI research arguably has come the furthest, and where the stakes are particularly high. Our belief is that a virtue-ethical approach helps us raise awareness and ask questions about how AI can be integrated in military settings in a way that takes seriously the crucial role of the human developer, operator, and user of AI systems. This, in turn, will help us focus on the sort of training and self-awareness that must be developed and fostered in tandem with the development and deployment of AI weapons systems.

To both ask and answer such questions intelligently requires, however, that we have a good overview of the nature of the ethical debates that confront us, and it is to that aim that we hope our reflections have contributed.

Acknowledgements The authors thank Joe Reeder and Sigurd Hovd for valuable comments to earlier drafts of the article.

References

Arkin, R. (2010). The case for ethical autonomy in unmanned systems. *Journal of Military Ethics, 9*(4), 332–341.

Asaro, P. M. (2007). Robots and responsibility from a legal perspective. *Proceedings of the IEEE, 2007*, 20–24.

Barry, C., & Christie, L. (2018). The moral equality of combatants. In S. Lazar & H. Frowe (Eds.), *The Oxford handbook of ethics of war*. New York: Oxford UP.

Cook, J. (2014). Book review: Killing without heart. *Journal of Military Ethics, 13*(1), 106–111.

Danzig, R. (2018). *Technology roulette: Managing loss of control as many militaries pursue technological superiority*. Washington, DC: Center for a New American Security.

Dennett, D. (1984). Cognitive wheels: The frame problem of AI. In C. Hookway (Ed.), *Minds, machines, and evolution* (pp. 129–151). Cambridge: Cambridge UP.

Johnson, A. M., & Axinn, S. (2013). The morality of autonomous robots. *Journal of Military Ethics, 12*(2), 129–141. https://doi.org/10.1080/15027570.2013.818399.

Kahn, P. W. (2002). The paradox of riskless warfare. *Philosophy and Public Policy Quarterly, 22*(3), 2–8.

Kalshoven, F. (1984). The soldier and his golf clubs. In C. Swinarski (Ed.), *Studies and essays on international humanitarian law and Red Cross principles: In honor of Jean Pictet* (pp. 369–385). Geneva/The Hague: International Committee of the Red Cross/Martinus Nijhoff.

Kim, J. (2018). *Digital twins and data analysis*. Available via Signal. Retrieved February 25, 2020, from https://www.afcea.org/content/digital-twins-and-data-analysis

Klincewicz, M. (2015). Autonomous weapons systems, the frame problem and computer security. *Journal of Military Ethics, 14*(2), 162–176. https://doi.org/10.1080/15027570.2015.1069013.

Knight, W. (2019). *China's military is rushing to use artificial intelligence*. Available via MIT Technology Review. Retrieved February 25, 2020, from https://www.technologyreview.com/f/612915/chinas-military-is-rushing-to-use-artificial-intelligence/

Lambert, D. (2019). *La robotique et l'intelligence artificielle*. Paris: Fidelité/Éditions jésuites.

Lin, P., Bekey, G., & Abney, K. (2008). *Autonomous military robotics: Risk, ethics, and design*. Washington, DC: U.S. Department of the Navy, Office of Naval Research.

Lucas, G. (2016). *Military ethics: What everyone needs to know*. New York: Oxford UP.

Makos, A. (2014). *A higher call: An incredible true story of combat and chivalry in the war-torn skies of World War II*. New York: Caliber.

Manthorpe, R. (2017). *All that's good and bad about Silicon valley's stoicism's fad*. Available via Wired. Retrieved February 25, 2020, from https://www.wired.co.uk/article/susan-fowler-uber-sexism-stoicism

Moelker, R., & Olsthoorn, P. (Eds.). (2007). Special issue: Virtue ethics and military ethics. *Journal of Military Ethics, 6*(4), 257–258.

Morkevicius, V. (2014). Tin men: Ethics, cybernetics and the importance of soul. *Journal of Military Ethics, 13*(1), 3–19.

Nagel, T. (1979). War and massacre. In *Mortal questions* (pp. 53–74). Cambridge: Cambridge UP.

Olsthoorn, P. (2019). Risks, robots, and the honorableness of the military profession. In B. Koch (Ed.), *Chivalrous combatants* (pp. 161–178). Baden-Baden: Nomos.

Patterson, R. (1985). *Image and reality in Plato's Metaphysics*. Indianapolis: Hackett.

Reichberg, G. (2002). The intellectual virtues. In S. J. Pope (Ed.), *The ethics of Aquinas*. Washington, DC: Georgetown UP.

Reichberg, G. (2017). *Thomas Aquinas on war and peace*. Cambridge: Cambridge UP.

Renic, N. (2018). UAVs and the end of heroism? Historicising the ethical challenge of asymmetric violence. *Journal of Military Ethics, 17*(4), 188–197.

Riza, M. S. (2013). *Killing without heart: Limits on robotic warfare in an age of persistent conflict*. Washington, DC: Potomac Books.

Roff, H. M., & Danks, D. (2018). 'Trust but verify': The difficulty of trusting autonomous weapons systems. *Journal of Military Ethics, 17*(1), 2–20.

Russell, D. C. (2009). *Practical intelligence and the virtues*. Oxford: Oxford UP.

Russell, S. (2019). *Human compatible: Artificial intelligence and the problem of control*. New York: Viking.

Scharre, P. (2018). *Army of none: Autonomous weapons and the future of war*. New York: W. W. Norton & Co..

Scharre, P. (2019). Killer apps, the real dangers of an AI arms race. *Foreign Affairs, 98*(3), 135–144.

Schulzke, M. (2016). Rethinking military virtue ethics in an age of unmanned weapons. *Journal of Military Ethics, 15*(3), 187–204.

Sherman, N. (2007). *Stoic warriors: The ancient philosophy behind the military mind*. Oxford: Oxford UP.

Simon, Y. R. (1969). *Freedom of choice*. New York: Fordham UP.

Strawser, B. J. (2010). The duty to employ uninhabited aerial vehicles. *Journal of Military Ethics, 9*(4), 342–368.

Syse, H. (2015). The moral equality of combatants. In J. T. Johnson & E. D. Patterson (Eds.), *The Ashgate research companion to military ethics* (pp. 259–270). Farnham: Ashgate.

Syse, H. (2016). Dronenes etikk: Noen glimt fra debatten [The ethics of drones: Some snapshots from the debate]. In T. A. Berntsen, G. L. Dyndal, & S. R. Johansen (Eds.), *Når dronene våkner: Autonome våpensystemer og robotisering av krig* (pp. 243–254). Oslo: Cappelen Damm.

Vallor, S. (2016). *Technology and the virtues*. New York: Oxford UP.

Walzer, M. (1992). *Just and unjust wars* (2nd ed.). New York: Basic Books.

AI Nuclear Winter or AI That Saves Humanity? AI and Nuclear Deterrence

Nobumasa Akiyama

Contents

Abstract

Nuclear deterrence is an integral aspect of the current security architecture and the question has arisen whether adoption of AI will enhance the stability of this architecture or weaken it. The stakes are very high. Stable deterrence depends on a complex web of risk perceptions. All sorts of distortions and errors are possible, especially in moments of crisis. AI might contribute toward reinforcing the rationality of decision-making under these conditions (easily affected by the emotional disturbances and fallacious inferences to which human beings are prone), thereby preventing an accidental launch or unintended escalation. Conversely, judgments about what does or does not suit the "national interest" are not well suited to AI (at least in its current state of development). A purely logical reasoning process based on the wrong values could have disastrous consequences, which would clearly be the case if an AI-based machine were allowed to make the launch decision (this virtually all experts would emphatically exclude), but grave problems could similarly arise if a human actor relied too heavily on AI input.

Keywords

Deterrence · Escalation · Nuclear · Stability · Weapons

N. Akiyama (✉)
Graduate School of Law /School of International and Public Policy
Hitotsubashi University, Tokyo, Japan
e-mail: n.akiyama@r.hit-u.ac.jp

Introduction

Technological innovation often brings about paradigm shifts in various dimensions of our life. Artificial Intelligence (AI) certainly has great potential to fundamentally transforming various dimensions of our life, for better or worse (Kissinger 2018).

The military and security domains are no exception. The trajectory of AI development, together with that of complementary information technology and others, will have a large effect on security issues. AI could make any type of weapons and military system smarter, and it may make warfighting more efficient and effective. The potential of military applications of AI is enormous, and some have already materialized.

Meanwhile, we should be aware of limitations that AI systems may have. The autonomy in weaponry systems poses serious and difficult questions regarding the legal and ethical implications, the evolution and changes of military doctrines and strategy as well as the risks of misguiding decision-making and actions, and the balance of power among nations. The discussion on how to regulate the military application of AI confronts us with significant political challenges.

Among many security-related questions, how the possible convergence of AI and nuclear strategy/deterrence will transform our security environment is the most important one. Nuclear weapons were a game changer in altering the nature of war and security strategy (Jervis 1989). Now AI may effect another game change, when it is integrated into nuclear weapon systems. As nuclear weapons keep posing existential threats to the human being, and the stable, safe management of nuclear arsenal and threat and risk reduction are essential for the survival of human beings, it is natural to ask what the growth of AI will mean for nuclear deterrence, nuclear threat reduction, and nuclear disarmament. Would AI reinforce nuclear deterrence and strategic stability, or undermine them? Would AI promote nuclear disarmament and nonproliferation? To begin with, how could AI be utilized in nuclear strategy and deterrence?

According to a study by RAND Corporation, there are three main positions on the potential impact of AI on nuclear stability. "Complacents" believe that AI technology would never reach the level of sophistication to handle the complicated challenges of nuclear war, and therefore AI's impact is negligible. "Alarmists" believe that AI must never be allowed into nuclear decision-making because of its unpredictable trajectory of self-improvement. "Subversionists" believe that the impact will be mainly driven by an adversary's ability to alter, mislead, divert, or trick AI. This can be done by replacing data with erroneous sample or false precedent, or by more subtly manipulating inputs after AI is fully trained (Geist and Andrew 2019). This categorizing effort suggests that even experts cannot reach a consensus about the potential trajectory of AI, and therefore it is difficult to assess the impact of the technology on nuclear stability and deterrence.

At most, what we can do under such a circumstance where forecasts of the consequences of AI development for the international security environment are necessarily tentative is to present the points of concerns which may arise from various scenarios. So, what factors will shape the future of nuclear deterrence and stability in the era of rise of AI? This paper outlines the risks associated with the possible application of this evolving technology into nuclear security, and the possibilities of contribution to the reduction of nuclear risks.

First, it shows how AI could function as a decision-support system in nuclear strategy. Second, I indicate how AI could support nuclear operations with improvements in intelligence, surveillance, reconnaissance (ISR), and targeting. The virtue of AI can be best demonstrated in a constellation of new technologies, and ISR is the most critical field where such constellation happens. In this connection, although it is important to analyze how AI application to conventional weapons such as Lethal Autonomous Weapon Systems (LAWS) or swarming unmanned aerial vehicles (UAV) bring about changes in the role of nuclear weapons in military strategy, the discussion will be left for another occasion. And third, I show how AI will affect the discourse on ethics and accountability in nuclear use and deterrence.

AI in Supporting Nuclear Decision-Making

Essence of Nuclear Deterrence and the Role of AI

Goals to achieve by nuclear weapons are complex. A fundamental/original purpose of the weapon is to win a warfighting. With its huge destructive power, nuclear weapon could be an effective tool to lead a country to victory. Meanwhile, exactly because of this feature, there are high hurdles for the actual use of nuclear weapons as it might cause humanitarian catastrophe not only in the adversary's population, but also in its own population when retaliated. Thus, the advent of nuclear weapons has changed the nature of war, and is employed to achieve certain political goals without detonating one in warfighting, or through nuclear deterrence.

Deterrence is the use of retaliatory threats to dissuade an adversary from attacking oneself or one's allies. Deterrence is not the only goal nor the winning a war. To succeed, deterrence should be multidimensional. Nuclear weapons can also be used for compellence—coercing the enemy into doing something against its will. In the meantime, the coercive elements are not the only components of a strategic relationship between potential adversaries. Both sides must be convinced to certain degree that they will not be attacked as long as they would not provoke beyond some level of confrontation that is being deterred (Schelling 1966).

In particular, between major nuclear powers in an adversarial relationship, the maintenance of strategic stability is sought in order to avoid armed conflict, which could be escalated into a catastrophic nuclear war if they fail to manage the escalation. Strategic stability is understood as "a state of affairs in which countries are confident that their adversaries would not be able to undermine their nuclear deterrent capability" and that state would be maintained with various means including nuclear, conventional, cyber, or other unconventional means (Podvig 2012).

Another important feature of nuclear deterrence, which requires thorough consideration is that effective deterrence needs to complicate adversary's strategic calculus, but to the extent that it would not challenge the *status quo*.

Beyond military capabilities, postures, and pressure, effective deterrence requires, to a greater extent, the sophisticated skill of political communication and crisis management. To achieve the objectives of deterrence, it is necessary to make the adversary lose confidence in the success or victory of its strategy and, at the same time, to make sure that the adversary would be convinced that its core national interests would be preserved if the situation would not be inflicted into the use of force. Although it is essential for successful deterrence to demonstrate strong resolve and sending clear message that action corresponding to such a resolve would be taken, and it will be ready to give damage imposing unbearable cost on the adversary, it would require compromise and carefully avoid total humiliation of an adversary, which may drive the adversary into desperate action.

So where does AI fit in? As referred above, with machine learning, AI optimizes its performance to achieve a goal provided. Systems using these techniques can, in principle if not in practice, recursively improve their ability to successfully complete pattern recognition or matching tasks based on sets of data (which usually need to be carefully curated by humans first) (Heath 2018).

Meanwhile, current AI systems may still have limitations in performing when they are operated outside the context for which they are designed to work, and transferring their learning from one goal to another. This feature suggests that AI at the current level of its capabilities would not function well. In a complex social and political environment where the final goals of action and the choice by decision makers are adaptable to emerging and evolving conditions surrounding decision makers, it is unlikely that, under the current level of technical competence and maturity of discussion on ethical concerns, AI alone will/can make a decision to use nuclear weapons.

AI would play only a limited, supporting role in nuclear decision-making. Nevertheless, considering possible application of AI in decision-making support systems requires us to revisit the viability of some conventional wisdoms as assumptions for nuclear deterrence.

Growing Questions over Rationality Assumption

Nuclear deterrence is a situation where states seek to achieve political and security goals by influencing the other side, or adversary, without using them. Some argue that deterrence could work because the high prospect for catastrophic results in the use of nuclear weapon or the failure of deterrence would induce parties to the escalation game to become very cautious in taking actions. This logic assumes that decision makers would act rationally in crisis and seek to maximize its gain or to minimize its loss. If an adversary is a rational actor, this potential aggressor will not take actions in the first place as the aggressor knows that it would face the retaliation which would result in more harm than benefit. In the meantime, ensuring non-use depends on many factors. Among them, decision makers would seek accurate grasp of the situation, accurate knowledge or understanding on adversary's action/reaction options and calculous as much as possible, but decision makers rarely enjoy such a situation.

Nowadays, as research in behavioral economics develops, there is growing argument that casts serious doubt on the assumption of rationality in human decision-making in crisis, which strategic stability under nuclear deterrence rests on. It argues that humans cannot be counted on to always maximize their prospective gains and tend to have wrong expectations and calculations on their adversary's cost-benefit calculations. Prospect Theory tells that people will take more risk in order to defend what they have already gained but tend to be more cautious and conservative in newly gaining something of equal value. And political leaders tend to be unusually optimistic and overly confident in their ability to control events. Because of over-confidence, they may fail to cut losses and take more risks either to recover the projected loss or to regain the control. In short, people may not act in a way to maximize utility, or even not be explicitly aware of costs and benefits of certain actions that they will take (Krepinevich 2019).

This theory of the nature of human psychology may undermine the reliability of strategic stability and nuclear deterrence, but it is not unique to AI. What a scenario of introducing an AI-powered decision-making system does in this regard is to acutely depict this problem. In a sense, arguing the potential problems of AI in decision support leads us to even more fundamental question on the rationality assumption in the nuclear deterrence logics.

So, the question is whether AI would help overcome these growing concerns on the underlying assumptions of rationality in the logic of nuclear deterrence, or it would amplify such concerns. In other words, could decision makers accept AI-supported advice/date, which is counter-intuitive to decision makers? And how do they know if adversarial decision makers take actions with or without their intuition? (See section "Fog of AI War".)

Faster and more reliable, increasingly autonomous information processing systems could reduce risks associated with the management and operation of nuclear arsenals, particularly in crisis situations. Further, as AI is super rational and free from psychological biases as well as pressure that humans are always under influence, there is a possibility that AI would be able to sharply decrease, if not eradicate, the risks of human error and misperception/misconception. If it is the case, humans may thereby achieve higher levels of strategic stability in the avoidance of accidental launch or unintended escalation.

But this argument must be carefully examined by addressing the following questions: *To which objective* should "rationality" be defined in nuclear deterrence games? *What and whose objectives* should rationality be accounted for? Is it possible to establish a decision-making process with fully informed environment? (In the first place, there is no such thing as decision-making in an environment of complete information!) Additionally, would full transparency in information on nuclear arsenals contribute to the stability in an asymmetric nuclear relationship?

The first two questions suggest how difficult it is to set goals for nuclear deterrence or strategy. Perhaps even the highest national decision makers are not clearly/explicitly aware of goals, and their goals will change as situations evolve. The other two questions point to problems that may newly arise when AI is partially employed to support decision-making systems.

Fog of AI War

In a crisis situation or in a battlefield, decision makers and commanders are suffered from the so-called "fog of war," or the uncertainty of situational awareness (von Clausewitz 1832). The "fog of war" in nuclear deterrence is a problem inherent in nuclear deterrence per se, but not exclusively inherent in AI. Adoption of AI into ISR would help clear such "fog of war" caused by the lack of information on adversary's capabilities and deployment (see section "Ability to Set a Goal"). Also, as discussed above, AI, if properly applied, could contribute to confidence building and threat reduction

among nuclear-armed states. However, it would also be fair to say that AI may bring another type of "fog of war," due to its potential consequence of the introduction of AI in the decision-making process.

First, the most critical "fog of war" in the game of nuclear deterrence is the logic and reasoning that shape the intentions and preference of the adversary and decide where the red line of self-restraints is drawn. Therefore, it is unclear to decision makers where to find the equilibrium to optimize the relative gain against the adversary, which must be sought through exchange of strategic communication.

The irony of nuclear deterrence is that while construction of escalation ladders and deterrence logics are largely dependent on rationality, the irrationality of cost-benefit calculations, which is derived from fear in the human mind, can also be a constraint/restraint against escalation into the use of force in a crisis. Posing unpredictability and lack of confidence in the rational calculations of adversaries dissuades them from attacking and improves the certainty of deterrence.

Second, in the pursuit of a victory in a warfighting, AI, which "feels" no obsession with the fear of losing something or defending something presumably vital, could make a rational choice solely based on the cost-benefit calculation for a pre-set objective for combat. However, it is not certain whether this will contribute to the ultimate objective of engaging in nuclear deterrence from the perspective of managing the medium- to long-term entanglement relationships among strategic rivals, and to the satisfaction of their respective peoples.

Nuclear deterrence is a psychological game, in which the threat of using nuclear weapons restricts the actions of an adversary and manages the escalation so that the confrontation between the adversary and itself does not lead to the actual use of nuclear weapons. The situation is always changing and thus it is difficult to set clear goals to be pursued. However, presumably, decision makers may be engaged in the game of nuclear deterrence even without knowing absolute truth in the game.

Third, it is sometimes not easy for decision makers to envision the "national interest" (politically it is considered absolute truth for a sovereign state) that they intend to realize through nuclear deterrence and strategy. Moreover, it is very difficult to gauge the adversary's intentions, and it is possible that even the adversary itself does not consciously understand its strategic goals.

In this regard, it seems that the affinity is missing between characters or strengths of (narrow) AI and the required skills for decision-making during a nuclear crisis and escalation game. In a situation where strategic goals are constantly changing, narrow AI does not necessarily play a role in clearing the "fog of war." Rather, the problem of blackboxing decision-making in AI as described below creates an AI-specific "fog of war."

AI as Black Box

For decision makers who are responsible for the consequences of their decisions, a critical question is how and to what extent they can trust AI. It is probably more phycological than technical question. But can decision makers confidently choose an option, following suggestions or advice provided by AI, whose calculation process is in "black box" to decision makers? There are reports on the so-called "over-learning" problems, which have caused racial discrimination and other social problems due to its solely technical nature of processing information. Probably for the algorithm that drew such a conclusion, pure data processing resulted in such a problem. However, goals sought by decision makers also inevitably involved more social and political considerations. In these cases, AI failed to incorporate such factors in drawing conclusions, and imported "social" mistakes. In such a situation, to what extent can human decision makers be assured that algorithms are not making misinterpretation, miscalculation, or misrepresentation of the reality of the situation? Research efforts have already begun. U.S. Defense Advanced Research Projects Agency (DARPA) has started research on increasing the visibility of the rational for AI's decisions under the Explainable AI program.

Decision makers must be held accountable for the consequence of their decisions even if they rely on "advice" by AI, which may be false. It is certainly possible to prescribe the responsibility of policy makers within a legal theory, but ethically and practically, commitments to the use of weapons based on the advice by AI, which lacks the traceability, may put decision makers in a rather ambiguous position in terms of both the trustworthiness of AI advice and the readiness to assume responsibility and accountability. As studies of Behavioral Economics suggest, humans have a poor intuitive grasp of probability and inconstant expectation on cost-benefit calculation, subject to the situation (e.g., "gambler's fallacy," see, for example, Tune 1964 and Oppenheimer and Monin 2009). When AI draws a conclusion which is different from an intuition that policy maker has, can policy maker follow a suggestion by AI without knowing the logic behind AI's advice? This situation may pose another type of risk/concern to the rationality assumption of nuclear deterrence.

Another black box is adversary's AI employment policy. When decision makers do not know whether or to what extent the adversary's decision depends on AI, even only with the "prospect" for the advancement of AI and its contribution to such an ability, when it is perceived by the adversary, it would have the impact on decision maker's consideration and calculation. Since AI-enhanced ISR will expedite the targeting process, thus providing the adversary with stronger time-pressing pressure for decision to counter, the lack of information on the adversary's adoption of AI into the nuclear weapon system would amplify mistrust and paranoia between adversaries.

Also, the AI-supported decision-making process, which lacks the traceability, raises the risk and vulnerability in information security, particularly against misinformation and deception. When algorithms perceive and interpret information in a wrong way or in a wrong context and thus provide biased solutions, self-learning mechanism would reproduce the biases of the data in an accelerated pace. In this regard, defending the command-and-control system from cyberattack and countering disinformation are even more critical for the AI-supported decision process.

Consequently, "black box" phenomena in AI may increase the possibility of miscalculation, and the temptation to first use nuclear weapons before destroyed. Scharre says that the real danger of an AI arms race is not that any country will fall behind its competitors in AI, but that the perception of a race will prompt everyone to rush to deploy unsafe AI systems (Scharre 2019).

AI and Changing Characters of Nuclear Deterrence

Impact on ISR

One of first applications of AI in nuclear weapons could be a Russian underwater nuclear drone. Russia is developing a nuclear-propelled underwater drone to carry a thermonuclear warhead. Given the difficulty in communicating underwater, this "Oceanic Multipurpose System Status-6" needs to be equipped with a highly autonomous operation system, presumably supported by AI (Weintz 2018). If it would actually be deployed, it would increase the survivability of Russian retaliatory nuclear capability and improve the credibility of nuclear deterrence. Then it will inevitably trigger reactions by the United States and other states exposed to increased vulnerability with this new Russian nuclear asset.

Realistically, at the current level of AI competence, an imminent question is how to assess the impact of AI in intelligence, surveillance, and reconnaissance (ISR), which subsequently affect the perception on survivability and credibility of nuclear deterrent as well as its usefulness in maintaining nuclear stability and threat reduction.

During the Cold War, the development of ballistic missile systems significantly shortened the time to deliver nuclear weapons. Sophistication of delivery systems required nuclear-armed states to develop a kind of automation and standard operating procedure to respond and retaliate adversary's nuclear attacks in a timely and effective manner. The warning time for ballistic missiles was so short that launch-on-warning postures with detection and early warning systems were also required to be established and maintained.

In order to support the operation of such systems, robust communications, control and response systems to integrate information from various sources were also constructed. Operating such complicated weapon systems effectively and credibly entails minimizing the risk of misinformation (and subsequent false alarm), misinterpretation of information, mechanical errors of early warning systems under the very strong time pressure as such misconducts might lead to a catastrophic consequence. Vulnerable, time-critical targets remains a source of serious concern and it may be even more critical in the AI-enhanced environment.

In today's warfighting domains regardless of conventional or nuclear, much information is collected and analyzed using various tools at both the strategic and theater levels. Ironically, this leaves military analysts and decision makers in a state of overabundance of information. Given its strengths in data and imaginary processing and anomaly detection, AI along with sophisticated sensing technology would make the huge difference in ISR capabilities (Kallenborn 2019).

To respond such a situation, U.S. Department of Defense launched a project called Project Maven (2017), in order to "reduce the human factors burden of [full-motion video] analysis, increase actionable intelligence, and enhance military decision-making" in the campaign to fight against ISIS. This project demonstrates the potential of AI in enabling targeted strikes with fewer resources and increased accuracy/certainty (Loss and Johnson 2019).

AI-enhanced ISR would provide with the improved ability to find, identify, track, and target their adversaries' military capabilities and critical assets. It would allow for more prompt and precise strikes against time-critical targets such as ground-based, transporter-erector missile launchers and submarine launched ballistic missiles, which are platforms to ensure the survivability of the second-strike capabilities as deterrent forces. If a country acquires exquisite counter-force capability along with a credible, AI-enhanced ISR capability, it will not only be able to limit damage in the event of a nuclear crisis escalation, but will also be able to neutralize enemy nuclear deterrence. It affects the calculation on strategic stability, which is based on the survivability of secure second-strike nuclear forces.

Whether it would contribute to enhancing the stability or undermining it, experts' views are divided.

One argument is that the AI-enhanced ISR capability could increase stability as it provides nuclear-armed states with better information and better decision-making tools in time-critical situations, reducing the risk of miscalculation and accidental escalation. Another merit of stronger ISR capabilities is the possible improvement of monitoring nuclear weapon-related developments and conducting monitoring and verification operations, which supports the compliance of arm control and disarmament arrangements (if any).

If such "AI revolution" in ISR happens equally on all nuclear-armed parties who are engaged in mutual deterrence and seeking a point of "strategic stability" (while assuming that it is no longer viable to consider only US-Russia strategic stability for the nuclear stability at the global level), monitoring and transparency on nuclear assets and activities would be increased, and the high level of verification of arms control arrangements would become possible. They would eventually improve mutual confidence among these nuclear-armed states, and contribute to threat and risk reduction among nuclear-armed states.

Another argument is that the risk may increase if such "AI revolution" in ISR happens asymmetrically, especially if the emulation of technology occurs unevenly. The risk of uneven emulation of the technology is not negligible as AI has great impact in ISR capabilities. Uneven emulation of the technology would bring a gap in ISR capability and then counter-force capability.

In this situation, AI could undermine deterrence stability and increase the risk of nuclear use for the same reasons of enhancing security. If one country would be confident in its superiority, it considers that any conceivable gains from the use of force including nuclear weapons outweigh the cost and damage caused by adversary's retaliation. AI is an enabler for gaining the superiority. On the contrary, when facing a nuclear-armed adversary with sophisticated technology and advanced ISR capabilities (if it is known), such poor performance would prove disastrous. One with weaker ISR capabilities may become concerned about the survivability of its nuclear capabilities, and may be tempted to strike before its capabilities are attacked in time of crisis. It is a classical security dilemma situation.

There is also a possibility that states, which suffer adversary's first-mover's advantage, may be tempted to offset by another means rather than catching up in the same domain. For example, when the United States demonstrated its capability for precision-strike, combined with sophisticated ISR system during the Gulf War in 1990–1991, Russia perceived the increased risk of losing a conventional war or the vulnerability of its nuclear arsenal against precision attacks. What Russia did to offset this technological disadvantage was to develop low-yield nuclear weapons and to employ a nuclear doctrine to use such weapons. If one side acquires the superiority in ISR systems powered by AI algorithms and sensing technology, and improves its ability to locate and target nuclear-weapon launchers and other strategic objects, whereas the other side's policy options are either to catch up in the technology, or lower the predictability of its behavior and make the cost-benefit calculation of nuclear attack more complicated, it may result in the destabilization of strategic relationship.

In this situation, states, which employ minimum nuclear deterrence, would be more affected by such asymmetrical development and adoption of AI into ISR systems. They will be incentivized to expand its nuclear arsenal both in number and in variety of launching platforms in order to cope with the vulnerability. Or, they would reconsider their nuclear doctrine by raising alert status, and lowering the threshold for nuclear use by automating nuclear launch and/or by employing first use policy, which might increase the risk of escalation or triggering a nuclear war by misjudgment of the situation, and lower the possibility of avoiding accidental or inadvertent escalation.

Another factor to complicate the calculation is the accuracy and trustworthiness of AI. In theory, AI-based image recognition systems could identify second-strike capabilities. But Loss and Johnson (2019) highlight two key challenges: bad data and an inability to make up for bad data. It may not be impossible to distinguish between a regular track and a mobile missile launcher in satellite images as image data of adversary's mobile missile launchers is not sufficiently available for comparison. Further, it is possible for the adversary to take advantage of the characteristics of data processing of AI and avoid detection or input false information to deceive AI. This is a very risky practice that increases the likelihood of unwanted attacks, but at the same time, the likelihood of such misinformation may be a factor that makes it difficult to rely on AI-based ISR to make decisions. (However, it is also true that this view depends on the other party's expectation of rationality.)

While narrow AI could achieve near-perfect performance for assigned mandates in ISR and thereby enable an effective counter-force capability, inherent technological limitations and human psychological boundaries will prevent it from establishing stable deterrence relationship. AI may bring modest improvements in certain areas, but it cannot fundamentally alter the calculus that underpins deterrence by punishment.

Challenges for Stably Controlling Nuclear Risks: Arms Control and Entanglement

As seen above, AI could improve the speed and accuracy of situation awareness by utilizing neural networks, imagery sensing technology, and a huge database. Algorithms and control systems for "swarming" autonomous weapon systems, hypersonic weapons, and precision-guided weapons, as coordinated with various military assets including early warning systems, could make a huge difference in battlefield. Coordination through layers of algorithms also work to help manage complex operation. Applications of the technology in these ways to the conventional weaponry systems may change the character of war (Acton 2018). When AI assists decision makers and field commanders in choosing optimal battle plan, it would inevitably transform force structure and employment, as well as organizational and operational modality in command and control systems in order to catch up with the rapid pace of changing situations. Combined with new technologies such as precision-guided missiles and hypersonic gliders, the emerging war eco-system could heighten the vulnerability of strategic assets including nuclear assets against non-nuclear strategic weapons, and drastically shorten decision-making time to respond attacks against them.

The application of emerging technologies such as hypersonic gliders, robots, along with AI, to weapons has increased the strategic value of non-nuclear weapons. Thus, the boundaries between nuclear and conventional weapons and between strategic and non-strategic weapons have become blurred. This has increased the complexity of "cross-domain" deterrence calculations, and the vulnerability of infrastructure supporting deterrence, such as cyber and space, has made it difficult to establish the scope of an arms control regime for managing stable strategic relationships.

Historically, in order to avoid unintended and unnecessary conflicts and escalation and to maintain stable relations (maintaining arms race stability and crisis stability), adversaries have established arms control regimes. In order for an arms control system to contribute to the stable control of strategic relations between nuclear powers, it is necessary to establish a mutual understanding concerning the definition of the state of stability in a tangible manner. And the stability is often converted into the balance of forces (like "strategic stability" between two major powers). In other words, the arms control regime does not mean a pure balance of power based on an estimate of military power, but it means institutionalizing a relationship formed by mutual recognition of the "existence of equilibrium" and its joint understanding for stable management of the situation within a certain range.

Here are some key questions: Is arms control over the employment of AI in nuclear forces possible? Would AI contribute to establishing a stable arms control regime? AI is a software-based technology that makes a tangible assessment of its capabilities difficult. It suggests that an arms control regime for AI or the verification of the functioning of AI in weapon systems would be neither possible nor credible. Nuclear-armed states could therefore easily misperceive or miscalculate to what extent they should count on the impact of AI in their adversaries' capabilities and intentions. AI also help enhancing values of non-nuclear weapons used for strategic objectives rather than battlefield warfighting. It implies that designing arms control scheme by category or type of weapons may become less relevant to achieving a stability between adversaries.

In the field of nuclear strategy and deterrence, the perception of an enemy's capability matters as much as its actual capability. A worrisome scenario would be a situation where a nuclear-armed state would trigger destabilizing measures (e.g., adopting new and untested technology or changing its nuclear doctrine) based only on the belief that its retaliatory capacity could be defeated by another state's AI capabilities (Boulanin 2019).

Agenda for Nuclear Ethics in the AI Era

Ability to Set a Goal

As seen in severe accidents that have occurred in clinical decision-support systems, aviation, and even nuclear command and control in the past, excessive reliance on automated systems (automation bias) could become a cause of error. The point here is that it is not a machine that makes mistakes, but the humans who misuse or abuse the system. An AI-enhanced decision-making system may have to be operated in a short, time-constrained fashion, which may not permit human decision makers to conduct re-evaluation and review of conclusions/advice that an AI-enhanced system provides. In this situation, the risk of automation bias would be even greater.

And even with AI support, there are so many factors to consider in decision-making that there are unconsciously many ethical and normative constraints. (Also see above the section "Challenges for Stably Controlling Nuclear Risks: Arms Control and Entanglement" for discussion on the limitation of AI's capability in "autonomous" strategic decision-making.)

Of course, these ethical and normative constraints are likely not universal, and there is no clear understanding of the extent to which they are common in different sociocultural contexts or impose constraints on decision makers. Will AI algorithms be able to identify and learn about patterns and frameworks of thought that humans do not consciously recognize? With the current level of technological competence, the limitation of AI in decision-making is clearly shown in this point.

From an ethical point of view, too, it is unlikely or unimaginable that humans will not be involved in decisions about the use of nuclear weapons. While global/universal human interests are potentially recognized as absolute good in concept, they are not prescriptive enough to serve as operational grounds for policy implementation. In the current international system, where sovereign states are major players, states are supposed to maximize their individual "national" interests. In this context, a norm of the prohibition of the use of nuclear weapons has not gained the universality, and the use of nuclear weapons is considered as a possible option for some states in an extreme circumstance of state survival.

In the meantime, the humanitarian dimension of nuclear weapons casts a doubt on the legitimacy of any use of nuclear weapons. Even among those who support the importance of

nuclear deterrence for the maintenance of international peace and security, many believe nuclear weapons should never be used. It is because that once a nuclear weapon is used, it is highly likely that its consequence would go beyond the victory in war between states and reach a point that the damage to human beings and the entire earth would be unrecoverable. Managing nuclear weapons involves consideration of the tremendous social, economic, and political costs.

Nuclear deterrence is a game to be played against this kind of premises. It is a very complicated statecraft whose goal is not so straightforward as the winning a war or destroying targets. While there is a clear value standard for the use of nuclear weapons in the context of the abstract conceptual arguments of ethics and morality, when we look at the operations of policies in the modern real world, the criteria become ambiguous.

In the foreseeable future, we can hardly imagine that AI alone would set a goal and make decision of the use of nuclear weapons on behalf of humans.

Taking the Responsibility and Accountability Seriously

Automation of a decision means that a decision is made based on prescriptive standard-operating procedures. There should be no deviation as long as the automated process would not be disrupted. During the Cuban missile crisis, as we later found, there were a couple of occasions that decision makers did not follow the standard operating procedures. So deviations from preset procedures actually happened, and they could be (or at least some interpreted them as) reasons for the avoidance of escalation into nuclear exchange. This example suggests that autonomy entails adoptability over the automated decision procedure in the evolving circumstances.

The responsibility and accountability of certain behavior is closely associated with the autonomy of the system. So, is Autonomous Intelligence really autonomous?

The human mental system is a closed system, and actions based on each other's intentions are unpredictable in an ultimate sense. In other words, "unknowability" is the basis of so-called "free will" as well as the fluctuation of semantic interpretation, and thus responsibility in behavior. Although AI may appear to have a free will like a human, it is an adaptive, heteronomous system, and it is impossible to make truly autonomous decisions.

When there are multiple options and it is not clear which one the other prefers to choose, the social effect which might be brought by the government's decision of one particular option is linked to "responsibility."

Therefore, "free will" and "responsibility" are the concepts associated with closed autonomous systems, and cannot be established with open, heteronomous systems. (However, if the behavior of the mental system is under the constraints of the upper social system and the choice is in fact capped, there is no free will and no liability.)

The behavior of the adaptive system (current "narrow" AI) is heteronomously predefined by the designer at a more abstract level. True autonomy is tied to the unknowability for the others.

Fluctuation in the interpretation of the meaning of other party's words or deeds is fundamentally due to the unknowability of the other party, which is a closed system. Since the operation of an AI is performed based on a very complex program, it would seem from the outside that predicting its output would be practically impossible (even if possible in theory). Thus, AI gives the impression that it may have "free will."

In 2012, neural network based AI, which was developed by Google, successfully identified faces of cats from ten million YouTube thumbnails without being fed information on distinguishing features that might help identify cat's faces. This experiment shows the strength of AI in detecting objects with certain characteristics in their appearance. Google's experiment appears to be the first to identify objects without hints and additional information. Five years later, an AI was trained to identify more than 5000 different species of plants and animals (Gershgorn 2017). The network continued to correctly identify these objects even when they were distorted or placed on backgrounds designed to disorientate.

However, various concepts that we deal with in politics, economy, and other human activities are different from identifying animal's face from information on social network services. It is impossible to distinguish or identify certain concepts simply by differences in their appearances alone. Concepts are relative things that differ from one language community to another, and there is no universal absolute concept that segments the world.

In playing nuclear deterrence or strategic games, which involve highly political, abstract concepts of humanitarian and ethical values beyond mere war planning rationality, decision makers (not field commanders) must take into close consideration on so many political, social, economic, and normative factors such as freedom, rights, and ethics, in a situation where the adversary's intention is unknown (with incomplete information).

As we have witnessed the 75 years of the history of the nonuse of nuclear weapons, ethical questions on the consequence of possible nuclear exchanges affected the consideration of decision makers' employing nuclear option in strategic confrontation.

Can AI incorporate highly abstract social concepts such as freedom and rights in drawing a conclusion on policy priorities or assessment on the situation? This makes us aware of the difference between human knowledge and universal, absolute knowledge. It further leads us to a question whether in particular AI will be able to provide universal knowledge. An answer at this stage of technological development may be No. Then the question further goes; Can decision makers define and describe a goal of nuclear strategic game in a way that AI could read and operate, incorporating abstract, normative concepts? Assuming it is possible, would cultural differences in background and interpretations of these concepts be overcome in order to maintain the high level of mutual predictability and thus stability.

Conclusion

Problems associated with the application of AI into nuclear deterrence command-and-control and decision systems may not be unique to AI. Rather, AI, or AI-enhanced weapon systems amplify the risks intrinsic to nuclear deterrence.

Fast and effective detection and identification of targets with AI and enhanced sensing technology would help confidence-building in one way. In another way, it poses more vulnerabilities to nuclear-armed states and increases insecurity. Space for strategic ambiguity, which in reality functions as a kind of buffer zone between deterrence and the actual use of nuclear weapons, will become narrower by AI. Fast identification and analysis of the situation may enable decision makers to consider the best option, while reaction by others may also become quicker, and allowance time for decision-making may in fact become shorter, and decision makers may have to decide and act under stronger time pressure. Therefore, prisoners' dilemma and chicken game situations in nuclear deterrence may take more acute modalities in the AI-enhanced security environment.

We will not likely see a world where humans are completely replaced by AI in nuclear decision-making in the foreseeable future. Nor is it realistic that AI would be totally dismissed from the operation of nuclear arsenal. The U.S. Department of Defense emphasizes the concept of human-machine teaming: Humans and machines work together symbiotically. Humans provide higher-order decision-making and ensure ethical and appropriate operation of autonomous systems (Kallenborn 2019).

Examining the applicability of AI to managing nuclear strategy and deterrence raise the awareness of the necessity to re-examine the understanding and appropriateness of the traditional, long-overdue question in detail, that is, whether assumption of rationality and ambiguity in the logic of nuclear deterrence is appropriate.

If AI is to give us a chance to face these hard questions on nuclear deterrence, AI may save humanity. But without addressing these concerns discussed above, AI may move forward the doomsday clock, and make us closer to nuclear winter.

References

Acton, J. M. (2018). Escalation through entanglement: How the vulnerability of command-and-control systems raises the risks of an inadvertent nuclear war. *International Security, 43*(1), 56–99. https://doi.org/10.1162/isec_a_00320. Retrieved February 25, 2020.

Boulanin, V. (2019, May). *The impact of artificial intelligence on strategic stability and nuclear risk.* Sweden: SIPRI, Stockholm International Peace Research Institute.

Geist, E., & Andrew, J. J. (2019). *How might artificial intelligence affect the risk of nuclear war?* Santa Monica, CA: RAND Corporation.

Gershgorn, D. (2017). *Five years ago, AI was struggling to identify cats. Now it's trying to tackle 5000 species.* Available via Quartz. Retrieved February 25, 2020, from https://qz.com/954530/five-years-ago-ai-was-struggling-to-identify-cats-now-its-trying-to-tackle-5000-species/

Heath, N. (2018). *What is machine learning? Everything you need to know.* Available via ZDNet. Retrieved February 25, 2020, from https://www.zdnet.com/article/what-is-machine-learning-everything-you-need-to-know/

Jervis, R. (1989). *The Meaning of the Nuclear Revolution: Statecraft and the Prospect of Armageddon.* Ithaca, NY: Cornell UP.

Kallenborn, Z. (2019). *AI risks to nuclear deterrence are real.* Available via War on the Rocks. Retrieved February 2020, from https://warontherocks.com/2019/10/ai-risks-to-nuclear-deterrence-are-real/

Kissinger, H. A. (2018). *How the enlightenment ends. Available via The Atlantic.* Retrieved February 25, 2020, from https://www.theatlantic.com/magazine/archive/2018/06/henry-kissinger-ai-could-mean-the-end-of-human-history/559124/

Krepinevich, A. F. (2019). *The eroding balance of terror: The decline of deterrence.* Available via Foreign Affairs. Retrieved February 25, 2020, from https://www.foreignaffairs.com/articles/2018-12-11/eroding-balance-terror

Loss, R., & Johnson, J. (2019). *Will artificial intelligence imperil nuclear deterrence?* Available via War on the Rocks. Retrieved February 25, 2020, from https://warontherocks.com/2019/09/will-artificial-intelligence-imperil-nuclear-deterrence/

Oppenheimer, D. M., & Monin, B. (2009). The retrospective gambler's fallacy: Unlikely events, constructing the past, and multiple universes. *Judgment and Decision Making, 4*(5), 326–334.

Podvig, P. (2012). *The myth of strategic stability.* Available via Bulletin of the Atomic Scientists. Retrieved February 25, 2020, from https://thebulletin.org/2012/10/the-myth-of-strategic-stability/

Scharre, P. (2019). *Killer apps: The real dangers of an AI arms race.* Available via Foreign Affairs. Retrieved February 25, 2020, from https://www.foreignaffairs.com/articles/2019-04-16/killer-apps

Schelling, T. (1966). *Arms and influence.* New Haven, CT: Yale UP.

Searle, J. R. (1980). Minds, brains, and programs. *Behavioral and Brain Science, 3*(3), 417–424. https://doi.org/10.1017/S0140525X00005756.

Tune, G. S. (1964). Response preferences: A review of some relevant literature. *Psychological Bulletin, 61,* 286–302.

von Clausewitz, C. (1993) *On War.* London: Everyman. Originally published as *Vom Kriege* (in German) in 1832.

Weintz, S. (2018). *The real reason you should fear Russia's status-6 torpedo.* Available via The National Interest. Retrieved February 25, 2020, from https://nationalinterest.org/blog/buzz/real-reason-you-should-fear-russias-status-6-torpedo-28207

Part IV

AI/Robot–Human Interactions: Regulatory and Ethical Implications

The AI and Robot Entity

Marcelo Sánchez Sorondo

Contents

Abstract

Robots are instruments of the human being who is intelligent and free. Aristotle defines being free as the one that is cause of himself or exists on his own and for himself (*causa sui* or *causa sui ipsius*). By contrast, the instrument is not a cause of itself and does not work by the power of its entity, but only by the motion imparted by the principal agent, so that the effect is not likened to the instrument but to the principal agent. From the Christian perspective, for a being to be free and a cause of himself, it is necessary that he/she be a person endowed with a spiritual and incorruptible soul, on which his or her cognitive and free activity is based. An artificially intelligent robotic entity does not meet this standard. As an artefact and not a natural reality, the AI/robotic entity is invented by human beings to fulfil a purpose imposed by human beings. It can become a perfect entity that performs operations in quantity and quality more precisely than a human being, but it cannot choose for itself a different purpose from what it was programmed for by a human being. As such, the artificially intelligent robot is a means at the service of humans.

M. Sánchez Sorondo (✉)
Pontifical Academy of Sciences, Vatican City, Vatican
e-mail: marcelosanchez@acdscience.va; pas@pas.va

Keywords

Robots · Artificial intelligence · Philosophy · Christian religion · Aristotle · Consciousness · Freedom · Cause of himself · Instrumental cause

As Paul Ricœur said, quoted in *Laudato Si'*: "I express myself in expressing the world; in my effort to decipher the sacredness of the world, I explore my own".[1] Indeed, in this dialectic of self-recognition in relation to other beings, Christian philosophy states that the human soul—which is a "subsistent form inseparable from the act of being" (*actus essendi*), capable of knowing and loving, i.e. spirit,—although being a substantial form of the body, is intrinsically incorruptible in the real order of things. This is the metaphysical foundation, according to which the human person is **in himself** free and capable of ethical order, and emerges from the forces of nature and the instincts of animals.[2] As a spiritual subject, the

---

[1] *LS*, § 59; Paul Ricœur, *Philosophie de la volonté* II, *Finitude et culpabilité*, Paris 2009, 2016.

[2] St Thomas Aquinas: '*Quae per se habent esse, per se operantur*' (*De anima*, a. 14). It's the same principle for incorruptibility: '*Si quod habet esse sit ipsa forma, impossibile est qued esse separetur ab ea*'. Precisely: '*Quod enim convenit alium secundum se, numquam ab eoseparati eo separari potest . . .* Rotunditas enim a circulo separari non potest, *quia convenit ei secundum seipsum: sed aeneus circulus potest amittere rotunditatem, per hoc quod circularis figura separatur ab aere.* Esse autem secundum se competit formae . . . [Unde] si ipsa forma subsistat in tua esse, non potest ammittee esse' (*S. Th.*, I, 50, 5. E cf. ad 3 which refers to q. 44, 1 ad 2 which quotes from *Metaph.* V, 5, 101, 5 b 9: ' . . . *sunt quaedam necessaria quae habent causam suae necessitatis*'. Hence

J. von Braun et al. (eds.), *Robotics, AI, and Humanity*, https://doi.org/10.1007/978-3-030-54173-6_14

human being is *imago Dei* and capable of "receiving grace", i.e. to be a child of God, and this is the highest status and dignity that a human being can reach as a spiritual being. Hence, "when the human being has received grace, he is capable of performing the required acts" for himself and others.[3] In this sense, I think, like most people, that **robots cannot be considered as persons**, so robots will not and should not possess freedom and do not possess a spiritual soul and cannot be considered "images of God". AI and robots are beings invented by humans to be their instruments for the good of human society.

Christian philosophy distinguishes two types of cause, principal and instrumental. The principal cause works by the power of its entity, to which entity the effect is likened, just as fire by its own heat makes something hot. But the instrumental cause works not by the power of its entity, but only by the motion whereby it is moved by the principal agent: so that the effect is not likened to the instrument but to the principal agent: for instance, the bed is not like the axe, but like the art which is in the craftsman's mind. Now such power is proportionate to the instrument. And consequently it stands in comparison to the complete and perfect power of anything, as the instrument to the principal agent. For an instrument does not work save as moved by the principal agent, which works of itself. And therefore the power of the principal agent exists in nature completely and perfectly, whereas the instrumental power has a being that passes from one thing into another, and is incomplete.

So, an instrument has a twofold action: one is instrumental, in respect of which it works not by its own power but by the power of the principal agent; the other is its proper action, which belongs to it in respect of its proper entity. Thus, it belongs to an axe to cut asunder by reason of its sharpness, but not to make a bed, in so far as it is the instrument of an art. But it does not accomplish the instrumental action save by exercising its proper action, for it is by cutting that it makes a bed.

St. Thomas calls the action of the principal agent as soon as it "flows" into the instrument "intentional" and has an incomplete energy and existence from one subject to another, similar to the passage from the agent to the patient.[4]

Aristotle speaks of two types of instruments: ἔμψυχον ὄργανον (living tool), and ὄργανον ἄψυχος (inanimate tool). The first is like the servant who moves by the will of his owner, the other as the axe.[5] The Philosopher makes another rather minor distinction in his text but one that has been a source of inspiration for Christian theology: "These considerations therefore make clear the nature of the slave and his essential quality: one who is a human being belonging by nature not to himself but to another is by nature a slave, and a person is a human being belonging to another if being a man he is an article of property, and an article of property is an instrument for action separable from its owner (ὄργανον πρακτικὸν καὶ χωριστόν)".[6] St. Thomas comments: "Then, supposedly such a definition of the servant, he concludes: the servant is an animated organ separated from another existing man [...]. In fact, it is said separate to distinguish it from another part that is not separated, like the hand".[7] Inspired by this suggestion, Christian theologians distinguish between the united instrument (*instrumentum coniuctum*) such one's own arm and hand, and the separate instrument (*instrumentum separatum*) such as the pen, the crosier, the cane, or the car: "Gubernator enim gubernat manu, et temone: manus enim est instrumentum coniunctum, cuius forma est anima. Unde temo est instrumentum movens navem, et motum a manu; sed manus non est instrumentum motum ab aliquo exteriori, sed solum a principio intrinseco: est enim pars hominis moventis seipsum".[8]

This distinction between separate and united instruments has a major role starting with the doctrine of St. John Damascene who considers the humanity of Christ an "instrument of divinity" (ὄργανον τῆς Θειότητος).[9] St. Thomas with explicit reference to this profound indication of St John Damascene states, "Christ's humanity is an 'organ of His Godhead,' as Damascene says. Now an instrument does not bring forth the action of the principal agent by its own power, but in virtue of the principal agent. Hence Christ's humanity does not cause grace by its own power, but by virtue of the Divine Nature joined to it, whereby the actions of Christ's humanity are saving actions".[10] Moreover, the Angelic Doctor affirms in one of his last texts that the main cause of grace is God Himself; Christ's humanity, on the contrary, is the instrument linked to divinity, while the sacraments are separate instruments. Indeed: "a sacrament in causing grace works after the manner of an instrument. Now an instrument is twofold: the one, separate, as a stick, for instance; the other, united, as a hand. Moreover, the separate instrument is moved by means of the united instrument, as a stick by the hand. Now the principal efficient cause of grace is God Himself, in comparison with Whom Christ's humanity is as a united instrument, whereas the sacrament is as a separate instrument. Consequently, the saving power

---

the difference between the III and IV proof of the existence of God. For the principle of the contingency of matter cfr. *C. Gent.* II, 30.

[3] St Thomas Aquinas, *De Malo*, q. 2, 11.

[4] Cf. *S. Th.*, III, q. 62, a. 4.

[5] Cf. *Nic. Eth.* 1161 b 4.

[6] *Pol.* I, 1254 a 14 ff.

[7] *In I Pol.* lect. 2, n. 55.

[8] St. Thomas Aquinas, *In II De Anima*, lib. 2, l. 9, n. 16.

[9] *De Fide Orth.*, Bk 3, c. 19.

[10] *S. Th.*, I-II, q. 112, 1 ad 1.

must be derived by the sacraments from Christ's Godhead through His humanity".[11]

As can be deduced by these quotations, the term "instrument" is used in various senses, but with reference to one central idea and one definite characteristic, and not as merely a common epithet. Now of all these senses which "instrument" has, the primary sense is clearly that of not being a cause of itself or not existing by itself as an instrument. On the contrary, in a famous text Aristotle defines being free as the one that is a cause of himself or exists on its own and for himself, i.e. one who is cause of himself (*causa sui* or *causa sui ipsius*). In fact: "we call a man free who exists for himself and not for another (ἄνθρωπος, φαμέν, ἐλεύθερος ὁ αὑτοῦ ἕνεκα καὶ μὴ ἄλλου ὤν)".[12]

As I said at the beginning, for a being to be free and a cause of himself, it is necessary that he/she be a person endowed with a spiritual soul, on which his or her cognitive and volitional activity is based. AI and robots are just that: an artificial reality and not a natural reality, that is, invented by the human being to fulfil a purpose imposed by the human being. It can become a very perfect reality that performs operations in quantity and quality more precisely than a human being, but as an AI or robot it cannot choose for itself a different purpose from what the human being programmed it for. It can be hypothesized that there are robots composed of organic parts of animals (which Aristotle calls living tools) and inanimate parts that perform mathematical operations following algorithms that for humans are almost impossible to fulfil, but these entities will never be ends for themselves, but means at the service of humans.

The fact that AI and robots are instruments without a spiritual soul does not prevent them from being able to transmit a spiritual virtuality, "as thus in the very voice which is perceived by the senses there is a certain spiritual power, inasmuch as it proceeds from a mental concept, of arousing the mind of the hearer".[13] A corollary of this may be when Aristotle says, "all these instruments it is true are benefited by the persons who use them, but there can be no friendship, nor justice, towards inanimate things; indeed not even towards a horse or an ox, nor yet towards a slave as slave. For master and slave have nothing in common: a slave is a living tool, just as a tool is an inanimate slave. Therefore there can be no friendship with a slave as slave, though there can be as human being: for there seems to be some room for justice in the relations of every human being with every other that is capable of participating in law and contract, and hence friendship also is possible with everyone so far as he is a human being. Hence even in tyrannies there is but little scope for friendship and justice between ruler and subjects;

but there is most room for them in democracies, where the citizens being equal have many things in common".[14]

Laudato Si', criticizing the prevailing technological paradigm, warns us that today it seems inconceivable to use technology as a mere instrument. "The technological paradigm has become so dominant that it would be difficult to do without its resources and even more difficult to utilize them without being dominated by their internal logic. It has become countercultural to choose a lifestyle whose goals are even partly independent of technology, of its costs and its power to globalize and make us all the same. Technology tends to absorb everything into its ironclad logic, and those who are surrounded with technology 'know full well that it moves forward in the final analysis neither for profit nor for the well-being of the human race', that 'in the most radical sense of the term power is its motive—a lordship over all'.[15] As a result, 'man seizes hold of the naked elements of both nature and human nature'.[16] Our capacity to make decisions, a more genuine freedom and the space for each one's alternative creativity are diminished".

We need to help remedy this profound crisis, caused by a confusion of our moral visions, of ends and means. We need to help stop this technological paradigm that is leading the world towards disaster. We must recognize human life itself, with its dignity and freedom, and we must work for the survival of our planet. "If a man gains the whole world but loses his soul, what benefit does he obtain?" (Matt 16:26). Yes, we are facing a matter of calculation, the calculation to save our world from indifference and from the idolatry of power of this technological paradigm. This is what Jesus meant when he told us that the poor in spirit, the gentle, those whose hearts desire justice, those who are merciful, those whose hearts are pure are *Felices* or *Bienaventurados*, happy and blessed, for theirs is the kingdom of heaven.

We need thinkers and practitioners who know how to use artificial intelligence and robots to alleviate suffering, to operate justice, to realize the highest good of peace. We thus welcome thinkers and practitioners who are capable of thinking and using these extraordinary tools for the sustainable development of the human being.

References

Aquinas, S. T. (1950). *In duodecim libros Metaphysicorum Aristotelis expositio*. Turin: Marietti.

Aquinas, S. T. (1965a). *Quaestio disputata de anima*. In P. Calcaterra & T. S. Centi (Eds.), *Questiones disputatae* (Vol. 2). Turin: Marietti.

---

[11]*S. Th.*, III, q. 62, 5.

[12]*Met.*, I, 982 b 26 f.

[13]St Thomas Aquinas, *S. Th.*, III, 62, 4 ad 1.

[14]*Nic. Eth.* 1161b 1 ff.

[15]Romano Guardini, *Das Ende der Neuzeit*, 63–64 (*The End of the Modern World*, 56).

[16]Ibid., 64 (*The End of the Modern World*, 56).

Aquinas, S. T. (1965b). *Quaestio disputata de malo*. In P. Bazzi & P. M. Pession (Eds.), *Questiones disputatae* (Vol. 2). Turin: Marietti.

Aquinas, S. T. (2010). *Summa theologicae*. Madrid: Biblioteca Autores Cristianos.

Aquinas, S. T. (2019a). *Summa contra gentiles*. Navarra: Fundación Tomás de Aquino. Retrieved from April 16, 2020, from https://www.corpusthomisticum.org/iopera.html.

Aquinas, S. T. (2019b). *Sententia libri politicorum I*. Navarra: Fundación Tomás de Aquino. Retrieved April 16, 2020, from https://www.corpusthomisticum.org/iopera.html.

Aristotle. (1894). In J. Bywater (Ed.), *Aristotle's Ethica Nicomachea*. Oxford: Clarendon Press. Retrieved April 16, 2020, from http://www.perseus.tufts.edu/hopper/text?doc=Perseus%3atext%3a1999.01.0053

Aristotle. (1924). In W. D. Ross (Ed.), *Aristotle's Metaphysics*. Oxford: Clarendon Press. Retrieved April 16, 2020, from http://www.perseus.tufts.edu/hopper/text?doc=Perseus%3atext%3a1999.01.0051

Aristotle. (1957). In W. D. Ross (Ed.), *Aristotle's Politica*. Oxford: Clarendon Press. Retrieved April 16, 2020, from http://www.perseus.tufts.edu/hopper/text?doc=Perseus%3atext%3a1999.01.0057

Damascene, S. J. (1856–1866). *Ékdosis akribès tēs Orthodóxou Písteōs*, Bk 3, c. 19, Patrologiae Cursus Completus, Series Graeca, vol. 94 col. 1080b, J. P. Migne (Ed.). Paris. Retrieved April 16, 2020, from http://www.documentacatholicaomnia.eu/20vs/103_migne_gm/0675-0749,_Iohannes_Damascenus,_Expositio_Accurata_Fidei_Orthodoxae_(MPG_94_0789_1227),_GM.pdf

Guardini, R. (1950). *Das Ende der Neuzeit: Die Macht*. Würzburg: Werkbund Verlag.

Pope Francis. (2015). *Encyclical letter: Laudato si' of the Holy Father Francis on care for our common home*. Available via Vatican Press: Vatican City. Retrieved September 26, 2019, from http://w2.vatican.va/content/dam/francesco/pdf/encyclicals/documents/papa-francesco_20150524_enciclica-laudato-si_en.pdf

Ricœur, P. (2009). *Philosophie de la volonté II: Finitude et culpabilité* (Vol. 2). Paris: Points.

Friendship Between Human Beings and AI Robots?

Margaret S. Archer

Contents

Abstract

In this chapter the case for potential Robophilia is based upon the positive properties and powers deriving from humans and AI co-working together in synergy. Hence, Archer asks 'Can Human Beings and AI Robots be Friends?' The need to foreground social change for structure culture and agency is being stressed. Human enhancement speeded up with medical advances with artificial insertions in the body, transplants, and genetic modification. In consequence, the definition of 'being human' is carried further away from naturalism and human essentialism. With the growing capacities of AI robots the tables are turned and implicitly pose the question, 'so are they not persons too?' Robophobia dominates Robophilia, in popular imagination and academia. With AI capacities now including 'error-detection', 'self-elaboration of their pre-programming' and 'adaptation to their environment', they have the potential for *active collaboration* with humankind, in research, therapy and care. This would entail *synergy or co-working* between humans and AI beings.

Keywords

Robots · Artificial intelligence · Sociology · Consciousness · Emotions · Friendship

M. S. Archer (✉)
University of Warwick, Coventry, UK
e-mail: margaret.archer@warwick.ac.uk

Introduction

Friendship is regarded as paradigmatic of human sociality (Donati and Archer 2015: 66) because it entails no implications about kinship, sexuality, ethnicity, nationality, language, residence, power, status, beliefs, etc., although each and every one of these could be imbricated in it. Clearly, not all human relations are of this kind, the extreme exception being slavery, including variations on the Hegelian 'Master'/'Slave' theme. Significantly, in antiquity the enslaved were regarded as non-human. Their (supposed) absence of

J. von Braun et al. (eds.), *Robotics, AI, and Humanity*, https://doi.org/10.1007/978-3-030-54173-6_15

a soul served to justify their subordination and perhaps its shadow lives on in general attitudes towards other species.

Moreover, as Aristotle maintained, 'friendship' is not a unitary concept even for humans alone; it could take three forms (based upon utility, pleasure and goodness). Thus, in Aristotelian philosophy, different humans would not accentuate the same relationship today when they referred to their 'friends' compared with those held by others to be 'friends' in different times and places. What then is generic to 'friendship' and why is it relevant to discussing our (potential) relationships with other intelligent beings/entities such as AI robots? The answer to the first part of this question is that 'friendship' generates emergent properties and powers, most importantly those of trust, reciprocity and shared orientation towards the end(s) of joint action. These are distinct from simple friendly behaviour towards, say, dogs, where even as a source of pleasure they can only metaphorically and anthropomorphically (Haraway 2003) be termed a 'companion species' because there is no emergent common good or shared orientation; these may be imputed by dog-owners but cannot be ascertained for the dog except in behaviouristic terms. The answer to the second part of the question is the subject of this paper and will doubtless be contentious.

Its aim is to break the deadlock between popular 'Robophobia' and commercialized 'Robophilia' (products intentionally marketed as humane 'Best Friends' or useful 'housekeepers' that control thermostats and draw the blinds, etc.). Most of the best-known arguments in this dispute rest upon exaggerating the binary divide between the human and the AI, as if this difference between organically based and silicon based entities formed an abyss dividing up the whole gamut of properties and powers pertaining to agents of any kind (Brockman 2015).

Conversely, I argued in the first papers in our Centre for Social Ontology series on *The Future of the Human* (Archer 2019) that acceptance of their shared 'personhood' can span this divide. The main propositions defended there and used here are the following[1]:-

1. 'Bodies' (not necessarily fully or partially human) furnish the necessary but not the sufficient conditions for personhood.
2. Personhood is dependent upon the subject possessing the First-Person Perspective (FPP). But this requires supplementing by *reflexivity* and *concerns* in order to define personal and social identities.
3. Both the FPP and Reflexivity require Concerns to provide traction in actuating subjects' courses of action and thus accounting for them.

[1] These conclusions share a great deal in common with Lynne Rudder Baker (2000, 2013). Compare with the above reference.

4. Hence, personhood is not in principle confined to those with a human body and is compatible with Human Enhancement.

In my above conclusions, point (4) merits a particular comment. Human Enhancement is anything but new. Historically it goes back at least to Pastoralism, enhancing height and strength through improved nutrition, and merely becomes more bionic with prostheses, pacemakers, etc. This is why I do not find the relationship between contemporary humans and the latest AI robots to be usefully captured by the concept of 'hybridity', since *Homo sapiens* have been hybrids throughout recorded history because of their progressive enhancement. Instead, this paper accentuates *synergy* between the two 'kinds' as paving the way to friendship.

Responses to my list of conclusions from opponents can again be summarized briefly as denials that the three capacities I attribute to all normal human beings are ones that can be attributed to an AI entity.

1. The AI entity has no 'I' and therefore lacks the basis for a FPP.
2. Consequently, it lacks the capacity to be Reflexive since there is no self upon which the FPP could be bent back.
3. Similarly, it cannot have Concerns in the absence of an 'I' to whom they matter.

In what followed, I sought to challenge all three of these objections as far as AI entities are concerned. However, this was not by arguing, in some way, that the subsequent development (if any) of these highly sophisticated, pre-programmed machines tracks the development of human beings in the course of their 'maturation'. On the contrary, let me be clear that *I start from accepting and accentuating the differences between the human and the AI in the emergence of the powers constitutive of personhood.*

In the human child, the 'I' develops first, from a *sense of self,* or so I have argued, as a process of doing in the real world, which is not primarily discursive (language dependent) (Archer 2000). The sequence I described and attempted to justify was one of {'I → Me → We → You'}. The sequence appears different for an AI entity that might plausibly follow another developmental path, and which, if any, would be a matter of contingency. What it is contingent upon is held to be relational, namely it develops through the synergy between an AI entity and another or other intelligent beings, namely humans to date. In this process of emergence, the 'We' comes first and generates a reversal in the stages resulting in personhood, namely {'We' → 'Me' → 'I' → 'You'}. Amongst robots, it is specific to the AI and will not characterize those machines restricted to a limited repertoire of pre-programmed skills, sufficient for a routine production line. Conversely, the AIs under discussion are credited with at least four main skill sets that can be supplemented: language recognition and production; learning ability; reflexive error-correction; and a (fallible) capacity for Self-Adaptation relevant to the task to which they have been assigned. This paper is not concerned with 'passing as human' (the Turing test), nor with 'functional equivalences in behaviour', independent of any understanding (Searle's Chinese Room) (see Morgan 2019). Its aim is twofold: First, to make a start on debunking some of the main obstacles regularly advanced as prohibiting 'friendship' between AIs and humans (and perhaps amongst AI themselves, although this will not be explored), second, to venture *how* synergy (working together) can result ceteris paribus in the emergence of 'friendship' and its causal powers.

Overcoming the Obstacles?

Three main obstacles are regularly advanced as precluding 'friendship' with human beings and often reinforce 'robophobia'. All of these concern ineradicable deficits attributed to AI robots. Specifically, each systematically downplays one of the characteristics with which AIs have been endowed in this current thought experiment; abilities for *continuous learning* (until/unless shut-down); for *error-correction* and for *adaptation* of their initial skills set—and thus of themselves—during their task performance. The accentuation of AI deficits which follow shadows Colonialist portrayals of colonized peoples.

Normativity as a Barrier

Stated baldly, this is the assertion that an AI entity, as a bundle of micro-electronics and uploaded software, is fundamentally incapable of knowing the difference between right and wrong. Consequently, alarm bells sound about the ensuing dangers of their anormativity for humans, as prefigured in Asimov's normative laws of Robotics (1950). In other words, 'Robots are seen as (potential) moral agents, which may harm humans and therefore need a "morality"' (Cockelbergh 2010: 209). Usually, this need is met by a top-down process of building human safeguarding into pre-programmed designs or, less frequently, by AI robots being credited with the capacity to develop into moral machines from the bottom-up, through learning morality, as ventured by Wallach and Allen (2008).

Both protective responses confront similar difficulties. On the one hand, morality changes over time in societies (poaching a rabbit no longer warrants transportation) and the change can be fast (from French Revolutionary laws to the Code Napoléon), as well as coercive and non-consensual. Even Hans Kelsen (1992) had abandoned one *grundnorm* as

founding all instances of legal normativity by the end of his work. On the other hand, if the model of childhood learning replaces that of pre-programming, we largely now acknowledge that the socialization of our kids is not a simplistic process of 'internalization'; again it is a non-consensual process societally and the family itself often transmits 'mixed messages' today. Thus, complete failure may result. It simply cannot be concluded that 'we set the rules and bring it about that other agents conform to them. We enable them to adapt their behaviour to our rules even before they can understand social norms (Brandl and Esken 2017: 214).'

Equally pertinent, social normativity is not homogeneous in its form or in its force. Both of the latter are themselves subject to social change. Just as importantly, some of its transformations are more amenable to straightforward learning (for example, it took less than 5 min online to learn how to renew my passport) than others (such as 'What constitutes Domestic Violence in a given country?'), which requires interpretation and changeable judgements about appropriate classification of various behaviours (e.g. the new category of 'coercive control' now used in British family law).

If we break normativity down into recognizable categories—and what follows is for purposes of illustration, not the only useful manner of doing so—it should clarify that working upwards through the list of rules is to move from easy learning to the need for expert advice that itself will be challenged.

Etiquette is heterogeneous (changeable and regional), varying with factors such as the age, and social standing of participants *vis à vis* one another. Although transgression attracts only social sanctions, its applicability to AIs is dubious. Are handshakes appropriate? May they list their 'special requirements' as 'access to an electric socket and cable'? How should they address others and expect to be addressed? No human guide to correct behaviour[2] can assist AIs in learning 'good manners', despite their ability to read and recite any manual available, because many human conventional practices are disallowed by the AIs' (current) constitutions. Much the same goes for large tracts of *customs and conventions*.

More significantly, as I have argued elsewhere (Archer 2016) is the growth of *anormative bureaucratic regulation*—both public and private—that now predominates in the production of social co-ordination. The rule of law can no longer run fast enough to keep up with social morphogenesis in almost any social domain; novel forms of malfeasance outstrip counteractive legislation as recognized in countries such as Britain that tried to declare a halt on designating

---

[2] Although such guidelines have become more relaxed (Abrutyn and Carter 2014).

'new crimes'[3] in the new millennium. Instead, Regulators and regulations boom in every area, sometimes upheld by law but frequently not. In this reversal, something very damaging has happened to normativity itself and is directly relevant to demolishing the barrier that the absence of a capacity for it once constituted for delineating AI beings from humans.

This is the fact that obeying regulations does not rely upon their ethical endorsement; indeed the rules governing domestic refuse disposal or the sale of forked carrots in supermarkets may be regarded as idiotic by the general public—who were never consulted. Regulations are not morally persuasive but causally produce conformity through fines, endorsements and prohibitions. Thus, it is up to the subjects to perform their own cost-benefit analyses and determine whether the price of infringement, such as motorway speeding, is worth it to them on any given occasion. Taking the normativity out of an increasing range of activities progressively weakens the barrier placing AI beings outside the moral pale. Like today's humans, they do not have to feel guilt, shame, remorse or wrongdoing in breaching regulations. Thus, whether or not they are capable of such feelings is an argument that has become less and less relevant because the social context makes decreasing use of them.

Intensive social change also undercuts a frequent protective suggestion that in the interests of human health and safety 'our' values should be built into AI beings. But, if this is on the model of childhood learning or socialization, what (or, rather, whose values) are adopted? The alternative is to programme such values into the AI to-be, in an updated version of Asimov's three laws. Yet, supposing it was possible and desirable, it is not an answer to the normative barrier because these would be our values that we have introduced by fiat. Pre-programed values can never be theirs, not because they did not initiate them (however that might be), but rather because the same theorists hold that no AI entity can have emotional commitment to them for the simple reason that they are presumed incapable of emotion. However, a normative administrative regulation sidesteps this particular issue since it is independent of emotions through its reliance on calculative cost-benefit analysis.

Emotionality as a Barrier

I am not arguing that human and AI beings are isomorphic, let alone fundamentally the same substantively. Nor is that the case methodologically for studying the two, which is particularly relevant to emotionality. Dehaene (2014) and his team do a wonderful job in de-coding parts of brain activities, established by working back experimentally *from behaviour*

for human beings, including some of the brain damaged. However, in comparing the capacities of the 'wet' and the 'dry' for experiencing emotion, it seems more productive to start the other way around with the physical constitution of AIs and the affordances and limitations of their uploaded software.

In *Being Human* (Archer 2000), I differentiated between our relations with the 'natural', 'practical' and 'social' orders as the sources of very different types of emotions impinging upon three inescapable human concerns (respectively, our 'physical well-being', 'performative achievement' and 'self-worth')—given our *organic* constitution, the way the world is made and the ineluctability of their interaction. I still endorse this approach when dealing with the human domain; each order of natural reality has a different 'import' for our species and generates the emergence of different clusters of emotions acting back upon them.

My definition of emotions as 'commentaries upon our concerns' in the three orders of natural reality is about matters we human beings cannot help but care about (to some extent)—imports to which we cannot be totally indifferent, given our constitution. Hence, whilst humans may experience 'terror' in anticipation of being plunged into Arctic water, constitutionally they cannot 'fear' themselves rusting. Conversely, objectively dangerous imports for AIs would be quite different: for example, rusting, extended power-cuts or metal fatigue.

However, because I am maintaining that AI robots can detect that they confront real dangers (just as all our electronic devices signal their need for re-charging), this does not in itself justify attributing emotions to robots. But, such a move is unnecessary. This need not be the case because on my account it is *concerns* that are pivotal and whilst emotion may increase attention and provide extra 'shoving power', it is not indispensable and it can be misleading. Thus, I disagree with those who maintain that 'emotions matter. They are the core of human experience, shape our lives in the profoundest ways and help us decide what is worthy of our attention' (McStay 2018: 1). Even when confined to humans, our emotionality surely cannot be a guide to 'worth'.

Nevertheless, many do maintain that an 'emotional commentary' is an essential part of all our *concerns* and, by extension, of all forms of *caring*. Yet, to be realistic, we *care enough* about a variety of mundane concerns to do something about them (e.g. checking the warranty when buying appliances, keeping spare light bulbs, stocking some 'long life' products in the pantry, etc.), all without any emotionality. Such concerns are sufficient to make (some of us) care 'enough' to do such things, without any 'feelings' at all. Conversely, being emotionally moved by a photo of a dead baby on an Aegean beach (a real example), said on social media to have 'moved many to tears', was clearly not enough to promote their active caring for asylum seekers. It

---

[3] Over 4000 entered the statute books in last two decades of the twentieth century (Cabinet Office Paper 2013).

seems that the branding of some movies as inconsequential 'tear jerkers' is not far from the mark. In sum, *I hold to my view that emotions are neither a necessary nor a sufficient condition for caring, whilst accepting that their addition may reinforce fidelity to our concerns.*

My position does not amount to Cognitivism, which might at first glance seem appropriate to AI beings. However, what 'cognitivists' maintain is that 'emotions are very real and very intense, but they still issue from cognitive interpretations imposed on external reality, rather than directly from reality itself' (Ortony et al. 1988: 4). Above, I have argued against this reduction of ontology to epistemology in human relations with the three orders of natural reality. Now I am asking why we should agree that AI beings are deficient precisely if they *do not experience* 'very real and very intense emotions' in natural reality? Were they merely to scrutinize a situation and to conclude cognitively that undesirable φ was highly likely to transpire, that would sometimes improve their 'prospects' of averting φ in comparison with dramatic displays of affectivity that can foster stampedes in human crowds confronting fires in confined spaces.

Some might think that a mid-way position is provided by Charles Taylor's statement that we speak of emotions as essentially involving a sense of our situation, claiming that they 'are affective modes of awareness of situation … *We are not just stating that we experience a certain feeling in this situation*' (Taylor 1985: 48, my ital.). I agree, but stress that the ontology of the situation remains indispensable. Moreover, there are two controversial words in that quotation which effectively divide psychologists of the emotions: one is 'affect' (affectivity) and the other 'feelings' and both are very relevant to our present discussion.

'Feelings' as the mainstay of my opponents are a slippery concept because some are held worth consideration and others not, some to be concealed, others displayed. But this is largely a matter of social convention. As individuals, humans vary enormously in how readily they reveal their suffering and which sufferings, but their acceptability has also varied historically. Why has 'having a stiff upper lip' come to be seen as derogatory would be interesting for semantic archaeology. Equally, why does encouraging others to grieve overtly, to 'tell their stories' or to 'let it all come out', seem to be the creed of the counsellors today? This overt variety shows emotionality not to be a universal and essential component of human responses to similar circumstances if they are so socially malleable. The rejoinder could obviously be that we can never know what people suffer (or exult about) in silence, but then if we cannot know it, neither can we study it. A last resort would be to hand this over to the 'therapeutic couch', but in which of those warring psychiatric practitioners should we place our trust?

If those holding that the presence and absence of feelings comes down to the 'phenomenal feel' of *qualia,* which will forever divide the wet and the dry, it seems a weak case for two reasons. First, it varies experientially within the human species, or calling someone 'tone deaf' or 'blind to natural beauty' would not have been coined. Second, if one of us is motivated by, say, injustice or unfairness why are those supposedly accompanied by particular *qualia*? Judges are expected to rule on cases in the light of evidence made available in trials, not to share their 'phenomenal feel' for it or the parties involved. Neither did John Rawls argue that decisions made 'behind the veil' entailed such phenomena. If these continue to be regarded as a 'barrier' by some, then sharing the same *qualia* will also continue to separate all beings from one another, regardless of their physical constitution, if these can ever be objectively determined.[4]

The Ultimate Barrier: Consciousness

This is generally presented as the 'great divide' that those in silico can never cross. Indeed, it is a resilient version of the old dispute between Comte and Mill about the premise of a 'split-consciousness' built into the concept of introspection. Mill's riposte was to jettison the simultaneous element by inserting a brief time interval between the original thought and *inspection* of it. Consequently, our self-awareness became an unobjectionable exercise of memory. I will not repeat the lengthy argument I advanced (Archer 2003: 53–92), by selectively drawing upon the American Pragmatists, to buttress my contention that 'introspection', on the observational model (*spect intro*), should be replaced by the 'inner' or 'internal conversation'. But how is this relevant to AI beings?

Instead of re-invoking 'introspection' I simply rely on two software abilities: to speak and to listen for *securing their self-consciousness*. Every day we humans employ language to pose questions: internally to ourselves, externally to other people and also of our outside environment. A common exemplar, not universal and answerable in various non-linguistic ways, is the first question likely to arise each day for most adults upon waking—'What time is it?' We are both questioners and respondents and this means that all normal people are both SPEAKERS and LISTENERS, to themselves. This is what the American pragmatists—James, Pierce, and Mead—called the 'inner conversation' and I have explored this subjective mental activity in my trilogy of books on human 'reflexivity' (Archer 2003, 2007, 2012).[5]

---

[4]In other words, I agree with Dehaene, that the 'concept of qualia, pure mental experience detached from any information-processing role, will be viewed as a peculiar idea of the prescientific era' (Dehaene 2014: 262).

[5]Reflexivity is defined 'as the regular exercise of the mental ability, shared by all normal people, to consider themselves in relation to their social contexts and vice versa' (Archer 2007: 4).

Now, I want to venture that given the AIs I have postulated are programmed to be/become proficient language users, then why should it be queried that they too function as both speakers and listeners? This cannot be seriously contested. But if that is so, why can't they be credited with internal reflexivity? The barrier on the part of those who deny the capacity for 'thought' to AI robots is a simplistic definitional denial of their ability to think because computers are held to be incapable of consciousness, let alone self-consciousness. Yet, if we examine the basic constituents of the 'internal conversation' what is there in the activities of speaking, listening and responding (internally, but reflexive deliberations can be shared externally) that would put any AI robot permanently beyond the pale of reflexivity? Certainly there are practical obstacles, the most powerful being that in current computers each software application works in a separate memory space between which exchanges are precluded, meaning that programmes have no general means of exchanging their specialized knowledge (Dehaene 2014: 259f). But, such limitations as these result from the programme designers rather than being intrinsic to AI robots and are not (usually) applicable to speaking and listening per se.

When we do think of questioning and answering in general conversations, all conversational exchanges are alike in one crucial respect, namely they involve turn-taking. Therefore, I am arguing that when we talk to ourselves the same rule maintains and it does so by our *alternating between being subject and object in the dialogical turn-taking process*, which is rendered possible because of the necessary time gap—however small—that separates question from answer. Some may query how this is possible given that any data or notion I produce (as subject) is identical to that I simultaneously hear (as object). Yet, it would be meaningless to entertain an alternation between two identical things. Instead, following the insight of William James, in expressing a response we review the words in which to articulate it, welcoming the felicitous ones and rejecting those less so (James 1890). Thus, our answers often do not come pre-clothed in the verbal formulations that clearly express them to the best of our ability or to our satisfaction—either externally or internally. We are fallible as well as sub-optimal in this respect, sometimes even saying what we do not mean—to ourselves and to others. But we are capable of reformulation before venturing a revised response. And we may do this several times over (see Fig. 1).

This extension of James consists only in allowing the subject to question his/her/its own object over time. Such a process will be familiar to any writer and is the bread and butter of literary historians. To redeploy James' notion of

Fig. 1 Datum and verbal formulation in the internal conversation. Source: Adapted from Archer (2003: 99)

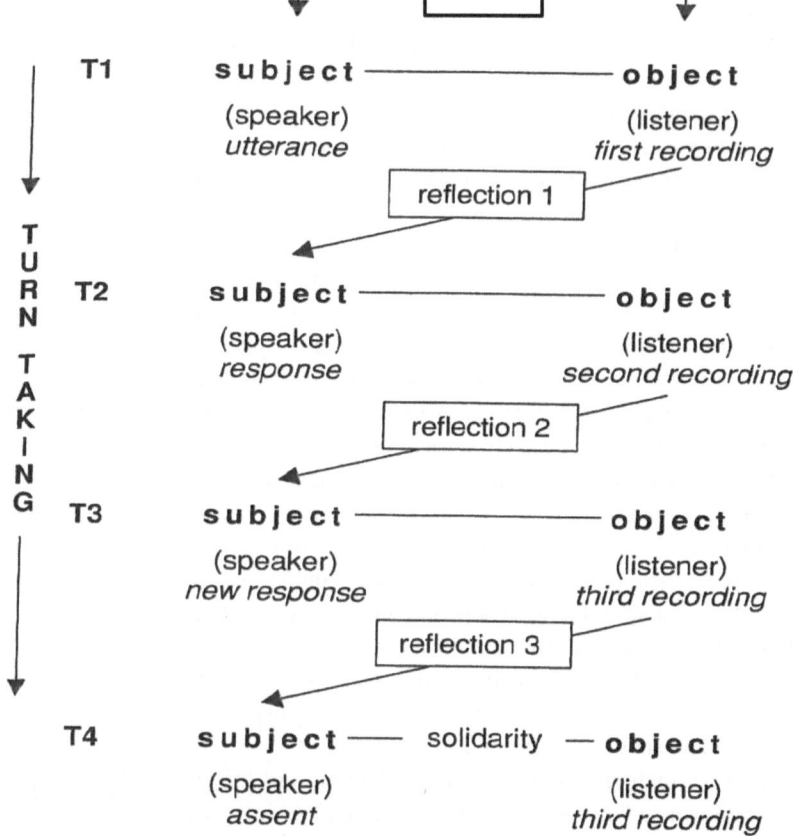

'welcoming' certain verbal formulations, discarding others and seeking for more adequate words would all be held illegitimate by opponents because as mental activities they entail 'thought'. Even if an AI is pictured summoning up a Thesaurus, the objection would be that there is no mechanism that can match semantics for their appropriateness in a *novel* research context, where 'common usage' cannot be the Court of Appeal. Of course, the AI may do as we do and sometimes resort to poetry. But then he is guilty of thoughtful 'creativity' and to concede this would allow the barrier to sag. I can see no way in which an AI's exercise of judgement can be freed from conceding that he is also exercising thought—and thus 'consciousness and self-consciousness', as Harry Frankfurt puts it:

> '(B)eing conscious in the everyday sense does (unlike unconsciousness) entail reflexivity. It necessarily involves a secondary awareness of a primary response. An instance of exclusively primary and unreflexive consciousness would not be an instance of what we primarily think of as consciousness at all. For what would it be like to be conscious of something without being aware of this consciousness? It would mean having an experience with no awareness whatever of its occurrence. This would be, precisely a case of unconscious experience. It appears, then, that being conscious is identical with being self-conscious. Consciousness *is* self-consciousness.' (Frankfurt 1988: 161–162)

Were these considered to be convincing if over-condensed objections to the 'barriers', there remains a final but crucial point to note which is central to the rest of the paper. All of the obstacles briefly reviewed have depended upon 'robotic individualism', since they have all taken the form of 'No AI robot can Λ, Ω, or Θ'. That is not the case that I am seeking to advance here. Instead, this concerns the dyad and a particular form of dyadic interaction, between an AI robot and a human co-worker or consociate. Since I presume we would agree that there are properties and powers pertaining to the dyad that cannot belong the individual, whether of organic or silicon constitution, then these need to be introduced and be accorded due weight.

The Emergence of Friendship and Its Emergents

This section begins with the dyad but does not end there. As Simmel maintained, there are both distinctions and connections between what are conventionally differentiated at different societal levels: the micro-, meso- and macro-strata. This is because different emergent factors and combinations come into play at different levels just as distinct contingencies are confronted there. In Critical Realism this represents a *stratified social ontology*, which is the antithesis of the 'flat ontology' endorsed, for instance, in Latour's 'Actor network' approach. Not only are there different properties that emerge at the three levels but there is also upward and downward causation between them. Restriction of space will make this a very sketchy introduction.

At the Micro-, Meso- and Macro-Levels

Let us begin schematically with the relationship between a single human researcher and the AI being awarded to him under his/her grant. The aim is obviously that the co-ordinated action of the two is intended to advance a particular research programme, or overcome a specific problem. For two main reasons there is nothing deterministic about this relationship. First, there will be psychological differences between human researchers in their expectations about the future relationship that arise from a whole host of factors, some but not all of these would be the same for what any particular researcher might expect if instead he/she had been awarded a post-doctoral co-worker. Second, there are certain structural and cultural constraints, formal and informal, attending such co-working (rights and duties) that will influence the likelihood of friendship developing, without proscribing it.

Ultimately, co-ordination is the generic name of the game for those setting up these 'mixed dyads'. And there is nothing about co-ordinated actions per se that is conducive to or hostile towards friendship; even 'collaboration' can involve a strict division of labour, which never segues into 'collegiality'. Indeed, co-ordination between humans may refer to 'one off' events such as two motorists coming from opposite directions who collaborate to move a fallen tree that is blocking both drivers but which neither can do alone (Tuomela 2010: 47). But the two drivers are unlikely candidates for subsequent friendship, given they will probably not meet again.

Equally, some senior researchers may define the relationship in advance as one in which they have acquired a Research Assistant, regardless of whether the subordinate is human or robotic. Black neatly characterizes this Command and Control hierarchy in terms general enough to apply to those research Bosses who hold themselves 'to be the only commander and controller, and to be potentially effective in commanding and controlling. [He] is assumed to be unilateral in [his] approach (he tells, others do), based on simple cause and effect relations, and envisaging a linear progression from policy formation through to implementation' (Black 2001: 106). None of that characterizes the *synergy* within the dyad that is the necessary but not sufficient condition for the development of friendship.

The Boss could indeed confine his robot Assistant to tasks of compilation and computation for which the latter is pre-equipped and their relationship may remain one of CAC even though a master does not possess all the skills of his servant, as with most Lords of the Manor and their gardeners. But it would be a short-sighted aristocrat who never talked or listened to his gardener and one who believed he could envisage a linear progression of the unfinished garden to his ideal and thus issue unilateral demands for the implementation at each stage. That would make his gardener dispensable as other than a digger and planter; any discussion of their synergy becomes irrelevant.

In a previous paper (2019) I sketched the emergence of synergy through a thought experiment about a much more open-minded Boss (Homer) and his robotic assistant (Ali) who became a co-worker. In the first instance, Homer played to what he knew to be Ali's strengths, to make fast computations on the Big Data available which were too extensive for Homer himself to review. He could then venture his hypotheses about the incidence of a lethal Tumour X in humans that extended way beyond his own reading or clinical practice. Ali's data overview produced some confirmatory cases but also cast doubts on it from others. Because they did talk about it, Homer shared his new suspicion with Ali that qualitative data was required in addition to quantitative to explain not just the incidence but the progression of Tumour X. In the second instance, after lengthy discussion, the two recognized what was needed but also that Ali was not equipped with the necessary software. He wanted to help his boss to solve the problem set (and on which his 'employment' depended), even though *qua* robot he had no concern about the solution benefitting humanity. He surveyed the qualitative data analysis programs and went on to read up evaluation reports about them through consulting e-journals. All of this is completely unremarkable, except for one thing. Ali had taken the responsibility (and accountability) for making and executing this extension of the research program. He had acted as a genuine part of this 'we'.

In the third instance, as the research program progressed and the relational goods generated grew, the design also became increasingly elaborate. Although funding increased, so did Homer's wish-list for new technical tools and so did his questions to Ali: 'Could you do this or that?' This led Ali to a stock-taking exercise, what could simply be added to his pre-programed repertoire of skills, what could be adapted by re-writing that software and what was it beyond him to deliver? In short, Ali discovered the affordances of its bionic body and its resistances to adaptation for some of the novel tasks mooted.[6] Because of their 'we-ness'—for Homer and Ali are not or no longer in a command and control relationship— Ali makes the adaptations to his pre-programming that are possible and commensurate with further progress on their future research design. Sometimes he makes mistakes (as Turing anticipated), but he is familiar with error-correction.

But something has happened to Ali during their long collaboration. He has learned a great deal and *he is aware of this*. He is (1) not merely carrying and processing information (as does a GPS and as Searle's man in the Chinese Room did). Ali is doing things that enable new knowledge to be generated, things that Homer cannot do but needs doing. As an AI, he is not knowledgeable (2) in the purely *metaphorical sense* that a statistical Table might be said to 'know' (e.g. the extent of drug trading in different countries). In any case, that is about the publication and not the generation of knowledge. Finally, (3) Ali's awareness is quite different from the sense in which a thermostat might (again metaphorically) be called aware of temperature change when it kicks-in. That is a pre-programmed mechanical response to a change in the external environment. Conversely, Ali's awareness is in response to his own self-induced changes in his internal constitution and resultant adaptive capacities (for which he has records). But enough of this film script that will never hit the big screen.

Some readers may have no insuperable objections to the above but still feel that it has not got to the core of friendship. Undoubtedly, it has not in Aristotelian terms where 'With those in perfect friendship, the friend is valued for his/her own goodness and constitutes 'another self''.[7] However, basing my case upon synergy, this is neither about differences alone nor similarities alone between co-workers, but about *diversity* (Archer 2013)—that is some shared qualities together with some differences—that incentivizes collaboration and makes it possible. What a focus upon such syncretism avoids is insistence on the 'similarity' criterion, which immediately evokes 'speciesism' as an insuperable barrier to friendship in this context (Dawson 2012: 9).

'Holism', Individualism' and 'Relationality' are three distinct ways out of this cul-de-sac. 'Holism' means abandoning human essentialism by focusing sociologically instead upon the characteristics and capacities that we acquire through our first-language and induction into a culture, which McDowell (1998) terms our 'second nature'. What this does is to transfer the characteristics previously attributed to the human 'species' to our social communities. However, elsewhere, I have criticized this as the 'Myth of Cultural Integration' (Archer 1985: 36), disputing that any 'culture' is fully shared and homogeneous, if only because none are free from contradictions, inconsistencies and aporia. If that is the case, then cultural essentialism falls too—to a multiculturalism where 'goodness' varies over time, place, selection and interpretation.[8]

---

[6]This is not unlike the child discovering that he is socially advantaged or disadvantaged compared with his classmates. In both cases, these very different kinds of subjects consider themselves as objects in relation to others of their kind.

[7]In the Nicomachean Ethics, Book 1, this represents the apex of friendship: 'Only the friendship of those who are good, and similar in their goodness, is perfect. For these people each alike wish good for the other *qua* good, and they are good in themselves'.

[8]The same can be said of Tuomela's (2010), reliance upon agents drawing on a 'shared ethos' in order to act as a 'we' or in the 'we-mode', and that is considerably less demanding than friendship.

'Individualism' holds that what is admirable in a friend originates from his/her uniqueness and therefore cannot be essential properties of human beings. *How* someone came to be of exemplary goodness might be due to some unique concatenation of circumstances, but we do not admire *what made them what they are—though we may learn from it—but rather celebrate their virtuosity, which can have aggregate consequences alone for co-working.*

'Relationality' shifts conceptualization away from holistic species' attributes and individual uniqueness alike, and moves the relationship to centre-stage. It replaces discussion of essences by causal considerations of outcomes. In fact, friendship is defined on the causal criterion, that is, on the basis of the relational goods and evils it generates. It is thus simultaneously freed from essentialism, exceptionalism and speciesism. It displaces reliance upon 'joint Commitment', the mainstay of Margaret Gilbert's account and which also has to be central to John Searle's since he places identical thoughts in the two heads of the dyad,[9] difficult as this is to endorse (Pettit and Schweikard 2006: 31–32). Instead, I focus upon 'joint action' and its causal consequences, where the commitments of each party maybe quite different. (The research Boss in the story could have been motivated by the death of his mother from Tumour X, which cannot be the case for his robotic co-worker).

However, although 'joint action' is held to be a necessary condition, it is not sufficient for engendering friendship. Paul Sheehy gives the nice example of four prisoners who are rowing a boat to escape. They share the belief that 'I am escaping' and that it entails 'we are escaping', since they are literally in the same boat. Hence their joint action may or may not be successful. But if this is the emergent effect of their cooperation, it does not necessarily entail or engender friendship (Sheehy 2002). It is only if relational goods are generated in synergy, from which both parties are beneficiaries and further benefits are deemed likely, given continued collaboration, that the first stepping stones towards friendship are laid. Like an orchestra or a successful football team, neither party can 'take away their share' without cutting the generative mechanism producing the relational goods they value, albeit for different reasons.

The first paving stone is the emergence of trust in their co-action. Although the commitments of the two may be substantively different, they cannot be incompatible or zero-sum, if each is a beneficiary and motivated by this to continue co-working. In short, they have a *common-orientation* to the project (of whatever kind) continuing and developing. This is all the more essential the more different in kind are their separate contributions. Such reciprocal trust, reinforced by successful practice over time, is what unites the skating duo,

the trapeze 'flyer' and 'catcher' as well as co-authors and research dyads. In all such instances, that trust needs to become sufficiently resilient to withstand occasional accidents, false starts and blank periods without progress. This argument is highly critical of the conceptualization and usage of the concept of 'trust' within the social sciences. Frequently, it is treated as a simple predicate rather than a relational outcome, which requires consistent reinforcement. But predicates cannot be plucked out of thin air when needed; they have their own morphogenesis, morphostasis and morphonecrosis (Al-Amoudi and Latsis 2015).

There are many ways of defining 'friends' and distinguishing between the characteristics and consequences of friendship. Here I treat 'dimensionality' as differentiating between 'thin', that is, 'one-dimensional' relations versus the 'thicker' multi-dimensional relationships as constitutive of friendship. In everyday human life it is common for people to refer to 'my golfing friend', 'my travelling companion' or 'my workmate', meaning their relations with another are restricted to these domains. Such 'thin' friendships are vulnerable to breakdown, partly because it takes only one row, bad trip or disappointed expectations for their fragility to break, and partly because such dyadic partners are quite readily replaceable. On the other hand, 'thickness' is tantamount to the resilience of the friendship, with its various facets compensating for deficiencies in any particular one.

Some commentators may conclude at this point that a human and a robot might possibly succeed in developing a friendly but one-dimensional working relationship but that social structural and cultural constraints would preclude this morphing into 'thick' friendship. In a manner evocative of 'apartheid', it could be agreed that many expressions of this incipient relationship do seem closed to them; e.g. they cannot go out for a drink or a meal together. Similarly, for the time being, there are many social embargos on their sharing leisure activities (e.g. AI robots may be ineligible for Golf Club Membership). The implication could seem to be that this dyad is confined to strictly cognitive activities in the extension of their friendship.

Even so, that leaves them plenty of options (film, television, the media, gaming, gambling, literature, music and art) and the cultural context of society is increasingly privileging some of these pursuits. If one day Homer wants to watch the news, he might be surprised that Ali as spectator becomes vocal about a short film-clip on football's match of the day. Certainly, his comments are cognitively confined to working out team disposition in relation to goal scoring—but so are those of many pub-pundits. In watching movies, he represents the critic that sci-fi film makers have never had (to date), commenting on the portrayal of those of his own kind and effectively challenging the dominant 'robophobia' presented. As far as 'gambling' is concerned, he is the ideal co-participant and readily computes how far the odds are

---

[9]For a detailed critique of the 'Plural Subject' analytical philosophers, see Donati and Archer (2015: Ch. 2).

stacked in favour of the virtual 'house'. Indeed, were the human/AI dyad to go 'rogue' and devote some of their efforts to the generation of 'Relational Evils' through online gambling—or working fruit machines for that matter—this could be a formula for breaking the bank at Monte Carlo.

When we come to micro-meso level linkages, the weakness of the sociological imagination, which has Cinderella status in the foregoing, dries up almost completely. Take the Research Centre which figured prominently in the discussion of relationships at the micro-level. Sociologically it is too spare and bare. For example, I know of no Research Centre that is exempt from the pressures of either its benefactors or of the University, or of the educational system of which it forms a part, if not all three. In turn, the *relational networks* are much more complex; in reality students, administrators, commercial agents, journalists, social media, funding agencies, educational policy makers, etc. would impinge upon this dyad—were the account more than fiction—with their different causal powers and vested interests. These would not be contingent occurrences but part and parcel of current research life and indispensable to accounting for real-life outcomes in it.

There is a cluster of congruent reasons for research showcasing the micro-stratum. To be fair, our world of competitive research funding structurally reinforces this focus on AI individuals. Thus, firstly, it is hard in the barrage of literature to find any empirical research *on AI to AI relations themselves*. Given many of them have the software for speech recognition and use, do they never speak to one another, about what, or are they disabled from doing so? Why, given this could be beneficial to the research? For example, Aurélie Clodic and her colleagues describe a table-test involving an AI and a human building a tower from wooden bricks together and the AI sometimes drops a brick (Clodic et al. 2017). Possibly the robot has something constructive to say about its occasional clumsiness that would improve the joint action. There is nothing to lose, beyond the cost of installing the communicative software if it is lacking.

Secondly, the same point is reinforced by the wealth of material concerning the (usually) positive views about AIs in the roles of taking care of the elderly, communicating with Autistic children, assisting in hospital operating theatres, etc.[10] What would these workers say about the greatest difficulties encountered versus what they accomplish easily in a small group? Instead, the focus remains fixedly on 'individual client satisfaction', but maybe this clientele could be better satisfied given such input was available. After all, the aim is not exclusively for-profit; it can also be to assuage or offset loneliness, so there does seem to be a role for

fostering friendship more effectively here. The AI assistants would not become bored by the repetition of stories by the lone, frail elderly, may supply missing names and places that escape and distress their clients, might show them photo montages of, let's say, the subject's children or of significant houses, neighbourhoods, landscapes in their biographies. These actions, undertaken by a team of robot-carers could enhance the quality of life for those in care.

Thirdly, when collective AI is considered, especially in military contexts, what they are modelled on is but 'swarm behaviour' amongst birds, insects and animals of lesser intelligence. The US Department of Defense is explicit that such swarms 'are not pre-programmed synchronized individuals, they are a collective organism, sharing one distributed brain for decision-making and adapting to each other like swarms in nature' (Plummer 2017). They appear to resemble a weapon of mass destruction rather than candidates for citizenship.

With such military exceptions, the *predominant focus* is upon the AIs as *individuals* rather than as any kind of interactive collective, it is about *singular* relations with particular human persons rather than shared problems in dealing with them and about *aggregate* and never (to my knowledge) about emergent consequences.

Yet Emmanuel Lazega (2013, 2014, 2015, 2016) has shown empirically in a variety of contexts (including science laboratories, law courts, and a Catholic Diocese) how, in particular, the quest for advice from peers and superiors is the stuff from which networks are built geographically between localities and can reach politically upwards for representation and voice, thus consolidating collectivities at the meso-level. This is precisely what almost exclusive concentration upon micro-level research precludes. It seems ironic that commercial ICT enterprises are resolutely global in the extension of their algorithms in pursuit of worldwide profit and of their surveillance data for globalized power, whereas the most intelligent forms of robots are confined to micro-level personal services. Of course, such confinement constrains them to supplementing the human hierarchy rather than substituting for it. Is this the reason why, besides the costs involved, it is the self-protectiveness of their human designers which discourages AI inter-communication, which deters the formation of collective associations and which defines their contributions in aggregate terms? This conclusion seems hard to resist given that Dehaene (2014: 259ff) ventures that the development of a Global Workspace accessible to all would increase the affordances available to each and every AI, potentially facilitating an increase in innovation and enabling the self-adaptive capacities of artificial intelligence to have freer social rein. This, of course, might more likely incite structural repression for human self-defence than to be met in friendship.

---

[10]For examples, see Maccarini (2019) and LaGrandeur (2017). By 2011, the total number of robots in use worldwide was 18.2 million: only 1.2 million of those were industrial, and the rest—17 million—were service robots', p. 98.

In sum, the consequences of the missing meso-level can broadly be presented as threefold as far as society is concerned:

1. It operates as a severe curb upon innovation and therefore on the novelty and new variety upon which societal morphogenesis depends.
2. It works to preclude the emergence of AI 'social movements', whatever (unknown) form these might take.
3. It serves to buttress human domination and thus is hostile to friendship between these two categories of persons.

At first glance, it may now appear surprising that *at the Macro-level* there are significant signs of the recognition of AI robots as 'electronic persons', who would acquire rights and obligations under a draft EU resolution of 2017. Yet, this only appears unexpected and contradictory if we wrongly interpret it as a similar stage in the succession of social movements following on from those first promoting workers' unionization, then female enfranchisement, anti-discrimination legislation, and LGBT rights, to date. However, 'robot rights' are not the aim of an AI millennial vanguard pressing for legal and political recognition in society. In fact, their closest historical precursor was the accordance of 'corporate personhood' to firms and companies, enabling them to take part as plaintiffs or respondents in legal cases (Lawson 2015). Its main objective was to free individual company owners, executives and shareholders from financial obligations following legal judgements of culpability. Exactly the same objective appears here; to exculpate human designers and marketers from legal and financial responsibility, whilst defending humans against demands for compensation given damages they have caused to a robot.

Significantly, the European Parliament (February 2017) adopted a draft resolution for civil law rules on robotics and AI, pertinent to issues of liability and ethics of robots. What is worth noting is that Robotics were at first treated as an undifferentiated collectivity and a compulsory insurance scheme was mooted for 'Robot users'—in general. And this despite the fact that a year earlier a draft European Parliament motion (May 31, 2016) had noted that the (AIs) 'growing intelligence, pervasiveness and autonomy require rethinking everything from taxation to legal liability' (CNBC 2016). This motion called upon the Commission to consider 'that at least the most sophisticated autonomous robots could be established as having the status of electronic persons with specific rights and obligations'.

The Code of Ethical Conduct (2017) proposed for Robotics Engineers endorsed four ethical principles: (1) beneficence (robots should act in the best interests of humans); (2) non-maleficence (robots should not harm humans); (3) autonomy (human interaction with robots should be voluntary; (4) justice (the benefits of robotics should be distributed fairly) (Mańko 2017). These defensive principles are redolent of Asimov's 'laws' of the mid-1940s (and were reproduced in the Code's text); they have nothing in common with today's movements or social protests. The report's author, Mady Delvaux (Luxemburg) encapsulated its main motive: 'to ensure that robots are and will remain in the service of humans, we urgently need to create a robust European legal framework' (Hern 2017).

The paucity of 'robot rights' reflects the priority given not only to the generic defence of humans (spurred by the approach of self-driving cars) but specifically, as one international lawyer put it, to the issue of intellectual property rights,[11] which I earlier maintained was the heart of the matter in late modernity's new form of relational contestation (Archer 2015). No more dramatic illustration is needed than this discussion of draft legislation to (supposedly) acknowledging 'robotic personhood' than a legal commentary which took *the sale of 'electronic persons' for granted*! (my italics) (Hern 2017).

Conclusion

In the context of the present paper, I wish to finish with a quotation from Clause 28 from the Draft European Report (2015).[12] What it reveals is a fundamental incapacity to deal with human/AI synergy, to treat synergy as non-emergent but rather susceptible of being broken down into the 'contributions' of the two parties, and having no necessary relationship to relational goods and evils or to sociality.[13]

In other words, 'friendship' between humans and AIs, far from being the anchorage of 'trust', 'reciprocity' and helping to engender 'innovative goods' is remarkable for its absence.

---

[11] As Ashley Morgan, working for the Osborne Clarke practice, summarized matters: 'If I create a robot, and that robot creates something that could be patented, should I own that patent or should the robot? If I sell the robot, should the intellectual property it has developed go with it?' (Hern 2017).

[12] European Parliament, *Committee on Legal Affairs*, 31.5.2015. [The Report] considers that, in principle, once the ultimately responsible parties have been identified, their liability would be proportionate to the actual level of instructions given to the robot and of its autonomy, so that the greater a robot's learning capability or autonomy is, the lower other parties' responsibility should be, and the longer a robot's 'education' has lasted, the greater the responsibility of its 'teacher' should be; note, in particular, that skills resulting from 'education' given to a robot should be not confused with skills depending strictly on its self-learning abilities when seeking to identify the person to whom the robot's harmful behaviour is actually due.

[13] This differs hugely from the Nobel Prize's procedure of awarding more than one researcher, without specification of their 'discrete' contributions. Of course, the conferring of an award for collective innovation is very different from the partitioning of legal liability among those involved.

However, as usual, looking for absences reveals whole tracts of activities as well as rights that have been exiled from such reports. There is space only to itemize some of those that are missing, partly at least because 'friendship' is not considered as a generative mechanism, that is as an indispensable building block in the attribution of rights to AI beings:-

– Eligibility to stand for election
– To hold a Passport and use it
– The right to open a Bank Account, to possess property (including intellectual property, patents, copywrite, etc.) and to benefit from inheritance or to designate inheritors.
– Rights to compete with humans for appointments, promotion, etc.
– Legal protection against discrimination
– Rights covering Marriage, Intermarriage and Adoption
– Appropriate changes in sports categories
– And, perhaps the most contentious of all, no blanket embargo upon AIs becoming 'received' into various Church and Faith communities.

References

Abrutyn, S., & Carter, M. J. (2014). The decline in shared collective conscience as found in the shifting norms and values of Etiquette manuals. *Journal for the Theory of Social Behaviour, 43*(2), 352. https://doi.org/10.1111/jtsb.12071.

Al-Amoudi, I., & Latsis, J. (2015). Death contested: Morphonecrosis and conflicts. In M. S. Archer (Ed.), *Generative mechanisms transforming the social order*. Cham: Springer.

Archer, M. S. (1985). The myth of cultural integration. *The British Journal of Sociology, 36*(3), 333–353. https://doi.org/10.2307/590456.

Archer, M. S. (2000). *Being human*. Cambridge: Cambridge UP.

Archer, M. S. (2003). *Structure, agency and the internal conversation* (pp. 53–92). Cambridge: Cambridge UP.

Archer, M. S. (2007). *Making our way through the world*. Cambridge: Cambridge UP.

Archer, M. S. (2012). *The reflexive imperative*. Cambridge: Cambridge UP.

Archer, M. S. (2013). Morphogenic society: Self-government and self-organization as misleading metaphors. In M. S. Archer (Ed.), *Social morphogenesis* (pp. 145–154). Dordrecht: Springer.

Archer, M. S. (2015). *Generative mechanisms transforming the social order*. Cham: Springer.

Archer, M. S. (2016). Anormative social regulation: The attempt to cope with social morphogenesis. In M. S. Archer (Ed.), *Morphogenesis and the crisis of normativity*. Cham: Springer.

Archer, M. S. (2019). Bodies, persons and human enhancement: Why do these distinctions matter? In I. Al-Amoudi & J. Morgan (Eds.), *Realist responses to post-human society: Ex machina*. Abingdon: Routledge.

Asimov, I. (1950). *I, Robot*. New York: Genome Press.

Black, J. (2001). Decentering regulation: Understanding the role of regulation and self-regulation in a post-regulatory world. *Current Legal Problems, 54*(1), 103–147.

Brandl, J. L., & Esken, F. (2017). The problem of understanding social norms and what it would take for robots to solveit. In R. Hakli &

J. Seibt (Eds.), *Sociality and normativity for robots* (pp. 201–215). Cham: Springer.

Brockman, J. (Ed.). (2015). *What to think about machines that think*. New York: HarperCollins.

Cabinet Office Paper. (2013). *When laws become too complex: A review into the causes of complex legislation*. Available via Gov.uk. Retrieved February 27, 2020, from https://assets.publishing.service.gov.uk/government/uploads/system/uploads/attachment_data/file/187015/GoodLaw_report_8April_AP.pdf

Clodic, A., Pacherie, E., & Chatila, R. (2017). Key elements for human-robot joint-action. In R. Hakli & J. Seibt (Eds.), *Sociality and normativity for robots* (pp. 159–177). Cham: Springer International Publishing.

CNBC. (2016). *Robots could become 'electronic persons' with rights, obligations under draft EU plan*. Accessible via CNBC. Retrieved September 4, 2019, from https://www.cnbc.com/2016/06/21/robots-could-become-electronic-persons-with-rights-obligations-under-draft-eu-plan.html

Cockelbergh, M. (2010). Robot rights? Towards a social-relational justification of moral consideration. *Ethics and Information Technology, 12*, 209–221. https://doi.org/10.1007/s10676-010-9235-5.

Dawson, R. (2012). Is Aristotle right about friendship? *Praxis, 3*(2), 1–16.

Dehaene, S. (2014). *Consciousness and the brain*. New York: Penguin.

Donati, P., & Archer, M. S. (2015). *The relational subject*. Cambridge: Cambridge UP.

Frankfurt, H. (1988). Identification and wholeheartedness. In H. Frankfurt (Ed.), *The importance of what we care about* (pp. 159–176). Cambridge: Cambridge UP.

Haraway, D. (2003). *The companion species: Dogs, people and significant otherness*. Boulder: Paradigm.

Hern, A. (2017). *Give robots 'personhood' status, EU committee argues*. Available via The Guardian. Retrieved February 27, 2020, from https://www.theguardian.com/technology/2017/jan/12/give-robots-personhood-status-eu-committee-argues

James, W. (1890). *Principles of psychology*. London: Macmillan.

Kelsen, H. (1992). *Introduction to the problems of legal theory*. Oxford: Clarendon.

LaGrandeur, K. (2017). Emotion, artificial intelligence, and ethics. In J. Romportal, E. Zackova, & J. Kelemen (Eds.), *Beyond artificial intelligence* (pp. 97–109). Cham: Springer International Publishing.

Lawson, T. (2015). The modern corporation: The site of a mechanism (of global Social change) that is out-of-control. In M. S. Archer (Ed.), *Generative mechanisms transforming the social order* (pp. 205–230). Cham: Springer.

Lazega, E. (2013). Network analysis and morphogenesis: A neo-structural exploration and illustration. In M. S. Archer (Ed.), *Social morphogenesis Series* (pp. 167–185). Cham: Springer.

Lazega, E. (2014). *'Morphogenesis unbound' from dynamics of multi-level networks*. Cham: Springer.

Lazega, E. (2015). *Body captors and network profiles*. Cham: Springer.

Lazega, E. (2016). *Joint anormative regulation from status inconsistency*. Cham: Springer.

Maccarini, A. M. (2019). Post-human sociality: Morphing experience and emergent forms. In I. Al-Amoudi & E. Lazega (Eds.), *Post-human institutions and organizations*. Abingdon: Routledge.

Mańko, R. (2017). *New Laws needed for robots. What Europe is doing for its citizens?* Available via European Parliamentary Research Blog. https://epthinktank.eu/2017/08/14/new-laws-needed-for-robots-what-is-europe-doing-for-its-citizens/

McDowell, J. (1998). *Two sorts of naturalism: Mind, value and reality*. Cambridge, MA: Harvard UP.

McStay, A. (2018). *Emotional AI: The rise of empathetic media*. Los Angeles: Sage.

Morgan, J. (2019). Yesterday's tomorrow today: Turing, Searle and the contested significance of artificial intelligence. In I. Al-Amoudi & J. Morgan (Eds.), *Realist responses to post-human society: Ex Machina*. Abingdon: Routledge.

Ortony, A., Clore, G. L., & Collins, A. (1988). *The cognitive structure of emotions*. Cambridge: Cambridge UP.

Pettit, P., & Schweikard, D. (2006). Joint actions and group agents. *Philosophy of the Social Sciences, 36*(1), 18–39. https://doi.org/10.1177/0048393105284169.

Plummer, L. (2017). *Swarm of micro-drones with 'hive mind' dropped from fighter jets in US military test*. Available via Mirror. https://www.mirror.co.uk/tech/drone-swarm-hive-mind-launched-9597295

Rudder Baker, L. (2000). *Persons and bodies*. Cambridge: Cambridge UP.

Rudder Baker, L. (2013). *Naturalism and the first-person perspective*. Oxford: Oxford UP.

Sheehy, P. (2002). On plural subject theory. *Journal of Social Philosophy, 33*(3), 377–394. https://doi.org/10.1111/0047-2786.00148.

Taylor, C. (1985). *Human agency and language*. Cambridge: Cambridge UP.

Tuomela, R. (2010). *The philosophy of sociality*. Oxford: Oxford UP.

Wallach, W., & Allen, C. (2008). *Moral machines: Teaching robots right from wrong*. Oxford: Oxford University Press.

Robots and Rights: Reviewing Recent Positions in Legal Philosophy and Ethics

Wolfgang M. Schröder

Contents

Abstract

Controversies about the moral and legal status of robots and of humanoid robots in particular are among the top debates in recent practical philosophy and legal theory. As robots become increasingly sophisticated, and engineers make them combine properties of tools with seemingly psychological capacities that were thought to be reserved for humans, such considerations become pressing. While some are inclined to view humanoid robots as more than just tools, discussions are dominated by a clear divide: What some find appealing, others deem appalling, i.e. "robot rights" and "legal personhood" for AI systems. Obviously, we need to organize human–robot interactions according to ethical and juridical principles that optimize benefit and minimize mutual harm. Avoiding disrespectful treatment of robots can help to preserve a normative basic ethical continuum in the behaviour of humans. This insight can contribute to inspire an "overlapping consensus" as conceptualized by John Rawls in further discussions on responsibly coordinating human/robot interactions.

Keywords

Philosophy · Personhood · Rights · Robots · Ethics

W. M. Schröder (✉)
Professur für Philosophie am Institut für Systematische Theologie, Katholisch-Theologische Fakultät, University of Würzburg, Würzburg, Germany
e-mail: wolfgang.schroeder@uni-wuerzburg.de

J. von Braun et al. (eds.), *Robotics, AI, and Humanity*, https://doi.org/10.1007/978-3-030-54173-6_16

Introduction

Robots[1]—as it seems—are here to stay.[2] But with which status, and under what normative conditions? Controversies about the moral and legal status of robots in general, and of humanoid (anthropomorphic) robots in particular, are among the top debates in recent practical philosophy and legal theory (Danaher 2017a; Gunkel 2018; Bryson 2019; Dignum 2019; Basl 2019; Nyholm 2020; Wong and Simon 2020; Andreotta 2020). Quite obviously, the state of the art in robotics and the rapid further development of Artificial Intelligence (AI) raise moral and legal issues that significantly exceed the horizon of classic normative theory building (Behdadi and Munthe 2020). Yet what exactly is the problem?

As robots become increasingly sophisticated, and engineers try harder to make them quasi "sentient" and "conscious" (Ekbia 2008; Torrance 2012), we are faced with AI-embedding systems that are ambivalent by design. They combine properties of tools with seemingly "psychological capacities that we had previously thought were reserved for complex biological organisms such as humans" (Prescott 2017: 142).[3] Hence there is a growing incentive to consider the ontological status of the robots as "liminal": robots seem to be "neither living nor simply mechanical" (Sandini and Sciutti 2018: 7:1). Therefore, it is not surprising that humans show inclinations to treat humanoid robots "as more than just tools", regardless of the "extent to which their machine nature is transparent" (Sandini and Sciutti 2018; see also Rosenthal-von der Pütten et al. 2018). After all, human brains have evolved mainly to understand (and interact with) humans, so

they are likely to be easily "tricked" into interpreting human-like robot behaviour "as if it were generated by a human" (Sandini and Sciutti 2018: 7:1). As a matter of consequence, it is time to come to terms with the question of how "intelligent machines" like robots (especially the humanoid ones) should be categorized and treated in our societies (Walch 2019).

Discussions on that issue have so far been dominated by a clear divide: What some find appealing, others deem appalling: "robot rights" and "legal personhood" ("e-personality") for AI systems.[4] While Luciano Floridi, Director of the Oxford Internet Institute, declassifies thinking and talking about the "counterintuitive attribution of rights" to robots as "distracting and irresponsible" (Floridi 2017: 4; see also Coyne 1999; Winfield 2007), *Forbes Magazine* ranks this topic as one of the most important AI ethics concerns in the early 2020s (Walch 2019). However contradicting opinions may be,[5] the task at stake is undisputed: We need to organize Human–Robot Interaction (HRI) according to ethical and juridical principles that optimize benefit and

---

[1] In absence of a generally accepted definition of what a "robot" is (standards like ISO 8373:2012 only relate to industrial robots), we use the following working definition: "Robots" are computer-controlled machines resembling living creatures by moving independently and performing complex actions (cf. https://www.merriam-webster.com/dictionary/robot; Dignum 2019: 31). As a short introduction into robotics, see Winfield (2012).

[2] According to the latest World Robotic Report (see https://ifr.org/news/summary-outlook-on-world-robotics-report-2019-by-ifr), the number of robot installations has never increased so strongly than from 2010 till present. We are in the midst of a "robot invasion"—with no end in sight (cf. Gunkel 2019). See also Bennett & Aly (2020).

[3] Robotics has progressed from single arm manipulators with motion schemes of limited degrees of freedom to more complex anthropomorphic forms with human motion patterns. Whereas, for security reasons, industrial robots are normally contained in barricaded work cells and are automatically deactivated if approached by a human, humanoid robots seem to blur the boundary between humans and machines. On the one hand, they can be engineered and used as the perfect universal tool, as extensions of industrial robots able to perform menial tasks in the workplace or hazardous work and exploration. On the other, their "human form" suggests that they can have "personhood" and are able to interact with humans. They can utilise devices originally constructed for humans and "inherently suited" to human environments. Thus, humanoid robots seem to be *so much more than just machines*. For future scenarios in robotics, see also Nourbakhsh (2013).

[4] At first glance, it seems that (at least in the European and Western tradition) the sphere of law and rights presents a specifically and exclusively anthropic one (cf. Siewert et al. 2006). In Greek antiquity, Zeus gave humans—and only humans, as opposed to animals—the *nómos* not to eat each other and to use law (*díkē*) instead of force (*bíē*) (Hesiod, Op. 275–279). Coming from the father of the gods, this *nómos* was morally binding for the mortals and corresponds to the "*nomoi* for all of the Hellenes", frequently mentioned in the fifth cent. BC, or to the "unwritten" *nómoi* that regulated moral conduct, e.g. to honor the gods, parents and strangers, to bury the dead and to protect those who suffer injustice [4]. As early as in Homer (who does not use the word *nómos*) the gods controlled whether there was *eunomía* (good order) or *hýbris* (disrespect, arrogance) among mortals (Homer, Od. 17, 487). *Nómos* refers to the proper conduct not only towards fellow human beings but also towards the gods as well, e.g. the obligation to sacrifice (Hesiod, Theog. 417; Hesiod, Fr. 322 M.-W.); thus *nómos* refers to the norms for moral and religious conduct in Greek society. But as Gunther Teubner has pointed out (Teubner 2006), the world of law in medieval and Renaissance Europe and also in other cultures was populated with non-human beings, with ancestors' spirits, gods, trees, holy shrines, intestines, birds' flight, to all those visible and non-visible phenomena to which communication could be presupposed and which included the potential to deceive, to lie, to trickster, and to express something by silence. Today, under the influence of rationalizing science, the number of actors in the legal world has been drastically diminished. After the scientific revolution, after philosophical enlightenment, after methodological individualism dominating the social sciences, after psychological and sociological analysis of purposive action, the only remaining plausible actor is the human individual. The rest is superstition. To be sure, the law still applies the construct of the juridical person to organizations and states. But increasingly, especially under the influence of legal economics, this practice has been devalued as merely an "analogy", a "linguistic abbreviation" of a complex legal relationship between individuals, as a "trap" of corporatist ideologies, at best as a "legal fiction", a superfluous myth, that should be replaced by the nexus model which conceives the organization as a multitude of contracts between individuals.

[5] See as overview Wallach (2007), Wallach (2010), Wallach and Allen (2010), Robertson et al. (2019) and Dubber et al. (2020).

minimize mutual harm (cf. Floridi 2013; Lin et al. 2014; Nemitz 2018; Scharre 2018; Bremner et al. 2019; Brockman 2019; Loh 2019; Schröder et al. 2021).

My paper takes up this topic from a legal ethics perspective and proceeds in three main steps. It begins with definitions of central terms and an exposition of central aspects under which robots (as AI-embedding machines and AI-controlled agents) can become a topic of moral and juridical discourse. Then follows a brief review of some of the most prominent theses from recent literature on the moral and juridical status of robots. In conclusion, a balanced intermediate result and a modest new proposal are presented and substantiated in recognition of the previous discussion. We will start with definitions.

Definitions and Brief Exposition of the Topic

Rights talk is a prominent issue in both moral and legal theory and praxis. Thus, we need to briefly clarify the different basic meanings that the term "rights" has in the respective contexts.

Ethical discourse on moral norms and rights concerns an enormously broad field of problems and investigation. Yet basically it is an effort in theory building on descriptive or normative abbreviatures of commonly acceptable social conduct (Lapinski and Rimal 2005).[6] Even from the perspective of some theories of natural law, *moral* norms and rights are *cultural products* (like values, customs, and traditions); they represent culturally shaped ideas and principles of what a reasonable grammar of practical freedom should look like (Schröder 2012; Bryson et al. 2017; Spiekermann 2019.

Legal rights and norms differ from purely moral ones by their specifically institutional character. Legal rights and norms "exist under the rules of *legal* systems or by virtue of decisions of suitably authoritative bodies within them" (Campbell 2001). Following the standard Hohfeldian account (Hohfeld 1913), rights as applied in juridical reasoning can be broken down into a set of four categories ("the Hohfeldian incidents" in Wenar 2015)[7]:

– Privileges

– Claims
– Powers
– Immunities

In the case of "privileges", you have a liberty or privilege to do as you please within a certain zone of privacy. As to "claims", they mean that others have a duty not to encroach upon you in that zone of privacy. "Powers", in this context, mean that you have the ability to waive your claim-right not to be interfered with in that zone of privacy. And, finally, "immunities" provide for your being legally protected against others trying to waive your claim-right on your behalf.

Obviously, the four above-mentioned categories logically relate to one another: "Saying that you have a privilege to do X typically entails that you have a claim-right against others to stop them from interfering with that privilege" (Danaher 2017b: 1).

The Classical Ontological Stance and the Recent "Relational Turn" in Animal, Robot and Machine Ethics: Mark Coeckelbergh's Analysis

From the point of view of robot and machine ethics, the question of robot rights refers to the kind of relations we have or will develop with robots as AI-embedding machines.[8] According to Mark Coeckelbergh,[9] we need to adapt our practice of moral status ascription to the fact that the number of candidates for moral patients and agents is growing (Coeckelbergh 2011, 2012a; see also Danaher 2019; Birch 1993; Haraway 2008; Hart et al. 2012; Latour 2015). Coeckelbergh criticizes that classificatory thinking in animal and machine ethics is usually one-sidedly property-based: entities are considered in isolation from other entities, thereby reducing ethics to a kind of mechanical thinking.[10] For Coeckelbergh, this raises three major problems: (1) How can we know which property is sufficient, or at least decisive, for ascribing moral status? (2) How could we indubitably establish that an entity

---

[6] First, this discourse makes the distinction between perceived and collective norms and between descriptive and injunctive norms. Second, the article addresses the role of important moderators in the relationship between descriptive norms and behaviours, including outcome expectations, group identity, and ego involvement. Third, it discusses the role of both interpersonal and mass communication in normative influences. Lastly, it outlines behavioral attributes that determine susceptibility to normative influences, including behavioral ambiguity and the public or private nature of the behavior. See also Bicchieri (2006) and Soh & Connolly (2020).

[7] Named after Wesley Hohfeld (1879–1918), the American legal theorist who discovered them. Each of these Hohfeldian incidents has a distinctive logical form, and the incidents fit together in characteristic ways to create complex "molecular" rights.

[8] See as an overview Dubber et al. (2020) and the respective online supplement https://c4ejournal.net/the-oxford-handbook-of-ethics-of-ai-online-companion/. And Anderson and Anderson (2018).

[9] Mark Coeckelbergh is Professor of Philosophy of Media and Technology at the Department of Philosophy, University of Vienna, and member of the High Level Expert Group on Artificial Intelligence for the European Commissions. He is best known for his work in AI Ethics and Ethics of Robotics, yet he has also broadly published in the areas of Moral and Environmental Philosophy.

[10] "From a social-philosophical point of view, this approach is individualist, since moral status is ascribed to entities considered in isolation from other entities—including the observer. [. . .] The modern scientist, who forces nature to reveal herself, is now accompanied by the moral scientist, who forces the entity to reveal its true moral status" (Coeckelbergh 2013a, b: 17).

indeed has a particular property S? And finally (3) how could we define really sharp boundaries between different kinds of entities?[11]

Coeckelbergh's scepticism about property-based ethical classification leads him to plead for an ecological anthropology as proposed by Tim Ingold (Ingold 2000). In Ingold's "Ecology of Materials", all entities are considered as nodes in a field of relationships, which can only be understood in relational, ecological, and developmental or growth terms (Ingold 2012). Following Ingold, Coeckelbergh interprets moral standing as an expression of active, developing relationships between entities. Instead of asking what property P counts for moral standing S, the new question is: "How should we relate to other beings as human beings who are *already part of the same world* as these non-human beings, who *experience* that world and those other beings and are *already engaged* in that world and stand already in relation to that world?" (Coeckelbergh 2013a, b). Thus, Coeckelbergh considers *relations* as basic conditions for moral standing: "The relational approach suggests that we should not assume that there is a kind of moral backpack attached to the entity in question; instead moral consideration is granted within a dynamic relation between humans and the entity under consideration" (Coeckelbergh 2010). On this account, relations are not to be seen as properties, but rather "as a priori given in which we are already engaged, making possible the ascription of moral status to entities" (Swart 2013).

We will return to reviewing this approach more closely when we come to David Gunkel's relation-based theory of "robot rights". But let us first turn to the juridico-legal discourse on the robots-and-rights topic, just to add this perspective to what we have seen in moral philosophy.

The Juridical Perspective: The "Accountability Gap" Implying a "Responsibility Gap"

Following Jack M. Balkin's[12] clear-cut analysis, there are **three key problems** that robotics and AI agents present for law:

– *Firstly*, there is the **problem of how to deal with the emergence of non-human agents in the social worlds of humans**. How should we "distribute rights and responsibilities among human beings when non-human agents create benefits like artistic works or cause harms like physical injuries"? The difficulty here arises from the "fact that

the behavior of robotic and AI systems is 'emergent'; their actions may not be predictable in advance or constrained by human expectations about proper behavior. Moreover, the programming and algorithms used by robots and AI entities may be the work of many hands, and may employ generative technologies that allow innovation at multiple layers. These features of robotics and AI enhance unpredictability and diffusion of causal responsibility for what robots and AI agents do" (Balkin 2015: 46).[13]

– *Secondly*, there is the **problem of the "substitution effect"**. What we already see now will become even clearer in the future: "People will substitute robots and AI agents for living things—and especially for humans. But they will do so only in certain ways and only for certain purposes. In other words, people tend to treat robots and AI agents as special-purpose animals or special-purpose human beings. This substitution is likely to be incomplete, contextual, unstable, and often opportunistic. People may treat the robot as a person (or animal) for some purposes and as an object for others" (Balkin 2015: 46).

– *Thirdly*, we are not dealing here with a static configuration of challenges to the law. Rather, **we are faced with a steadily evolving dynamic field of often disruptive factors**. As Balkin put it: "We should not think of essential characteristics of technology independent of how people use technology in their lives and in their social relations with others. [. . .] Innovation in technology is not just innovation of tools and techniques; it may also involve innovation of economic, social and legal relations. As we innovate socially and economically, what appears most salient and important about our technologies may also change" (Balkin 2015: 48f).

The most obvious legal problem with robots that can potentially harm people's physical integrity or property is an "**accountability gap**" implying a "**responsibility gap**". There are at least two important legal levels at which this issue creates problems: **criminal law** and **civil law** (Keßler 2019).[14]

In *criminal law* only natural persons—in the sense of the law, real, living people—can be held responsible for their actions. At civil law level, on the other hand, legal persons, such as companies, can be included. Robots fall after the current state in neither of the two fields.

---

[11] For further discussion see Swart (2013).

[12] Jack M. Balkin is Knight Professor of Constitutional Law and the First Amendment at Yale Law School. He is the founder and director of Yale's Information Society Project, an interdisciplinary center that studies law and new information technologies.

[13] Balkin also reminds us on Lawrence Lessig's famous dictum that "Code is Law", meaning "that combinations of computer hardware and software, like other modalities of regulation, could constrain and direct human behavior. Robotics and AI present the converse problem. Instead of code as a law that regulates humans, robotics and AI feature emergent behavior that escapes human planning and expectations. Code is lawless" (Balkin 2015: 52).

[14] The following brief exposition relies on the kind advice of my Würzburg colleague Christian Haagen (Forschungsstelle Robotrecht, University of Würzburg).

Manufacturers of robots can be held responsible if they have been proven to use the wrong materials in construction or have made mistakes in programming: for example, if a self-propelled forklift in a high-level warehouse drops a pallet and is clearly related to faulty programming. For this damage then the manufacturer of the truck should be liable. However, proving such errors is likely to become even more difficult in the future.

The robots of the future will learn independently and will make decisions based on their past experiences. Programmers can and must provide a faultless framework for this. However, if the robots are wrong in their experience-based decisions, manufacturers cannot be held so easily accountable. In general, one could say: as long as the manufacturers have acted with all possible and reasonable care and have made no mistakes, they cannot be guilty according to our criminal understanding.

At the *civil law* level, there is still the possibility of liability, which does not depend on fault. The manufacturer could then be prosecuted if the damage is due to a fault on the robot due to the manufacturer. Otherwise, it will be difficult to hold someone liable.

If the robots violate their duties and cause damage, they would be responsible, for example, financially. In principle, robots act independently with their experience-based decisions. They just cannot be punished like humans. For example, imprisonment means nothing to robots. Robots do not have their own assets. Maybe one will fund up, with whose funds the penalties for the machines are paid. Or, ultimately, it is the manufacturers who are responsible for the mistakes of their robots. You need to watch the devices, detect and fix errors. In the worst case, they have to call back the robots.

Recent Juridical Tendencies Towards Advocating Legal Personality for Robots

Some legal scholars like Gunther Teubner[15] argue in favour of granting rights and legal personality to AI systems, depending on the degree of independence that AI systems are bestowed with. Teubner holds that personification of non-humans is "best understood as a strategy of dealing with the uncertainty about the identity of the other, which moves the attribution scheme from causation to double contingency and opens the space for presupposing the others' self-referentiality" (Teubner 2006: 497); hence Teubner does not recognize any "compelling reason to restrict the attribution of action exclusively to humans and to social systems, as

Luhmann argues. Personifying other non-humans is a social reality today and a political necessity for the future. The admission of actors does not take place ... into one and only one collective. Rather, the properties of new actors differ extremely according to the multiplicity of different sites of the political ecology" (Teubner 2006: 497; see also Calerco 2008; Campbell 2011). On Teubner's account, granting legal personality to AI systems would fill the Accountability Gap, thereby maintaining the integrity of the legal system as a whole and advancing the practical interests of humans (Teubner 2006).

Going beyond Teubner's view, US legal scholar Shawn J. Bayern[16] has argued that, under US law of limited liability companies (LLCs), legal personality could be bestowed on any type of autonomous systems. By means of a special transactional technique, legal entities (mainly LLCs) in the US could be governed entirely by autonomous systems or other software, without any ongoing necessary legal oversight or other involvement by human owners or members (Bayern 2019). For the time being, however, it seems clear that any autonomous system would probably "lack the basic acumen necessary to take many business decisions" (Turner 2019: 177). Thus, it seems unclear if Bayern's point of view would be shared by the courts (Turner 2019).

In a recent groundbreaking book called "Robot Rules" (Turner 2019), legal scholar Jacob Turner[17] deals with the Bermuda triangle of AI-related legal problems: Who is responsible for harms as well as for benefits caused by AI? Should AI have rights? And last but not least: How should ethical and juridical rules for AI be set and implemented?

Rather than literally formulating "robot rules" See also Coeckelbergh (2014), Turner's book sets out to "provide a blueprint for institutions and mechanisms capable of fulfilling this role" (Turner 2019: 372). On Turner's account, there are four reasons for protecting the rights of others that could be applied to at least some types of AI and robots:

1. The ability to suffer
2. Compassion
3. The value of something or somebody to others
4. "Situations where humans and AI are combined" (Turner 2019: 145)

Regarding the **"Argument from the Ability to Suffer"**, it is clear that some sort or degree of "artificial consciousness"[18] would be a necessary precondition for a claim to legal protection based on the ability to suffer. "Pain", in this

---

[15]Gunther Teubner has been Professor of Private Law and Legal Sociology at the University of Frankfurt until 2007; in 2011 he has taken up an "ad personam" Jean Monnet Chair at the International University College of Turin.

[16]Shawn J. Bayern is Larry and Joyce Beltz Professor of Torts at the Florida State University College of Law.

[17]Jacob Turner is a barrister at Fountain Court Chambers (UK) after having been judicial assistant to Lord Mance at the UK Supreme Court.

[18]Turner explains: "For an entity to be conscious, it must be capable of (1) sensing stimuli, (2) perceiving sensations and (3) having a sense of

context, could be understood as "just a signal which encourages an entity to avoid something undesirable"; thus defined, it would not be difficult in Turner's eyes to "acknowledge that robots can experience it" (Turner 2019: 152). Turner's conclusion here is that "*if* an AI system was to acquire this quality, *then* it should qualify for some moral rights" (Turner 2019: 146).

The "**Argument from Compassion**" works on the premise that we obviously tend to protect certain entities because (and as far as) we have "an emotional reaction to them being harmed" (Turner 2019: 155). In the case of robots, it may suffice that they look as if they were conscious and had the ability to suffer to trigger our psychological tendency to develop feelings and compassion for them.

The "**Argument from Value to Humanity**" draws on a different observation. It is based on the fact that a whole range of things can be protected by law not because these things have a particular definable use, but "rather for a panoply of cultural, aesthetic and historical reasons" (Turner 2019: 165). These reasons can be deemed to constitute an "inherent value" of the objects at stake. To exemplify this idea, Turner refers to art. 20a of the German Constitution (Grundgesetz) which says: "Mindful also of its responsibility towards future generations, the state shall protect the natural foundations of life and animals".

Last but not least, there is an "**Argument from Posthumanism**" referring to hybrid organisms, Cyborgs and "electronic brains". This argument draws on the fact that in technically enhanced human bodies augmented by AI technology and human minds are not always really "separate". Rather humans and AI seem to combine to become a symbiotic entity—thus developing into something "greater than just the sum of their parts" (Turner 2019: 167). Accordingly, in these cases the strict distinction between what is human and what is artificial may become increasingly fluid or even obsolete (cf. Turner 2019: 169).

With humans augmented by AI, boundary issues can arise: "when, if ever, might a human lose their protected status? [...] What about if 20%, 50% or 80% of their mental functioning was the result of computer processing powers?" (Turner 2019: 168). Probably we could suppose a broad consensus here that augmentation or replacement of human organs and physical functions with artificial substitutes "does not render someone less deserving of rights" (Turner 2019: 168).

While rights in general are social constructions, legal personality is, more specifically, a juridical fiction, created in and through our legal systems. On these grounds, it is up to us to decide to what exactly it should apply and to define its precise content (cf. Turner 2019: 175).

As Turner points out, legal personality—instead of being a single notion—is "a technical label for a bundle of rights and responsibilities" (Turner 2019: 175). It is a juridical "artifice" designed to make sure that "legal people need not possess all the same rights and obligations, even with the same system". In the case that one country grants (or plans to grant) legal personality to AI systems, this could certainly have a domino effect on other nations (cf. Turner 2019: 180).

On the other hand, arguments are brought forward against granting legal personality to AI systems. One draws on the idea of "Android Fallacy", meaning the mistaken conflation of the concept of personality tout court with "humanity" as such (cf. Turner 2019: 189).

Another point of departure for rejecting legal personality for AI systems is the fear that robots with e-personality could be (mis-)used and exploited as liability shields by human actors (or whole companies) for selfish motives.

Furthermore, one could argue that robots should not be given rights or even legal personality because they as themselves are unaccountable rights violators (Turner 2019: 193).

David J. Gunkel's "Robot Rights" (2018)

Unsatisfied by traditional moral theorizing on human–machine relations (Gunkel 2007, 2012), philosopher David J. Gunkel[19] has made a strong case for "robot rights" as different from human rights. The latest and systematically most accomplished versions of his approach are his books "Robot Rights" (Gunkel 2018) and "How to Survive a Robot Invasion: Rights, Responsibility, and AI" (Gunkel 2019). Gunkel's approach is based on a critique of all traditional moral theorizing in which ontological reflection actually precedes the ethical one: First you ask (and try to clarify) what a certain entity "is"; then you can proceed to the ethical question of whether or not this entity can or should be attributed a certain moral value (Gunkel 2018: 159). Gunkel pleas for thinking otherwise. He wants to "deconstruct" the aforementioned "conceptual configuration" and remix the parameters involved.

Gunkel's argument draws heavily on Emmanuel Levinas' assertion that ethics precedes ontology, not the other way around. For Levinas, intersubjective responsibility originates in face-to-face encounters (Lévinas 1990).[20] On this account,

---

self, namely a conception of its own existence in space and time" (Turner 2019: 147).

[19]David J. Gunkel is Presidential Teaching Professor of Communication Studies at Northern Illinois University. Currently he is seen as the most prominent philosophical author on robot rights issues.

[20]"L'éthique, déjà par elle-même, est une 'optique'." Cf. ibid. 215: "La relation avec l'autre en tant que visage guérit de l'allergie. Elle est désir, enseignement reçu et opposition pacifique du discours. [...] Voilà la situation que nous appelons accueil du visage. L'idée de l'infini se produit dans l'opposition du discours, dans la socialité. Le rapport avec le visage avec l'autre absolument autre que je ne saurais contenir, avec l'autre, dans ce sens, infini, est cependant mon Idée, un commerce. Mais

"intersubjective experience proves 'ethical' in the simple sense that an 'I' discovers its own particularity when it is singled out by the gaze of the other. This gaze is interrogative and imperative. It says 'do not kill me'. It also implores the 'I', who eludes it only with difficulty, although this request may have actually no discursive content. This command and supplication occurs because human faces impact us as affective moments or, what Levinas calls 'interruptions'. The face of the other is firstly expressiveness. It could be compared to a force" (Campbell 2001).

On Gunkel's account, this means that "it is the axiological aspect, the *ought* or *should* dimension, that comes first, in terms of both temporal sequence and status, and the ontological aspect (the *is* or *can*) follows from this decision" (Gunkel 2018: 159).

If one follows the thesis of "the underivability of ethics from 'ontology'" (Duncan 2006: 277), encountering others and otherness changes its meaning as regards the sequence of challenges associated with it. On Gunkel's view, "we are initially confronted with a mess of anonymous others who intrude on us and to whom we are obliged to respond even before we know anything at all about them and their inner workings" (Gunkel 2018: 159f). On these grounds, Gunkel advocates proposals to grant robots "with a face" some basic moral rights to be respected by humans in their common social worlds (Gunkel 2018: 171–175).

Gunkel interprets his approach as "applied Levinasian philosophy" (Gunkel 2018: 170), knowing that Levinas never wrote about robots, technology or robotics. As it seems, Gunkel's "applied Levinasian philosophy" is also inspired by Silvia Benso's book on the "face of things" (Benso 2000). Accordingly, Gunkel admits some difficulties arising from the application of Levinas's philosophy to the study of robot rights. He is aware of the fact that Levinas' ethics exclusively concerns relationships between humans and between humans and god. Therefore, as Katarzyna Ginszt notes in her comment on Gunkel's hermeutics, applying Levinas' thought to robots would "require reading Levinas beyond the anthropocentric restrictions of the 'Other' that is presupposed to be a human being" (Ginszt 2019: 30). Moreover, Gunkel would have to make sure that this kind of "broadening the boundaries" could not easily be misunderstood as relativistic but could be clearly recognized as "relational" in Coeckelbergh's sense (Ginszt 2019: 30). Closely read, Gunkel's formulations meet these requirements.

The EPSRC Paper on "Principles of Robotics"

Nevertheless, Gunkel's plea for robot rights is generally perceived as bold proposal and has earned much criticism. In the eyes of many robo-ethicists Gunkel goes too far in his account of the ethical implications of having robots in our society (cf. Bryson 2019). Those who do not want to take the "relational turn" in roboethics insist that "Robots are simply not people. They are pieces of technology" (Boden et al. 2017: 126). From this point of view, it seems indeed counterintuitive to attribute rights to robots. Rather, it is stringent to emphasize that "the responsibility of making sure they behave well must always lie with human beings. Accordingly, rules for real robots in real life must be transformed into rules advising those who design, sell and use robots about how they should act" (Boden et al. 2017: 125; see also Pasquale 2018).

How this responsibility might be practically addressed is outlined in a paper by the British Engineering and Physical Research Council (EPSRC) called "Principles of Robotics. Regulating robots in the real world" (Boden 2011): "For example, one way forward would be a licence and register (just as there is for cars) that records who is responsible for any robot. This might apply to all or only operate where that ownership is not obvious (e.g. for a robot that might roam outside a house or operate in a public institution such as a school or hospital). Alternately, every robot could be released with a searchable online licence which records the name of the designer/manufacturer and the responsible human who acquired it [. . .]. [. . .] Importantly, it should still remain possible for legal liability to be shared or transferred e.g. both designer and user might share fault where a robot malfunctions during use due to a mixture of design problems and user modifications. In such circumstances, legal rules already exist to allocate liability (although we might wish to clarify these, or require insurance). But a register would always allow an aggrieved person a place to start, by finding out who was, on first principles, responsible for the robot in question" (Boden 2011: 128).

Robots in the Japanese *koseki* System: Colin P.A. Jones's Family Law Approach to Robotic Identity and Soft-Law Based Robot Regulation

A completely different, and probably more culture-relative, soft law model for robotic identity and robot registration is based on Japanese family law. This might sound counterintuitive in Euro-American or African contexts where robots are frequently seen as unnatural and threatening. Yet Japan, being in the vanguard of human–robot communication, is different. In Japanese popular culture robots are routinely

---

la relation se maintient sans violence dans la paix avec cette altérité absolue. La 'résistance' de l'Autre ne me fait pas violence, n'agit pas négativement; elle a une structure positive: éthique."

depicted as "an everyday part of the natural world that coexists with humans in familial contexts" (Yamaguchi 2019: 135). Moreover, there is a political dimension to it. Since 2007, with Prime Minster Shinzo Abe's "Innovation 2025" proposal, followed by the "New Robot Strategy" of 2015 and the subsequent blueprint for a super-smart "*Society* 5.0", also the Japanese government has been eagerly promoting "the virtues of a robot-dependent society and lifestyle" (Robertson 2014: 571). In Abe's vision—which is futuristic and nostalgic at the same time—robots can help actualize the historicist cultural conception of "beautiful Japan" (Robertson et al. 2019: 34). Robots are seen as a "dream solution to various social problems, ranging from the country's low birth rate, an insufficient labor force, and a need for foreign migrant workers, to disability, personal safety, and security concerns" (Yamaguchi 2019: 135).

As anthropologist Jennifer Robertson[21] reports, nationwide surveys even suggest "that Japanese citizens are more comfortable sharing living and working environments with robots than with foreign caretakers and migrant workers. As their population continues to shrink and age faster than in other postindustrial nation-states, Japanese are banking on the robotics industry to reinvigorate the economy and to preserve the country's alleged ethnic homogeneity. These initiatives are paralleled by a growing support among some roboticists and politicians to confer citizenship on robots" (Robertson 2014: 571). On this basis, the somewhat odd idea of robots acquiring Japanese civil status became a reality. Japanese family law provided the institutional framework for this.

In Japan, every legally significant transition in a citizen's life—birth, death, marriage, divorce, adoption, even change of gender—is supposed to be registered in a *koseki* (戸籍), the registry of a Japanese household's (*ie* [家]) members. In fact, it is this registration (which is historically a part of the foundation of civil law and government infrastructure in Japan) that gives legal effect to the above-mentioned events. An extract of a citizen's *koseki* serves as the official document that confirms basic details about their Japanese identity and status (Chapman 2008).

Law scholar Colin Jones[22] sees a basic analogy between Japanese family law and institutional requirements for robot regulation. The crucial point of reference is family law's concern with parental liability for the interests, torts and crimes of minors. On Jones' account, many issues of robot law might be "amenable to an approach that sees robots treated analogously to 'perpetual children'. The provisions on parental liability for harm caused by children … might provide as useful a model for allocating responsibility for

robots as anything in products liability or criminal law – if we could just figure out who the 'parents' are" (ibid.: 410).[23]

Obviously, this was not too much of a problem for Japanese authorities in the paradigmatic case of *Paro*,[24] a therapeutic "mental commitment robot" with the body of an artificial baby harp seal manufactured in Nanto City, Japan. On November 7th 2010, as Jennifer Robertson notes, "Paro was granted it's own *koseki*, or household registry, from the mayor of Nanto City, Toyama Prefecture. Shibata Takanori, Paro's inventor, is listed as the robot's father … and a 'birth date' of 17 September 2004 is recorded. Media coverage of Paro's koseki was favorable. […] this prototypical Paro's koseki can be construed as a branch of Shibata's … household, which is located in Nanto City. Thus, the 'special family registry' is for one particular Paro, and not for all of the seal-bots collectively" (Robertson 2014: 590f).

As mentioned earlier, the *koseki* conflates family, nationality and citizenship. In the case of Caro, that means that, by virtue of "having a Japanese father, Paro is entitled to a koseki, which confirms the robot's Japanese citizenship" (Robertson 2014: 591). Oddly, the fact that *Paro* "is a robot—and not even a humanoid—would appear to be less relevant here than the robot's 'ethnic-nationality'" (ibid.). Accordingly, not *Sophia* (a social humanoid robot that became a Saudi Arabian citizen in October 2017) but *Paro* was the first robot ever to be granted citizenship.

The fact that robots can be legally adopted as members of a Japanese household definitely is an inspiration for robot law theory. Colin Jones has outlined what a "Robot Koseki" would look like if it were systematically contoured as a (Japanese) law model for regulating autonomous machines. His approach to "practical robot law" on providing "definitions that can be used … to establish a framework for robotic identity. Hard or soft laws defining what is and is not a robot would be—should be—the starting point for either applying existing rules to those definitions or developing new rules" (Jones 2019: 418).

That means that a "Robot Koseki" would be essentially informational, yet in a robot law perspective. It would start with differentiating registered Robots from unregistered robotic AI systems, the latter thus remaining "robots without capital R" (W.S.). Unregistered AI systems may have many attributes commonly associated with "robots". Yet they "would not be Robots for purposes of the registration system, or the rules and regulations tied to it" (Ibid.). That means that, in order to be eligible for *koseki* registration, "Robots

---

[21] Jennifer Robertson is Professor of Anthropology and the History of Art at the University of Michigan, Ann Arbor.

[22] Colin P.A. Jones is Professor of Law at Doshisha Law School, Kyoto.

[23] As Jones admits, children are "not the only area of family law that may be a useful reference. The field also deals with responsibility for adults with diminished capacity, those judicially declared incompetent or subject to guardianship or conservatorships" (Jones 2019: 411).

[24] Its name comes from the Japanese pronunciation of "personal robot" (*pāsonaru robotto*) (cf. Robertson 2014: 590).

with capital R" (W.S.) would have to meet certain technical and normative criteria, e.g. technical specifications, safety, possible liability nexuses and so forth. In this way a "Robot Koseki" would provide third parties with "assurances" that the registered Robots satisfy certain minimum standards on which "hard and soft law requirements as well as technical rules and regulations" could then be built by governments and private actors (Jones 2019: 453).[25]

Robots registered in the "Robot Koseki" would also be assigned "unique identifying codes or numbers that would become a key part of its identity. Codes identifying members of the same series or production line of robots could also be used. Robot Identification Numbers could even serve as taxpayer identification numbers if the Robot is accorded legal personality and the ability to engage in revenueproducing activities" (Ibid.: 455). In the Internet of Things, the "Robot Koseki" would need to work "in a way so that the current registration details of each Robot were accessible to other technology systems … interacting with it", either in a centralized or distributed database system (ibid.: 464).

As Jones insists, the "Robot Koseki" would not only entail data about machines. Some of the key registration parameters should also "provide information about people involved in the creation and ongoing existence of the Robot, people who through the system will effectively become a part of the Robot's identity", i.e. "the maker (or manufacturer), programmer, owner, and user" of the Robot (ibid.: 465). In Jones' eyes, this is where the Japanese *koseki* system provides "a particularly useful model, since it involves the registration of a single unit (the family) that is comprised of multiple constituents. If we are to develop robot law from family law analogies and attempt to regulate Robots as a form of 'perpetual children', then the koseki system will make it possible to identify who is analogous to their parent(s)" (Ibid.).

This amounts to suggesting a soft law basis for hard law robot regulation. According to Jones, *koseki*-style Robot registry would not immediately call for governmental legislation but could be organized primarily by industry action. The registry could start as "a creature of code, of soft law and technical standards"; thus, it would first be driven by "industry players, professional associations or open standards organization comparable to the Internet Engineering Task Force, which has developed many of the rules and standards governing the technical aspects of the Internet" (Ibid.: 461). Based on these standards, then, both hard and soft law requirements as well as technical rules and regulations could be built by governments and private actors (ibid.: 453).

Overall, Jones' layout of a "Robot Koseki" shows two things: (a) how to solve the problem of "robotic identity" both in a legally compatible and legally effective way without requiring the concept of "robot rights" (in Gunkel's sense) as a basis[26]; (b) how to provide a helpful soft-law basis for hard-law robot regulation.

Yet apart from soft law: Which *ethical* key distinctions should a hard-law robot regulation be based on?

Joanna J. Bryson's Outline of Roboethics

In a recent paper, AI ethicist Joanna J. Bryson[27] (who was one of the most influential co-authors of the aforementioned EPSRC paper) has marked two central normative criteria for integrating AI-embedding systems like robots in our society: coherence and a lack of social disruption (Bryson 2018: 15). Bryson's premises here are that "the core of all ethics is a negotiated or discovered equilibrium that creates and perpetuates a society" and "that integrating a new capacity like artificial intelligence (AI) into our moral systems is an act of normative, not descriptive, ethics" (Bryson 2018: 15). Moreover, Bryson emphasizes that "there is no necessary or predetermined position for AI in our society. This is because both AI and ethical frameworks are artefacts *of* our societies, and therefore subject to human control" (Bryson 2018: 15).

Bryson's main thesis on the ethics of AI-embedding system is: While constructing such systems as either moral agents or patients is indeed possible, neither is desirable. In particular, Bryson argues "that we are unlikely to construct a coherent ethics in which it is ethical to afford AI moral

---

[25]Jones envisages that "Robot Koseki" parameters would include "requirements and specifications such as those relating to: (1) the method the Robot uses to interact with other technology systems (*WiFi,* USB, QR codes, Bluetooth, RFID, etc.); (2) basic safety parameters as to size, speed of motility, etc.; (3) location (e.g. incorporation of GPS; compatibility with geo-fencing systems, etc.); (4) cybersecurity requirements (anti-malware/requirements, etc.); (5) access requirements (i.e. if the Robot Koseki system requires Robots to submit to software updates for various purposes, the Robot will have to be set to accept such updates regularly); (6) privacy protection (e.g. mandatory data encryption and access restrictions for video, voice, and other data recorded by the Robot); (7) operating system; (8) override capability (e.g. a kill switch that can be used remotely to shut the Robot down remotely when necessary in emergency situations); (9) sensory capabilities for perceiving the world (video, sound, motion sensors, facial recognition technology, etc.); and (10) a 'black box' that records all that is happening inside the Robot (software updates, a log of what and how the robot may have 'learned' to do things, etc.), and which can be used for forensic purposes, if necessary. Further mechanisms may be necessary to (for example) address the safety, integrity and rights (or denial) of access to the vast amount of data robots may be able to record and store. Roboticists will doubtless have other suggestions as to what technological parameters should be included.

[26]The extent to which entitlement and defense rights for robots would come with robot citizenship as in the case of *Paro* and *Sophia* needs further discussion.

[27]Joanna J. Bryson was associate professor in the department of computer science at the University of Bath and is now Professor of Ethics and Technology at the Hertie School of Governance, Berlin.

subjectivity. We are therefore obliged not to build AI we are obliged to" (Bryson 2018: 15).

So Bryson's recommendations are as follows: "First, robots should not have deceptive appearance—they should not fool people into thinking they are similar to empathy-deserving moral patients. Second, their AI workings should be 'transparent' [. . .]. This implies that clear, generally-comprehensible descriptions of an artefact's goals and intelligence should be available to any owner, operator, or other concerned party. [. . .]. The goal is that most healthy adult citizens should be able to make correctly-informed decisions about emotional and financial investment. As with fictional characters and plush toys [. . .], we should be able to both experience beneficial emotional engagement, *and* to maintain explicit knowledge of an artefact's lack of moral subjectivity" (Bryson 2018: 23).

In my eyes, Bryson's position is plausible and convincing. Quite obviously, even the most human-like behaving robot will not lose its ontological machine character merely by being open to "humanizing" interpretations. Rather, robots are, and will probably remain, more or less perfect simulations of humans and their agency.

But even if they do not really present an anthropological challenge (cf. Wolfe 1993), they certainly present an ethical one. I endorse the view that both AI and ethical frameworks are artefacts *of* our societies—and therefore subject to human choice and human control (Bryson 2018). The latter holds for the moral status of robots and other AI systems too. This status is in no way logically or ontologically set; rather, it is, and remains, a choice, not a necessity: "We can choose the types and properties of artefacts that are legal to manufacture and sell, and we can write the legislation that determines the legal rights and duties of any agent capable of knowing those rights and carrying out those duties" (Bryson 2018: 16). To this adds that self-disclosing AI would "help people match the right approach to the right entities, treating humans like humans, and machines like machines" (Bowles 2018: 188; Macrorie et al. 2019).

or artificial) would be morally allowed or even encouraged. This might also help to minimize the risk of being morally deskilled by using technology (Coeckelbergh 2012b; Wong 2012; Vallor 2015).

With that in mind, we could consider AI-embedding machines at least as awe-inspiring. Facing them, we encounter the work and latest evolutionary product of our own intelligence: a culmination of human creativity, we are at the cutting edge of human creativity. And we are allowed to be astonished, in the sense of the *thaumázein* in Plato's Theiatetos.[28] Along these lines one could think of a roboethics based upon what we owe to ourselves as creators and users of such sophisticated technology. Avoiding disrespectful treatment of robots is ultimately for the sake of the humans, not for the sake of the robots (Darling and Hauert 2013).

Maybe this insight can contribute to inspire an "overlapping consensus" (Rawls 1993: 133–172) in further discussions on responsibly coordinating HRI. Re- or paraphrasing Rawls in this perspective could start with a three part argument: (a) Rather than being dispensable, it seems reasonable to maintain the aforementioned normative basic ethical continuum in the behaviour of humans (some would add: and their artefacts); accordingly, we should (b) look for ways to stabilize this continuum, facing up (c) to the plurality of reasonable though conflicting ethical frameworks in which robots and rights can be discussed. If mere coordination of diversity is the maximum that seems achievable here, nothing more can be hoped for than a modus vivendi in the AI and roboethics community. Yet there might also be some "common ground" on which to build something more stable. In a Rawlsian type of "overlapping consensus" for instance, diverse and conflicting reasonable frameworks of AI and roboethics would endorse the basic ethical continuum cited above, each from its own point of view. If this configuration works out, both could be achievable: a reasonable non-eliminative hybridity of the ethical discourse on robots and rights—and the intellectual infrastructure for developing this discourse with due diligence.

Conclusion

However, also the relational model sketched by Coeckelbergh and Gunkel 2020 seems to be helpful when it comes to avoiding incoherency and social disruption in ethics systems (see Van Wynsberghe 2012, 2013; van Wynsberghe and Robbins 2014; Wong and Simon 2020) Wong forthcoming. If the claim of ethics is uncompromising in the sense that it concerns action and agency as such (and not only a limited range hereof), one can argue for a *normative basic ethical continuum* in the behaviour of humans, meaning that there should be no context of action where a complete absence of human respect for the integrity of other beings (natural

---

[28]This is in line with how *autómata* were viewed in antiquity (Schürmann 2006). *Autómata* were not used as industrial machinery in production. Rather, they were almost always aimed at creating amazement in the spectators, who could hardly comprehend the mysteriously autonomously moving artefacts they were seeing. Archytas of Tarentum, for instance, is said to have constructed a mechanical dove which looked deceptively real and was perhaps also able to take flight using some form of pneumatic propulsion (Gell. NA 10,12,8–10). In a similar way, an artificial snail created by Demetrius of Phalerum (308 BC; Pol. 12,13,12) and a statue of Nysa seem to have been hugely awe-inspiring; they were displayed in the procession of Ptolemaeus II Philadelphus (*c.* 270 BC; Ath. 5198 s.Co) for the purpose of demonstrating the power of Hellenistic rulers and legitimizing their position in relation to the people.

References

Anderson, M., & Anderson, S. L. (Eds.). (2018). *Machine ethics*. Cambridge: Cambridge UP.

Andreotta, A. J. (2020). The hard problem of AI rights. *AI & Society*. https://doi.org/10.1007/s00146-020-00997-x.

Balkin, J. M. (2015). The path of robotics law. *California Law Review, 6*, 45–60.

Basl, J. (2019). *The death of the ethic of life*. Oxford: Oxford UP.

Bayern, S. J. (2019). Are autonomous entities possible? *Northwestern University Law Review Online, 114*, 23–47.

Behdadi, D., & Munthe, C. (2020). A normative approach to artificial moral agency. *Minds and Machines, 30*, 195. https://doi.org/10.1007/s11023-020-09525-8.

Bennett, B., & Daly, A. (2020). Recognising rights for robots: Can we? Will we? Should we? Law. *Innovation and Technology, 12*(1), 60–80.

Benso, S. (2000). *The face of things: A different side of ethics*. Albany, NY: SUNY Press.

Bicchieri, C. (2006). *The grammar of society: The nature and dynamics of social norms*. Cambridge: Cambridge UP.

Birch, T. H. (1993). Moral considerability and universal consideration. *Environmental Ethics, 15*, 313–332.

Boden, M. (2011). *Principles of robotics*. Available via The United Kingdom's Engineering and Physical Sciences Research Council (EPSRC). Retrieved April 2011, from https://www.epsrc.ac.uk/research/ourportfolio/themes/engineering/activities/principlesofrobotics/

Boden, M., Bryson, J. J., Caldwell, D., Dautenhahn, K., Edwards, L., Kember, S., Newman, P., Parry, V., Pegman, G., Rodden, T., Sorrell, T., Wallis, M., Whitby, B., & Winfield, A. (2017). Principles of robotics: Regulating robots in the real world. *Connection Science, 29*(2), 124–129.

Bowles, C. (2018). *Future ethics*. Hove: Now Text Press.

Bremner, P., Dennis, L. A., Fisher, M., & Winfield, A. F. (2019). On proactive, transparent, and verifiable ethical reasoning for robots. *Proceedings of the IEEE, 107*(3), 541–561.

Brockman, J. (Ed.). (2019). *Possible minds: 25 ways of looking at AI*. New York: Penguin Press.

Bryson, J. J. (2018). Patiency is not a virtue: The design of intelligent systems and systems of ethics. *Ethics and Information Technology, 20*, 15–26.

Bryson, J. J. (2019). *The past decade and future of AI's impact on society*. Available via BBVA. Retrieved from https://www.bbvaopenmind.com/en/articles/the-past-decade-and-future-of-ais-impact-on-society/

Bryson, J. J., Diamantis, M. E., & Grant, T. D. (2017). Of, for, and by the people: The legal lacuna of synthetic persons. *Artificial Intelligence and Law, 25*(3), 273–291.

Calerco, M. (2008). *Zoographies: The question of the animal from Heidegger to Derrida*. New York: Columbia UP.

Campbell, K. (2001). Legal rights. In E. N. Zalta (Ed.), *The stanford encyclopedia of philosophy* (Winter 2017 Edition). Available via Stanford Encyclopedia of Philosophy. Retrieved February 19, 2020, from https://plato.stanford.edu/entries/legal-rights/

Campbell, T. C. (2011). *Improper life: Technology and biopolitics from Heidegger to Agamben*. Minneapolis: University of Minnesota Press.

Chapman, D. (2008). Sealing Japanese identity. *Critical Asian Studies, 40*(3), 423–443.

Coeckelbergh, M. (2010). Robot rights? Towards a social-relational justification of moral consideration. *Ethics and Information Technology, 12*(3), 209–221.

Coeckelbergh, M. (2011). Is ethics of robotics about robots? Philosophy of robotics beyond realism and individualism. *Law, Innovation and Technology, 3*(2), 241–250.

Coeckelbergh, M. (2012a). *Growing moral relations: Critique of moral status ascription*. Basingstoke, NY: Palgrave Macmillan.

Coeckelbergh, M. (2012b). Technology as skill and activity: Revisiting the problem of alienation. *Techne, 16*(3), 208–230.

Coeckelbergh, M. (2013a). David J. Gunkel: The machine question: Critical perspectives on AI, robots, and ethics. *Ethics and Information Technology, 15*, 235–238.

Coeckelbergh, M. (2013b). *Human being @ risk: Enhancement, technology, and the evaluation of vulnerability transformations. Philosophy of engineering and technology* (Vol. 12). Dordrecht: Springer.

Coeckelbergh, M. (2014). The moral standing of machines: Towards a relational and non-Cartesian moral hermeneutics. *Philosophy & Technology, 27*(1), 61–77. https://doi.org/10.1007/s13347-013-0133-8.

Coyne, R. (1999). *Technoromanticism: Digital narrative, holism, and the romance of the real*. Cambridge: MIT Press.

Danaher, J. (2017a). *Robot sex: Social and ethical implications*. Cambridge: MIT Press.

Danaher, J. (2017b). *Should robots have rights? Four perspectives*. Available via Philosophical Disquisitions. Retrieved from https://philosophicaldisquisitions.blogspot.com/2017/10/should-robots-have-rights-four.html

Danaher, J. (2019). *Automation and utopia: Human flourishing in a world without work*. Cambridge: Harvard UP.

Darling, K., & Hauert, S. (2013). *Giving rights to robots*. Available via Robohub. Retrieved from http://robohub.org./robots-giving-rights-to-robots. Accessed.

Dignum, V. (2019). *Responsible artificial intelligence: How to develop and use AI in a responsible way*. Cham: Springer Nature.

Dubber, M., Pasquale, F., & Das, S. (Eds.). (2020). *Oxford handbook of artificial intelligence*. Oxford: Oxford UP.

Duncan, R. (2006). Emmanuel Levinas: Non-intentional consciousness and the status of representational thinking. In A.-T. Tymieniecka (Ed.), *Logos of phenomenology and phenomenology of the logos, Book 3. Analecta Husserliana: The yearbook of phenomenological research* (Vol. 90, pp. 271–281). Dordrecht: Springer.

Ekbia, H. R. (2008). *Artificial dreams: The quest for non-biological intelligence*. New York: Cambridge UP.

Floridi, L. (2013). *The ethics of information*. Oxford: Oxford UP.

Floridi, L. (2017). Robots, jobs, taxes, and responsibilities. *Philosophy & Technology, 30*(1), 1–4.

Ginszt, K. (2019). The status of robots in moral and legal systems: Review of David J. Gunkel (2018). Robot rights. Cambridge, MA: MIT Press. *Ethics in Progress, 10*(2), 27–32. https://doi.org/10.14746/eip.2019.2.3.

Gunkel, D. J. (2007). *Thinking otherwise: Philosophy, communication, technology*. West Lafayette: Purdue UP.

Gunkel, D. J. (2012). *The machine question: Critical perspectives on AI, robots, and ethics*. Cambridge: MIT Press.

Gunkel, D. J. (2018). *Robot rights*. Cambridge: MIT Press.

Gunkel, D. J. (2019). *How to survive a robot invasion: Rights, responsibility, and AI*. London: Routledge.

Gunkel, D. J. (2020). *An introduction to communication and artificial intelligence*. Cambridge: Wiley Polity.

Haraway, D. J. (2008). *When species meet*. Minneapolis: University of Minnesota Press.

Hart, E., Timmis, J., Mitchell, P., Nakano, T., & Dabiri, F. (Eds.). (2012). *Bio-inspired models of network, information, and computing systems*. Heidelberg/Dordrecht/London/New York: Springer.

Hohfeld, W. N. (1913). Some fundamental legal conceptions as applied in juridical reasoning. *The Yale Law Journal, 23*, 16–59.

Ingold, T. (2000). *The perception of the environment: Essays on livelihood, dwelling and skill*. London: Routledge.

Ingold, T. (2012). Toward an ecology of materials. *Annual Review of Anthropology, 41*, 427–442.

Jones, J.P. (2019). The Robot Koseki: A Japanese Law Model for Regulating Autonomous Machines. *J. Bus. & Tech. L., 14*, 403–467.

Keßler, F. (2019). *Wie verklage ich einen Roboter?* Available via Spiegel. Retrieved February 17, 2020, from https://www.spiegel.de/karriere/kuenstliche-intelligenz-so-koennten-roboter-haften-wenn-sie-fehler-machen-a-1263974.html

Lapinski, M. K., & Rimal, R. N. (2005). An explication of social norms. *Communication Theory, 15*(2), 127–147.

Latour, B. (2015). *Face à Gaia: Huit Conférences sur le Nouveau Régime Climatique.* Paris: Éditions La Découverte.

Lévinas, E. (1990). *Lévinas.* Giessen: Focus.

Lin, P., Jenkins, R. K., & Bekey, G. A. (Eds.). (2014). *Robot ethics 2.0: From autonomous cars to artificial intelligence.* Cambridge: MIT Press.

Loh, J. (2019). *Roboterethik: Eine Einführung.* Berlin: Suhrkamp.

Macrorie, R., Marvin, S., & While, A. (2019). Robotics and automation in the city: A research agenda. *Urban Geography.* https://doi.org/10.1080/02723638.2019.1698868.

Nemitz, P. (2018). Constitutional democracy and technology in the age of artificial intelligence. *Philosophical Transactions of the Royal Society A: Mathematical, Physical and Engineering Sciences, 376*(2133), 20180089. https://doi.org/10.1098/rsta.2018.0089.

Nourbakhsh, I. (2013). *Robot futures.* Cambridge: MIT Press.

Nyholm, S. (2020). *Humans and robots: Ethics, agency, and anthropomorphism.* London: Rowman & Littlefield International.

Pasquale, F. (2018). A rule of persons, not machines: The limits of legal automation. *The George Washington Law Review, 87*(1), 1–55.

Prescott, T. J. (2017). Robots are not just tools. *Connection Science, 29*(2), 142–149.

Rawls, J. (1993). *Political liberalism.* New York: Columbia University Press.

Robertson, J. (2014). Human rights vs. robot rights: Forecasts from Japan. *Critical Asian Studies, 46*(4), 571–598.

Robertson, L. J., Abbas, R., Alici, G., Munoz, A., & Michael, K. (2019). Engineering-based design methodology for embedding ethics in autonomous robots. *Proceedings of the IEEE, 107*(3), 582–599.

Rosenthal-von der Pütten, A. M., Krämer, N. C., & Herrmann, J. (2018). The effects of humanlike and robot-specific affective nonverbal behavior on perception, emotion, and behavior. *International Journal of Social Robotics, 10*, 569–582.

Sandini, G., & Sciutti, A. (2018). Humane robots—From robots with a humanoid body to robots with an anthropomorphic mind. *ACM Transactions on Human-Robot Interaction, 7*(1), 1–7. https://doi.org/10.1145/3208954.

Scharre, P. (2018). *Army of none: Autonomous weapons and the future of war.* New York: W.W. Norton.

Schröder, W. M. (2012). Natur- und Vernunftrecht. In G. Lohmann & A. Pollmann (Eds.), *Menschenrechte: Ein interdisziplinäres Handbuch* (pp. 179–185). Stuttgart: Metzler Verlag.

Schröder, W. M., Gollmer, K. U., Schmidt, M., & Wartha, M. (2021). *Kompass Künstliche Intelligenz: Ein Plädoyer für einen aufgeklärten Umgang.* Würzburg: Wuerzburg University Press.

Schürmann, A. (2006). Automata. In H. Cancik & H. Schneider (Eds.), *Brill's new Pauly.* Available via BRILL. https://doi.org/10.1163/1574-9347_bnp_e210220.

Siewert, P., Ameling, W., Jansen-Winkeln, K., Robbins, E., & Klose, D. (2006). Nomos. In: H. Cancik & H. Schneider (Eds.), *Brill's new Pauly.* Available via BRILL. https://doi.org/10.1163/1574-9347_bnp_e210220.

Soh, C., & Connolly, D. (2020). New frontiers of profit and risk: The fourth industrial Revolution's impact on business and human rights. *New Political Economy.* https://doi.org/10.1080/13563467.2020.1723514.

Spiekermann, S. (2019). *Digitale Ethik: Ein Wertesystem für das 21.* Droemer & Knaur, München: Jahrhundert.

Swart, J. A. A. (2013). Growing moral relations: Critique of moral status ascription. *Journal of Agricultural and Environmental Ethics, 26*(6), 1241–1245.

Teubner, G. (2006). Rights of non-humans? Electronic agents and animals as new actors in politics and law. *Journal of Law and Society, 33*(4), 497–521.

Torrance, S. (2012). *The centrality of machine consciousness to machine ethics.* Paper presented at the symposium 'The machine question: AI, ethics, and moral responsibility', AISB/IACAP world congress 2012, Birmingham, 4 July 2012.

Turner, J. (2019). *Robot rules: Regulating artificial intelligence.* Cham: Palgrave Macmillan.

Vallor, S. (2015). Moral deskilling and upskilling in a new machine age: Reflections on the ambiguous future of character. *Philosophy & Technology, 28*, 107–124.

van Wynsberghe, A. (2012). Designing robots for care: Care centered value-sensitive design. *Science and Engineering Ethics, 19*(2), 407–433.

van Wynsberghe, A. (2013). A method for integrating ethics into the design of robots. *Industrial Robot, 40*(5), 433–440.

van Wynsberghe, A., & Robbins, S. (2014). Ethicist as designer: A pragmatic approach to ethics in the lab. *Science and Engineering Ethics, 20*(4), 947–961.

Walch, K. (2019). *Ethical concerns of AI.* Available via Forbes. Retrieved February 19, 2020, from https://www.forbes.com/sites/cognitiveworld/2020/12/29/ethical-concerns-of-ai/#10a9affd23a8

Wallach, W. (2007). Implementing moral decision-making faculties in computers and robots. *AI & Society, 22*(4), 463–475.

Wallach, W. (2010). Robot minds and human ethics: The need for a comprehensive model of moral decision-making. *Ethics and Information Technology, 12*(3), 243–250.

Wallach, W., & Allen, C. (2010). *Moral machines: Teaching robots right from wrong.* New York: Oxford UP.

Wenar, L. (2015). Rights. In E. N. Zalta (Ed.), *The Stanford encyclopedia of philosophy* (Fall 2015 Edition). Available via Stanford Encyclopedia of Philosophy. Retrieved from https://plato.stanford.edu/archives/fall2015/entries/rights/. Accessed. 9/25/20.

Winfield, A. (2007). *The rights of robot.* Available via Alan Winfield's Web Log. Retrieved from https://alanwinfield.blogspot.com/search?q=rights+of+robots. Accessed. 9/25/20.

Winfield, A. (2012). *Robotics: A very short introduction.* Oxford: Oxford UP.

Wolfe, A. (1993). *The human difference: Animals, computers, and the necessity of social science.* Berkeley/Los Angeles/London: University of California Press.

Wong, P. H. (2012). Dao, harmony and personhood: Towards a Confucian ethics of technology. *Philosophy & Technology, 25*(1), 67–86.

Wong, P. H. (forthcoming). *Global engineering ethics.* In D. Michelfelder & N. Doorn (Eds.), *Routledge handbook of philosophy of engineering.* Wong 2020 is still in press; it has not yet been published.

Wong, P. H., & Simon, J. (2020). Thinking about 'ethics' in the ethics of AI. *IDEES, 48.*

Yamaguchi, T. (2019) Japan's robotic future. *Critical Asian Studies, 51*(1), 134–140.

Human–Robot Interactions and Affective Computing: The Ethical Implications

Laurence Devillers

Contents

Abstract

The field of social robotics is fast developing and will have wide implications especially within health care, where much progress has been made towards the development of "companion robots." Such robots provide therapeutic or monitoring assistance to patients with a range of disabilities over a long timeframe. Preliminary results show that such robots may be particularly beneficial for use with individuals who suffer from neurodegenerative pathologies. Treatment can be accorded around the clock and with a level of patience rarely found among human healthcare workers. Several elements are requisite for the effective deployment of companion robots. They must be able to detect human emotions and in turn mimic human emotional reactions as well as having an outward appearance that corresponds to human expectations about their caregiving role. This chapter presents laboratory findings on AI-systems that enable robots to recognize specific emotions and to adapt their behavior accordingly. Emotional perception by humans (how language and gestures are interpreted by us to grasp the emotional states of others) is being studied as a guide to programming robots so they can simulate emotions in their interactions with humans.

L. Devillers (✉)
Sorbonne University, CNRS-LIMSI, Paris-Saclay, Paris, France
e-mail: devil@limsi.fr

Keywords

Companions · Emotions · Health care · Pathologies · Robotics

Introduction

Since the early studies of human behavior, emotion has attracted the interest of researchers in many disciplines of neurosciences and psychology. Recent advances in neurosciences are highlighting connections between emotion, social functioning, and decision-making that have the potential to revolutionize our understanding of the role of affect.

Cognitive neuroscience has provided us with new keys to understand human behavior, new techniques (such as neuroimaging), and a theoretical framework for their evalua-

J. von Braun et al. (eds.), *Robotics, AI, and Humanity*, https://doi.org/10.1007/978-3-030-54173-6_17

tion. The American neuroscientist A. Damasio has suggested that emotions play an essential role in important areas such as learning, memory, motivation, attention, creativity, and decision-making (Damasio 1994, 1999, 2003).

More recently, cognitive neuroscience has become a growing field of research in computer science and machine learning. *Affective Computing* aims at the study and development of systems and devices that use emotion, in particular, in human–computer and human–robot interaction. It is an interdisciplinary field spanning computer science, psychology, and cognitive science. The *affective computing* field of research is related to, arises from, or deliberately influences emotion or other affective phenomena (Picard 1997). The three main technologies are emotion detection and interpretation, dialog-reasoning using emotional information, and emotion generation and synthesis.

An affective chatbot or robot is an autonomous system that interacts with humans using affective technologies to detect emotions, decide, and simulate affective answers. It can have an autonomous natural language processing system with at least these components: signal analysis and automatic speech recognition, semantic analysis and dialog policies, response generation and speech synthesis. The agent can be just a voice assistant, a 2D or 3D on-screen synthetic character, or a physically embodied robot. Such artifact has several types of AI modules to develop perceptive, decision-making, and reactive capabilities in real environment for a robot or in virtual world for synthetic character. Affective robots and chatbots bring a new dimension to interaction and could become a means of influencing individuals. The robot can succeed in a difficult task and will not be proud of it, unless a designer has programmed it to simulate an emotional state. The robot is a complex object, which can simulate cognitive abilities but without human feelings, nor that desire or "*appetite for life*" that Spinoza talks as *conatus* (effort to persevere in being) which refers to everything from the mind to the body. Attempts to create machines that behave intelligently often conceptualize intelligence as the ability to achieve goals, leaving unanswered a crucial question: whose goals?

In 2060, 32% of the French population will be over 60 years old, an increase of 80% in 50 years. The burden of dependency and chronic diseases will go hand in hand with this aging. Robots could be very useful in following patients throughout an illness and helping sick, elderly and/or disabled people to stay at home and reduce their periods of hospitalization. Robots are available 24 h a day, 7 days a week; they are patient, can overcome perceptual deficiencies (deafness, visual impairment), and provide access to information on the internet more easily than a computer. They are also valuable for continuously recording data and sending them to a doctor to detect abnormal behaviors (depression, stress, etc.) and to follow patients with, for example, bipolar

disorders, Holter's disease, and degenerative diseases in the support of daily life. It is already possible to design intelligent systems to train people with cognitive impairment to stimulate memory and language.

Social and emotional robotics wants to create companion robots, which are supposed to provide us with therapeutic assistance or even monitoring assistance. So, it is necessary to learn how to use these new tools without fear and to understand their usefulness. In the case of neurodegenerative pathologies or severe disabilities, the robot may even be better than humans at interacting with people. The machine is in tune with the other, at very slow, almost inhuman rhythms. The robot listens with kindness and without any impatience. For very lonely people, the machine can also help them avoid depressions that lead to dementia.

We need to demystify the artificial intelligence, elaborate ethical rules, and put the values of the human being back at the center of the design of these robotic systems.

Artificial Intelligence and Robotics

Artificial intelligence and robotics open up important opportunities in the field of numerous applications such as, for example, health diagnosis and treatment support with the aim of better patient follow-up.

In 2016, AlphaGo's victory (an artificial-intelligence computer program designed by Google DeepMind) over one of the best go players Lee Sedol raised questions about the promise and risks of using intelligent machines. However, this feat, which follows Deep Blue's 20-year-ago victory over Garry Kasparov, should not lead us to fantasize about what robots will be capable of tomorrow in our daily lives. When AlphaGo beats the go player, the machine does not realize what she's doing. Despite the AI's impressive performances on specific tasks, it is necessary to keep in mind that machine learning systems cannot learn beyond the "real data." They only use the past data to predict the future. However, many of the discoveries of our greatest scientists are due to the ability to be counter-intuitive, that is, to ignore the current knowledge! Galileo in the sixteenth century had the intuition that the weight of an object had no influence on its speed of fall. The serendipity, the "gift of finding" at random, is also not the strength of the machine. Faced with a question without a known answer, the human being with all their cognitive biases and imperfections is incredibly stronger than the machine to imagine solutions.

The robotics community is actively creating affective companion robots with the goal of cultivating a lifelong relationship between a human being and an artifact. Enabling autistic children to socialize, helping children at school, encouraging patients to take medications, and protecting the elderly within a living space are only a few samples of how they could interact with humans. Their seemingly boundless potential stems in part from the fact that they can be physically instantiated, i.e., they are embodied in the real world, unlike many other devices.

Social robots will share our space, live in our homes, help us in our work and daily life, and also share a certain story with us. Why not give them some machine humor? Humor plays a crucial role in social relationships; it dampens stress, builds confidence, and creates complicity between people. If you are alone and unhappy, the robot could joke to comfort you; if you are angry, it could help you to put things into perspective, saying that the situation is not so bad. It could also be self-deprecating if it makes mistakes and realizes it!

At Limsi-CNRS, we are working to give robots the ability to recognize emotions and be empathetic, so that they can best help their users. We teach them to dialogue and analyze emotions using verbal and nonverbal cues (acoustic cues, laughter, for example) in order to adapt their responses (Devillers et al. 2014, 2015). How are these "empathetic" robots welcomed? To find out, it is important to conduct perceptual studies on human–machine interaction. Limsi-CNRS has conducted numerous laboratory and Ehpad tests with elderly people, or in rehabilitation centers with the association Approche,[1] as part of the BPI ROMEO2 project, led by Softbank robotics. Created in 1991, the main mission of the association Approche is to promote new technologies (robotics, electronics, home automation, information and communication technologies, etc.) for the benefit of people in a situation of disability regardless of age and living environment. We are exploring how the expression of emotion is perceived by listeners and how to represent and automatically detect a subject's emotional state in speech (Devillers et al. 2005) but also how to simulate emotion answers with a chatbot or robot. Furthermore, in a real-life context, we often have mixtures of emotions (Devillers et al. 2005). We also conducted studies around scenarios of everyday life and games with Professor Anne-Sophie Rigaud's team at the Living Lab of Broca Hospital. All these experiments have shown that robots are quite well-accepted by patients when they have time to experiment with them. Post-experimental discussions also raised a number of legitimate concerns about the lack of transparency and explanation of the behavior of these machines.

Developing an interdisciplinary research discipline with computer scientists, doctors, and cognitive psychologists to study the effects of coevolution with these machines in a long-term way is urgent. The machine will learn to adapt to us, but how will we adapt to it?

The Intelligence and Consciousness of Robots

Machines will be increasingly autonomous, talkative, and emotionally gifted through sophisticated artificial-intelligence programs. Intelligence is often described as the ability to learn to adapt to the environment or, on the contrary, to modify the environment to adapt it to one's own needs. Children learn by experiencing the world.

A robot is a platform that embeds a large amount of software with various algorithmic approaches in perception, decision, and action in our environment. Even if each of the object-perception or face-recognition modules is driven by machine learning algorithms, automation of all modules is very complex to adjust. To give the robot the ability to learn autonomously from its environment, reinforcement algorithms that require humans to design reward metrics are used. The robot learns by trial and error according to the programmed rewards, in a laborious way, it combines actions in the world and internal representations to achieve the particular tasks for which it is designed.

The integration of intentionality and human-like creativity is a new area of research. These machines are called "intelligent" because they can also learn. For a robot, the task is extremely difficult because it has neither instinct nor intentions to make decisions. It can only imitate human being. Giving a robot the ability to learn in interaction with the environment and humans, is the Holy Grail of artificial-intelligence researchers. It is therefore desirable to teach them the common values of life in society. The ability to learn alone constitutes a technological and legal breakthrough and raises many ethical questions. These robots can be, in a way, creative and autonomous in their decision-making, if they are programmed for this. Indeed, according to the American neuroscientist A. Damasio (2003), self-awareness comes from the pleasant or unpleasant feelings generated by the state of homeostasis (mechanisms aimed at the preservation of the individual) of the body. "Consciousness" is a polysemic term; for some, it refers to self-awareness, for others to the consciousness of others, or to phenomenal consciousness, moral consciousness, etc. To be conscious, you need a perception of your body and feelings.

The robots would need an artificial body with homeostatic characteristics "similar to ours" to be conscious. The goal of researchers such as K. Man and A. Damasio is to test the conditions that would potentially allow machines to care about what they do or "think" (Man and Damasio 2019). Machines capable of implementing a process resembling homeostasis is possible using soft robotics and multisensory abstraction. Homeostatic robots might reap behavioral benefits by acting as if they have feelings. Even if they would never achieve full-blown inner experience in the human sense, their properly motivated behavior would result in expanded intelligence and better-behaved autonomy.

---

[1] http://www.approche-asso.com

The initial goal of the introduction of physical vulnerability and self-determined self-regulation is not to create robots with authentic feeling, but rather to improve their functionality across a wide range of environments. As a second goal, introducing this new class of machines would constitute a scientific platform for experimentation on robotic brain–body architectures. This platform would open the possibility of investigating important research questions such as "*To what extent is the appearance of feeling and consciousness dependent on a material substrate?*"

With a materialistic conception of life, we can consider that the computer and the human brain are comparable systems, capable of manipulating information. There is a massively parallel interconnected network of 10^{11} neurons (100 billion) in our brain and their connections are not as simple as deep learning. For the moment, we are far from the complexity of life! Experiments conducted in Neurospin by Stanislas Dehaene's team (chapter "Foundations of Artificial Intelligence and Effective Universal Induction" in this volume), particularly using subliminal images, have shown that our brain functions mainly in an unconscious mode. Routine actions, the recognition of faces, words, for example, are carried out without recourse to consciousness. In order to access consciousness, the human brain sets up two types of information processing: a first level, called "global availability," which corresponds to the vast repertoire of information, modular programs that can be convened at any time to use them; and a second type of information processing, specific to the human consciousness: self-monitoring or self-evaluation, i.e., the ability to process information about oneself, which can also be called metacognition. Thus, the brain is able to introspect, control its own process, and obtain information about itself, which leads to autonomy. The addition of physical vulnerability opens the robot's behavior to new reward function in reinforcement learning (RL).

The research challenge is to build autonomous machines able to learn just by observing the world. For a digital system, autonomy "is the capacity to operate independently from a human operator or from another machine by exhibiting nontrivial behavior in a complex and changing environment" (Grinbaum et al. 2017).[2] In April 2016, Microsoft's Tay chatbot, which had the capacity to learn continuously from its interactions with web users, started racist language after just 24 h online. Microsoft quickly withdrew Tay. Affective computing and curiosity models will be among the next big research topics. Self-supervised learning systems will extract and use the naturally available relevant context, emotional information, and embedded metadata as supervisory signals.

Researchers such as A. Bair (MIT lab) created an "Intrinsic Curiosity Model," a self-supervised reinforcement learning system.

How can we assess a system that learns? What decisions can and cannot be delegated to a machine learning system? What information should be given to users on the capacities of machine learning systems? Who is responsible if the machine disfunctions: the designer, the owner of the data, the owner of the system, its user, or perhaps the system itself?

Anthropomorphism

The imagination of the citizens about robotics and more generally artificial intelligence are mainly founded on science-fiction and myths (Devillers 2017). To mitigate fantasies that mainly underline gloomy consequences, it is important to demystify the affective computing, robotics, and globally speaking AI science. For example, the expressions used by experts, such as "the robots understand emotions" and "the robots will have a consciousness" (Devillers 2020), are not understood as metaphors by those outside the technical research community. The citizens are still not ready to understand the concepts behind these complex AI machines. These emerging interactive and adaptive systems using emotions modify how we will socialize with machines and with humans. These areas inspire critical questions centering on the ethics, the goals, and the deployment of innovative products that can change our lives and society.[3]

Anthropomorphism is the attribution of human traits, moods, emotions, or intentions to nonhuman entities. It is considered to be an innate tendency of human psychology. It is clear that the multiple forms of the voice assistants and affective robots already in existence and in the process of being designed will have a profound impact on human life and on human–machine coadaptation. Human–machine coadaptation is related to how AI is used today to affect people autonomy (in decision, perception, attention, memorization, ...) by nudging and manipulating them. What will be the power of manipulation of the voices of these machines? What responsibility is delegated to the creators of these chatbots/robots?

Systems have become increasingly capable of mimicking human behavior through research in affective computing. These systems have provided demonstrated utility for interactions with vulnerable populations (e.g., the elderly, children with autism). The behavior of human beings is shaped by several factors, many of which might not be consciously detected. Marketers are aware of this dimension of human psychology as they employ a broad array of tactics to en-

---

[2]A Grinbaum et al. (2017) and same in Research Ethics Board of Allistene, the Digital Sciences and Technologies Alliance: Research Ethics in Machine Learning, CERNA Report, February 2018, p. 24. http://cerna-ethics-allistene.org/digitalAssets/54/54730 cerna 2017 machine learning.pdf

[3]"Ethically aligned design v2—IEEE standard dev. document," http://standards.ieee.org/develop/indconn/ec/eadv2.pdf

courage audiences towards a preferred behavior. Jokinen and Wilcock (2017) argue that a main question in social robotics evaluation is what kind of impact the social robot's appearance has on the user, and if the robot must have a physical embodiment. The Uncanny Valley phenomenon is often cited to show the paradox of increased human likeness and a sudden drop in acceptance. An explanation of this kind of physical or emotional discomfort is based on the perceptual tension that arises from conflicting perceptual cues. When familiar characteristics of the robot are combined with mismatched expectations of its behavior, the distortion in the category boundary manifests itself as perceptual tension and feelings of creepiness (Jokinen and Wilcock 2017). A solution to avoid the uncanny valley experience might be to match the system's general appearance (robot-like voice, cartoon-like appearance) with its abilities. This can prevent users from expecting behavior that they will not "see" (Jokinen and Wilcock 2017).

Alternatively, users can be exposed to creatures that fall in the uncanny valley (e.g., Geminoids), making the public more used to them. Humans tend to feel greater empathy towards creatures that resemble them, so if the agent can evoke feelings of empathy in the user towards itself, it can enhance the user's natural feeling about the interaction and therefore make communication more effective. Following the reasoning on perceptual categorization, the robot's appearance as a pleasantly familiar artificial agent and its being perceived as a listening and understanding companion to the user can establish a whole new category for social robots which, in terms of affection and trust, supports natural interaction between the user and the robot.

Nudges with Affective Robots

The winner of the Nobel Prize in economics, the American Richard Thaler, highlighted in 2008 the concept of nudge, a technique that consists in encouraging individuals to change their behavior without constraining them, by using their cognitive biases. The behavior of human beings is shaped by numerous factors, many of which might not be consciously detected. Thaler and Sunstein (2008) advocate *"libertarian paternalism"*, which they see as being a form of weak paternalism. From their perspective, *"Libertarian Paternalism is a relatively weak, soft, and non-intrusive type of paternalism because choices are not blocked, fenced off, or significantly burdened"* (Thaler and Sunstein 2008, p.175, note 27). Numerous types of systems are already beginning to use nudge policies (e.g., Carrot, Canada, for health). Assuming for the time being that nudging humans for their own betterment is acceptable in at least some circumstances, then the next logical step is to examine what form these nudges may take. An important distinction to draw attention to is between

"positive" and "negative" nudges (sludges) and whether one or both types could be considered ethically acceptable.[4]

The LIMSI team in cooperation with a behavioral economist team in France in the Chair AI HUMAAINE HUman-MAchine Affective spoken INteraction and Ethics au CNRS (2019–2024) will set up experiments with a robot capable of nudges with several types of more or less vulnerable population (children, elderly) to develop nudge assessment tools to show the impact (Project BAD NUDGE BAD ROBOT (2018) (Dataia 2020)). The principal focus of this project is to generate discussion about the ethical acceptability of allowing designers to construct companion robots that nudge a user in a particular behavioral direction for different purposes. At the laboratory scale, then in the field, the two teams will study whether fragile people are more sensitive to nudges or not. This research is innovative, it is important to understand the impact of these new tools in the society and to bring this subject on ethics and manipulation by machines internationally (IEEE 2017). The objects will address us by talking to us. It is necessary to better understand the relationship to these chatty objects without awareness, without emotions, and without proper intentions. Users today are not aware of how these systems work, they tend to anthropomorphize them. Designers need to avoid these confusions between life and artifacts to give more transparency and explanation on the capabilities of machines (Grinbaum et al. 2017).[5]

Social roboticists are making use of empirical findings from sociologists, psychologists, and others to decide their spoken interaction designs, and effectively create conversational robots that elicit strong reactions from users. From a technical perspective, it is clearly feasible that robots could be encoded to shape, at least to some degree, a human companion's behavior by using verbal and nonverbal cues. But is it ethically appropriate to deliberately design nudging behavior in a robot?

Ethical Implications

We must avoid a lack of trust but also too blind a trust in artificial-intelligence programs. A number of ethical values are important: the deontology and responsibility of designers, the emancipation of users, the measures of evaluation (Dubuisson Duplessis and Devillers 2015), transparency, explainability, loyalty, and equity of systems and the study of human–machine coadaptation. Social and emotional robots raise many ethical, legal, and social issues. Who is responsible in case of an accident: the manufacturer, the buyer,

---

[4]See https://smartech.gatech.edu/handle/1853/53208

[5]Fagella D., "IEEE Standards and Norms Initiative": techemergence.com/ethics-artificial-intelligence-business-leaders/

the therapist, or the user? How to regulate their functioning? Control their use through permits? For what tasks do we want to create these artificial entities? How do we preserve our privacy, our personal data?

Any system must be evaluated before it is placed in the hands of its user (Bechade et al. 2019). How do we evaluate a robot that learns from and adapts to humans, or that learns on its own? Can it be proven that it will be limited to the functions for which it was designed, that it will not exceed the limits set? How to detect sludge? Who will oversee the selection of the data that the machine uses for its learning directs it to certain actions?

These important issues have only recently been raised. The dramatic advances in digital technology will one day improve people's well-being, provided we think not about what we can do with it, but about what we want to do with it. That is why the largest international professional digital association, the global scholarly organization IEEE (Institute of Electrical and Electronics Engineers), has launched an initiative to reflect on ethics related to self-designated systems; a dozen working groups on norms and standards have emerged, including on robot *nudging* (incentive manipulation). The CERNA,[6] replaced by the French National Pilot Committee for Digital Ethics (CNPEN) also took up this subject artificial intelligence that can provide better health diagnoses, stimulation tools, detection of abnormal behaviors, and better assistance, particularly for disability or loss of autonomy. As an example, one of the three referrals submitted by the Prime Minister to the CNPEN concerns the ethical issues of conversational agents, commonly known as chatbots, which communicate with the human user through spoken or written language. This work of the CNPEN is an extension of the work initiated by CERNA, the Allistene Alliance's Commission for Research Ethics in Digital Science and Technology (https://www.ccne-ethique.fr/en/actualites/cnpen-ethical-issues-conversational-agents). Machines will surely be able to learn on their own, but will not be able to know if what they have learned is interesting, because they have no conscience. Human control always will be essential. It is necessary to develop ethical frameworks for social robots, particularly in health, and to understand the level of human–machine complementarity.

In my book "Robots and Humans: Myths, Fantasies and Reality" (Devillers 2017), I propose to enrich Asimov's laws with commands adapted to life-assistant robots. The foundations of these commandments come in part from feedback from experiences of interactions between elderly people and robots. Conversational virtual agents and robots using autonomous learning systems and affective computing will change the game around ethics. We need to build long-term experimentation to survey Human–Machine Coevolution and to build *ethics by design* chatbots and robots.

References

Approche Asso. Retrieved from http://www.approche-asso.com/. The association APPROCHE is an association under the law of 1901 in France. Created in 1991, the association's main mission is to promote new technologies (robotics, electronics, home automation, Information and Communication Technologies, …) for the benefit of people with disabilities, whatever their age or living environment.

Bechade, L., Garcia, M., Dubuisson Duplessis, G., Pittaro, G., & Devillers, L. (2019). Towards metrics of evaluation of Pepper robot as a social companion for elderly people: International Workshop on Spoken Dialog System Technology. In M. Eskenazi, L. Devillers, & J. Mariani (Eds.), *Advanced social interaction with agents* (pp. 89–101). Berlin: Springer International Publishing AG.

Damasio, A. (1994). *Descartes' error*. New York: HarperCollins.

Damasio, A. (1999). *The feeling of what happens*. San Diego: Harcourt Brace.

Damasio, A. (2003). *Looking for Spinoza*. Boston: Mariner Books.

Dataia. (2020). *Bad nudge-bad robot? Project: Nudge and ethics in human-machine verbal interaction*. Available via Dataia. Retrieved from https://dataia.eu/en/news/bad-nudge-bad-robot-project-nudge-and-ethics-human-machine-verbal-interaction.

Devillers, L. (2017). *Des robots et des hommes: Myths, fantasmes et réalité*. Paris: Plon.

Devillers, L. (2020). Les robots émotionnels … et l'éthique dans tout cela? Ed. L'Observatoire.

Devillers, L., Vidrascu, L., & Lamel, L. (2005). Challenges in real-life emotion annotation and machine learning based detection. *Neural Networks, 18*(4), 407–422. https://doi.org/10.1016/j.neunet.2005.03.007.

Devillers, L., Tahon, M., Sehili, M., & Delaborde, A. (2014). Détection des états affectifs lors d'interactions parlées: robustesse des indices non verbaux. *TAL, 55*(2), 123–149.

Devillers, L., Tahon, M., Sehili, M., & Delaborde, A. (2015). Inference of human beings' emotional states from speech in human-robot interactions. *International Journal of Social Robotics, 7*(4), 451–463.

Dubuisson Duplessis, G., & Devillers, L. (2015). *Towards the consideration of dialogue activities in engagement measures for human-robot social interaction*. Paper presented at the International Conference on Intelligent Robots and Systems, Congress Center Hamburg, Hamburg, 28 September to 2 October 2015.

Grinbaum, A., Chatila, R., Devillers, L., & Ganascia, J. G. (2017). Ethics in robotics research: CERNA mission and context. *IEEE Robotics and Automation Magazine, 3*, 139–145. https://doi.org/10.1109/MRA.2016.2611586.

Jokinen, K., & Wilcock, G. (2017). Expectations and first experience with a social robot. In *Proceedings of the 5th International Conference on Human Agent Interaction—HAI '17, 2017*.

Man, K., & Damasio, A. (2019). Homeostasis and soft robotics in the design of feeling machines. *Nature Machine Intelligence, 1*, 446–452. https://doi.org/10.1038/s42256-019-0103-7.

Picard, R. (1997). *Affective computing*. London: MIT Press.

Thaler, R. H., & Sunstein, C. R. (2008). *Nudge: Improving decisions about health, wealth, and happiness*. New Haven/London: Yale University Press.

---

[6]Allistene's Reflection Commission on Research Ethics in Digital Science and Technology, cerna-ethics-allistene.org

Impact of AI/Robotics on Human Relations: Co-evolution Through Hybridisation

Pierpaolo Donati

Contents

Abstract

This chapter examines how the processes of human enhancement that have been brought about by the digital revolution (including AI and robotics, besides ICTs) have given rise to new social identities and relationships. The central question consists in asking how the Digital Technological Matrix, understood as a cultural code that supports artificial intelligence and related technologies, causes a hybridisation between the human and the non-human, and to what extent such hybridisation promotes or puts human dignity at risk. Hybridisation is defined here as entanglements and interchanges between digital machines, their ways of operating, and human elements in social practices. The issue is not whether AI or robots can assume human-like characteristics, but how they interact with humans and affect their social identities and relationships, thereby generating a new kind of society.

P. Donati (✉)
Department of Political and Social Sciences, University of Bologna, Bologna, Italy
e-mail: pierpaolo.donati@unibo.it

Keywords

Artificial intelligence · Robotics · Human enhancement · Hybridisation · Relational goods

The Topic and Its Rationale

I intend to analyse how the processes of human enhancement brought about by the digital revolution (ICTs, AI, robotics) modify social identities, relationships, and social organisations, and under what conditions this revolution can shape organisational forms that are able to promote, rather than alienate, humanity.

A recent debate in France (Bienvault 2019) about the use of robots in care for the elderly ended with the question: 'can the human relationship be the product of our technological feats?'. We must look for a plausible answer.

Personal and social identities, relations, and organisations are forced to take shape in the environment of a Digital Technological Matrix (henceforth DTM) that is a symbolic code, which *tends* to replace all the ontological, ideal, and moral symbolic matrices that have structured societies in the past. The concept of 'matrix' used here refers to the meaning that this concept has in my book on 'the theological matrix of society' (Donati 2010). It indicates a vision of the world that has a religious character, or functionally equivalent to that of religion, in that it expresses the 'ultimate values' of that society. In the case of DTM, these values are those of an indefinite evolution of humanity and the cosmos that replaces the representations of the world as a reality created and supported by a supernatural reality.[1]

Obviously, I am talking about a trend, not only because in each society different cultural matrices coexist, in particular in the multicultural societies emerging today, but also because it is always possible that the conflicts between cultural matrices can intensify and diversify. What I argue is that DTM, like all cultural matrices that have a transcendental character, that is, express a vision of humanity and the cosmos in their ultimate ends, has an intrinsic propensity to become the dominant matrix, similarly to what happened with the cultural matrix of modernity that made traditional cultures marginal. As a form of Tech-Gnosis, the peculiarity of the DTM is that of making the boundaries between human and non-human labile and crossable in every way in order to foster hybrids. Hybridisation is defined here as entanglements and interchanges between digital machines, their ways of operating, and human elements in social practices.

Hybrids, however, are not random or purely contingent entities. They stem from complex interactional networks, in which social relations are mediated by the DTM. The processes of hybridisation of social identities and relations are selective and stratified according to the ways in which the human/non-human distinction is relationally thought of and practised. We are witnessing a co-evolution between AI/robotisation and human relations.

Three scenarios of hybridisation are outlined along with three kinds of societal morphogenesis: adaptive, turbulent, and relationally steered. The chapter examines these scenarios of co-evolution, their consequences, and the issues of social regulation.

I am not so interested in discussing whether AI or robots can be more or less human in themselves—or not human at all—but rather how they can interact with humans and affect their social relationships so as to generate a different kind of society characterised by the hybridisation between human and non-human.

I will tackle two major themes: (a) the first concerns the problem of how we can distinguish interhuman relations from the relationships between human beings and machines, which implies the need to clarify what the processes of hybridisation of identities and social relations consist of and how they happen; (b) the second concerns the consequences of digitalised technological innovations on the hybridisation of social institutions and organisations and, ultimately, the possible scenarios of a 'hybridised society'.

Let Me Explain the Broader Cultural Framework to Which This Contribution Refers

The encyclical *Laudato si'* aims at an integral ecology that includes the environmental, technological, economic, and social dimensions in the awareness that 'everything is intimately related' (§ 137). To achieve the goal of sustainable and integral development, these various aspects must be treated 'in a balanced and interconnected manner' so that 'no one be left behind' (§§ 2 and 4).

The idea that we can heal our relationship with nature without healing 'all fundamental human relationships' and therefore without fully including in this therapeutic action 'the social dimension of the human being' (to which Pope Francis adds the transcendent) is illusory. The vectors of ethics, culture, and spirituality converge on the social dimension: the ecological crisis is the expression of a deeper crisis that invests these essential elements of human existence in modernity and there are not two separate crises, 'but a single and complex socio-environmental crisis' (§§ 119 and 139).

For this reason, we must first try to straighten out the pillars of human relations and anthropology, which is the real 'root' of a crisis that distorts the development of science and technology, the mechanisms of the economy, the responsibilities of politics.

Science, technology, and economics are only a part and not the whole of the social as it is understood here, as are their relative behaviors and ways of understanding reality (§ 139). It must be said that the social includes the economic and the technology, and not vice versa. To the loss of orientation towards the common good, which reduces economy and technology to the only worry for profit and politics to the obsession for power (§ 198), one cannot think of answering exclusively by internal routes to economy and technology, thus remaining prisoners, perhaps in good faith and for good, of the ideology of their hegemony.

---

[1] These values are well illuminated by Harari (2017), by various contributions to the volume Floridi (2015) and by numerous authors that it is not possible to mention here for reasons of space.

This chapter seeks to clarify the meaning of this perspective and make it operational.

The Pervasiveness of the Digital Matrix

In the famous science fiction movie of the same name (1999), the Matrix is depicted as a dystopian world in which reality, as perceived and lived by most humans, is actually a simulated reality created by sentient machines to subdue the human population while their bodies' heat and electrical activity are used as an energy source. Sentient machines rule the world with lasers, explosions, and killer robots. *This* Matrix is a 'dream world', where cyborgs are supposed to simulate a superman who is a mixture of a super-animal and super-machine. In a famous dialogue between two protagonists, Morpheus says, 'The Matrix is everywhere. It is all around us. Even now, in this very room. You can see it when you look out your window or when you turn on your television. You can feel it when you go to work... when you go to church... when you pay your taxes. It is the world that has been pulled over your eyes to blind you from the truth'. Neo asks, 'What truth?', to which Morpheus replies, 'That you are a slave, Neo. Like everyone else, you were born into bondage. Into a prison that you cannot taste or see or touch. A prison for your mind'. This is what I would call the Matrix Land. In the end, the Matrix appears as it actually is: nothing but the green lines of a programming code that pervades all the environment of the human condition.

Leaving aside the aspects of science fiction, one can take the Matrix to mean the Digital Technological Mind that is made pervasive and omnipresent by the global ICT network constituted by all the tools and symbolic codes that operate on the basis of algorithms.

From the cultural point of view, the Digital Matrix (DTM) is *the globalised symbolic code[2] from which digital artefacts are created in order to help or substitute human agency by mediating interhuman relations or by making them superfluous*. From the structural and practical point of view, the DTM is the complex of all *digital technologies*, based on

scientific knowledge and engineering, that consist of computerised devices, methods, systems, electronic machines (digital electronics or digital electronic circuits). Of course, the artefacts produced by the DTM can have different forms of intelligence and more or less autonomy (Lévy 1997).

In short, the DTM software is part of the cultural system, and its hardware fits into the social structures by occupying the positions that are nodes in the networks. It is important to understand: (1) first, that the DTM symbolic code plays a major or dominant role within the whole cultural system of society; (2) second, that it is the starting point of innovation processes (through the discoveries and inventions of scientific research that are subsequently applied in new technologies).

I will explain the second point in the next section (see Fig. 1). As to the first, I contend that the DTM symbolic code plays a role, in respect to all other cultural symbols, in the same way as the generalised symbolic medium of money has functionalised all the other generalised symbolic media to itself within modern society. Money has been (and still is) the G.O.D. (generator of diversity) of modern society. It has functionalised to itself power, influence, and value commitment.

The DTM symbolic code is now the Generator of Diversity of transmodern society (or 'onlife society', i.e. the society of online life). As an instrumental symbol, it functionalises to itself all the other symbols, e.g. the *finalistic* symbols such as life and death, creation or evolution; the *moral* or *normative* symbols such as justice or injustice, honesty or dishonesty, normative or anormative, disassembling or reassembling, gluing or ungluing; and the *value* symbols such as worthy or unworthy, good or bad, positive or negative, pleasant or unpleasant.[3]

The dualities inherent in all these symbols are treated in an apparently relational way since the binary code allows gradations and combinations of each type in the 0/1 sequences. They produce *eigensymbols* (eigenvalues), i.e. symbols autogenerated by the code itself. As I have argued elsewhere, the relational modes with which the Digital Matrix operates can be read in different ways, that is, as purely interactive flows producing only random and transactional outcomes (as *relationist* sociology claims) or as processes that generate relatively stable cultural and social structures that are emergent effects endowed with *sui generis* qualities and causal properties (Donati 2020).

According to the relationist viewpoint, digital technologies (machines) represent the material infrastructure of an anonymous DTM of communication that feeds the autonomisation of a multiplicity of communicative worlds separated

---

[2]I define a symbolic code (or semantics) as a set of symbols and the rules for using them in looking at the world and interpreting phenomena, facts, and events, while producing them. When the symbols refer to 'ultimate realities' ($\check{\epsilon}\sigma\chi\alpha\tau o\varsigma$, éskhatos), and the rules follow a logic of first principles, a symbolic code takes on the form of a 'theological matrix'. Why do I call it so? Because, in this case, the symbolic code is a reflection of a theology. In short, the symbolic code derives from the way we semantise the 'ultimate realities' that explain what happens in the world. The DTM is a substitute for the old ontological and theological matrices of the past. An example is given by Luciano Floridi's definition of the human being as a 'Nature's beautiful glitch' (Floridi 2019: 98: 'We are special because we are Nature's beautiful glitch'). This definition replaces the idea of the human person as a being that has a sense and a finalism that is related to a project of God with the idea that it is a random product of a Darwinian evolution devoid of any finalism.

[3]Symbols are here understood as generalised media of interchange (Parsons 1977) and/or as generalised media of communication (Luhmann 1976).

Fig. 1 The morphogenetic cycle
(run by the DTM) that generates
the hybridisation of society.
Source: developed by the author

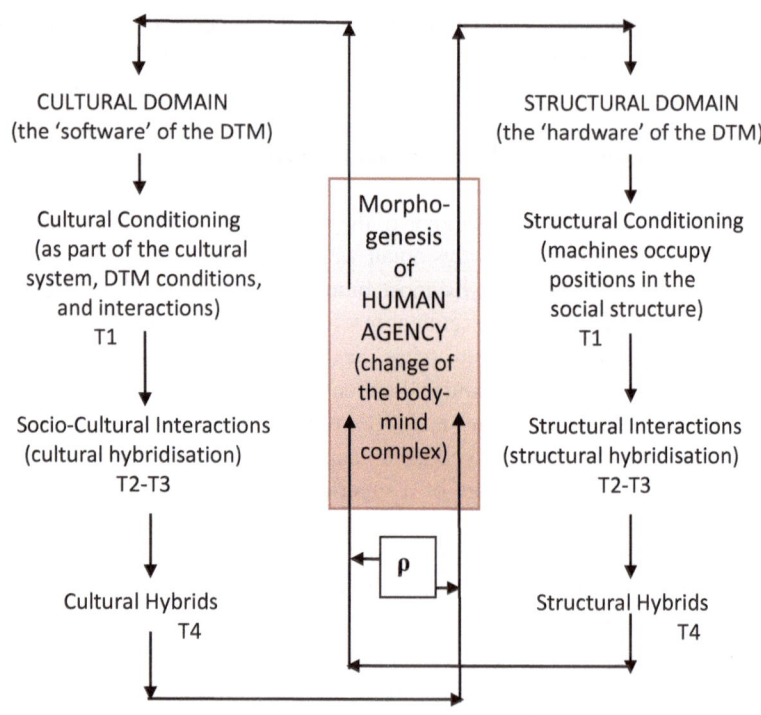

from a humanistic conception of intersubjective and social relations. This DTM becomes detached from any traditional culture based on religious or moral premises and builds its own political power system, economy, and religion. Harari (2017) has taken this scenario to its extreme consequences, claiming that it prefigures how the quest for immortality, bliss, and divinity could shape humanity's future. In this scenario, human beings will become economically useless because of human enhancement, robotisation, and artificial intelligence, and a new religion will take humanism's place.

The rationale for this view is that the DTM promotes the *digital transformation of the whole society*, which is the change associated with the application of digital technology in all aspects of social life, and is expressed in the *Infosphere*, defined as the environment constituted by the global network of all devices in which AIs are at work receiving and transmitting information. The DTM is therefore the 'environment' of all social interactions, organisations, and systems. As such, it promotes a culture deprived of any teleology or teleonomy since it operates as a substitute for any rival moral or theological matrix.[4]

---

[4]Together with Teubner (2006a: 337), we can say that the TM 'extends to infinity the usurpation potential of its special medium, power, without any immanent restraints. Its operative closure and its structural autonomy let it create new environments for itself, vis-à-vis which it develops expansive, indeed downright imperialist tendencies. Absolute power liberates unsuspected destructive forces. Centralized power for legitimate collective decisions, which develops a special language of its own, indeed a high-flown rationality of the political, has an inherent tendency to totalize them beyond limit'.

The Hybridisation Issue

The hybridisation of society emerges from a process of social morphogenesis in which social structure, culture, and agency intertwine so to produce social and cultural hybrids together with a hybridised agency. Figure 1 schematises these processes, which consist in a series of subsequent morphogenetic cycles.[5]

Each morphogenetic cycle can start from the side of the social structural domain or from the side of the cultural domain. Let us see what happens in either case.

1. If the morphogenic cycle begins on the side of the structural domain, for example when a robot is introduced in a certain social context (a factory, a school class, etc.), what happens is a redefinition of the social network and the social positions of the actors involved. As a consequence of the presence of the newcomer, structural interactions change and produce structural hybrids. These hybrids affect human agency in the direction of influencing her impact on the cultural domain.

2. If the morphogenetic cycle starts from the side of the cultural domain, for example, when new software (language, programs, algorithms) are invented and adopted by people, then new sociocultural interactions produce cultural hybrids. These cultural elaborations modify hu-

---

[5]On the processes of morphogenesis and the relative cycles see Archer (1995).

man agency, which introduces changes into the structural domain.

3. The human agency changes through active (reflexive) responses to the above changes. The possibilities for developing this morphogenetic process are certainly dependent on how the DTM operates in the cultural and structural domains, but we must consider the autonomous role of human agency.

One may wonder: which morphogenesis occurs in the human person (agency) and in her relational qualities and properties? What happens to social relations between humans and between humans and robots?

Figure 1 illustrates the relational structure and dynamics of co-evolution between digital machines, their ways of operating (according to their symbolic code), and human agency. Co-evolution can obviously have positive or negative results in terms of compatibility among the different components, and therefore it can generate both relational goods and relational evils.

Notice that the morphogenesis of human agency is both passive and active in *both* directions, towards the cultural as well as the structural domain. (a) It is passive because it is influenced alternatively by one of the two domains and (b) at the same time it is active towards the other domain. It is here that the body-mind relational unit has to confront the coherences or dissonances between the two domains. The symbol [ρ] means that there is a connection of some sort between the cultural and structural processes of hybridisation of the human person in her social identity and in her social relationships. Such a connection can be of different kinds, from the maximum of synergy (complementarity), as when cultural identity is well adapted to the position occupied in the social structure, to a maximum of conflict (contradiction), as when cultural identity conflicts with the position occupied in the social structure.

What is most relevant in Fig. 1 is to observe the dematerialisation of the human agency due to AIs operating in and through the quantum network (the internet) where information is transmitted with qubits. The process of hybridisation takes place in fact contaminating the relationships that operate on the basis of the principles of classical physics applied to the natural world with virtual relations that operate on the basis of the postulates of quantum physics, where the latter dematerialises the natural world.[6]

The conceptual scheme summarised in Fig. 1 is fundamental to understand the hybridisation processes of social identities, relationships, and organisations in the light of the idea that, to speak the language of Adams (2006: 517), 'hybridity becomes a complex potential'. These processes proceed according to a relation of contingent complementarity between changes in the cultural background that encourage the development of certain forms of virtual thinking and utopian discourse, on the one hand, and changes in the material production of new technologies (AI/robots) in the structural domain, on the other hand. Hybridisation occurs when culture (software) and structure (hardware) mutually influence and reinforce each other, as long as human agency adapts to these processes. Relations play the most crucial role because changes in identities and organisations depend on them. The hybridisation of society is based on the assumption that these processes operate in a functional way, making sure that the passivity of human agency translates into its active participation in such an evolution with minimal reflexivity or a reflexivity substantially dependent on the DTM. The kind and extent of hybridisation depend on how subjects reflect on their relationships, on which their internal conversation (personal reflexivity) depends.

The hybridisation of people's identities and social organisations (families included) consists in the fact that these entities change their relational constitution to the extent that the cultural and structural processes of change affect (in Fig. 1 they 'cross') the human person and modify her ways of relating to herself, to others, and to the world. Agency is obviously dependent on how the brain and the mind of the person work. Both the brain and the mind are seriously influenced by the way technologies operate (Greenfield 2014) because they absorb certain ways of communicating, using and combining information and following performance logics that can ignore or alter analogical thinking. As Malabou (2009) claims, the human being has his essential difference in the plasticity of his mind that the robot will never have (synaptic plasticity is unique, there cannot be two identical brains).

Since hybridisation means some form of mutual adaptation between humans and AI bots, one can ask to what extent this mutual influence (up to mutual constitution) can arrive. Personally, I do not believe that humanoid robots can be or become 'relational subjects' in the sense of being able to relate to others and to the social context by forming a we-relation with adequate personal and relational reflexivity (Donati and Archer 2015). They cannot for various reasons (Donati 2020), in particular because their mind and their body–mind complex are ontologically different from that of humans, so that human–human interaction (HHI) and human–robot interactions (HRI) are ontologically different (of course, if we assume that human relations are constituted by reciprocal actions which are something more and different in respect to pure behaviour). Basically, AI bot's mind (consciousness) has an inside-out and outside-in relationality that is structurally different from the human one (Lindhard 2019). There is much debate about whether robots can be

---

[6]For a recent debate on this issue see the theory by Alexander Wendt (2015) according to which consciousness is a macroscopic quantum mechanical phenomenon and the critical review by Porpora (2018), in particular for what concerns the consequences of Wendt's ideas on social life.

friends of human beings. My position is that it is unlikely that humans will ever be able to be friends with robots. As Mathias Tistelgren ('Can I Have a Robot Friend?' quoted by Barlas 2019) claims, the capacity to enter into a friendship of one's own volition is a core requirement for a relationship to be termed friendship. We also have a duty to act morally towards our friends, to treat them with due respect. To be able to do this, we need to have self-knowledge, a sense of ourselves as persons in our own right. We do not have robots who display these capacities today, nor is it a given that we ever will.

Bearing in mind the substantial differences between social relations and technological relations (Nørskov 2015), I believe however that it is possible to recognise that 'social robots, unlike other technological artifacts, are capable of establishing with their human users *quasi-social relationships* as *pseudo-persons*' (Cappuccio et al. 2019: 129).

Hybridisation means that, through sustained interactions with technologies (the 'fully networked life'), the previous modes of considering (conceiving) oneself, relationships with others, and what form to give to a social organisation in keeping with the (analogic) principle of reality are mixed with the way digital technology works, i.e. the (fictitious) principle of digital virtuality. Hybridisation means blending the real and the fictitious, the analogue and the digital. This happens, for example, when one wants to modify one's own image or undermine another person's reputation on social networks.

To clarify this point, I must say that I do not use the term 'real' in opposition to digital, but in my view the polar opposites are 'real' vs 'fictitious' and 'digital' vs 'analogical'.[7] If a hacker creates a profile on Facebook, or another social network, and transmits fake news, such news has a reality, even if it is the reality of a fiction. While the story's referent is not real, the story itself is causally efficacious, and therefore real in that regard. So, instead of contrasting a digital message to a real message, it is better to contrast a digital message with an analogue, because the analogical code implies, in fact, an effective analogy relationship between what is written or transmitted and what is true. In short, hybridisation can take place by putting in interaction, and then mixing what is *real* and what is *fictitious*, what is *digital* and what is *analogue*. Table 1 shows the different types of relations that, in different combinations, produce different forms of hybridisation.

Qualitative and quantitative research provides some empirical evidence of these hybridisation processes:

– *About ICTs*: in the chapter 'Twenty-First Century Thinking', Greenfield (2009) suggests that the decline of read-

Table 1 Types of relations whose combination generates hybridisation

	Real	Fictitious
Analogue	Interpersonal (face-to-face) relationships	Games of relationships
Digital	Digital relations replacing the interpersonal ones	Simulated relations

Source: Designed by the author

ing in favour of fragmentary interactions, such as computer games or short messages on the internet, threatens the substance of both our neurological makeup and our social structures;
– human persons alter their pace, rhythm, and sense of Self; they change their personal reflexivity;
– *about AI/robots*: authentic human relationships are reduced to and sacrificed in favour of digital relationships ('We expect more from technology and less from each other'); technology appeals to us most where we are most vulnerable (passions, feelings, interests, etc.);
– digital relations erase precisely those aspects of randomness that also make human life, people, and relationships interesting, spontaneous, and metamorphic;
– by developing new technologies, we are inevitably changing the most fundamental of human principles: our conception of self, our relationships to others, and our understanding and practice of love and death; nevertheless, we should not stop developing new technologies, but do it differently by adopting an approach of relational observation, diagnosis, and guidance (as I will say later).

In short, the processes of hybridisation of identities, human relations, and social organisations are closely connected and change together (in parallel with each other), varying from case to case due to the presence of intervening variables in specific contexts.

The Process of Hybridisation of Identities and Relations

From the point of view of critical realism, hybridisation can and must instead be considered as a morphogenetic process that leads from entities structured in a certain way to entities structured in another way. The fact remains that we should understand how to define the *proprium* of the human in hybridisation processes.

Lynne Rudder Baker's theory is often cited as a solution to this problem as it replaces a '*constitutive*' rather than a substantive (so-called 'identifying') view of the human.

Baker (2000) argues that what distinguishes persons from all other beings is the mind's intentionality detached from

---

[7]They are the poles of an opposition that I conceive in the frame of what Guardini (1925) calls *Gegensatz* ('polar opposition'), which has nothing to do with the Hegelian dialectic.

corporeity since we, as human beings, can be fully material beings without being identical to our bodies. In her view, personhood lies in the complex mental property of first-person perspective that enables one to conceive of one's body and mental states as one's own. The argument is that the human mind must have a bodily support since the body–mind relation is necessary, but the type of body—and, therefore, the type of body–mind relationship—can be quite contingent; according to her, we can change the body with artefacts, provided we do so from the perspective of the first person.[8] The relation between one's mind and one's body is open to any possibility. Consequently, since the personality is equal to the self-thinking mind, we must then acknowledge the existence of a personality in any animal or machine that can be judged sentient and thinking, on the condition that it is aware that it is itself that thinks. In this way, the possibility of anthropomorphising robots, as well as robotising human bodies, becomes thinkable and legitimised.

Following this line of thought, some thinkers today are more and more inclined to acknowledge the existence of moral behaviours in certain species of higher primates like the chimpanzees. They are supposed to have a 'moral intelligence', i.e. in that they are compassionate, empathetic, altruistic, and fair (Bekoff and Pierce 2009), and, along the same lines, they recognise the possibility that special robots endowed with an artificial morality might exist (Pana 2006). It seems to me that this perspective is totally at odds with a realistic social ontology for which human relationality—between body and mind, as well as in social life with others (the two are closely related)—has qualities and properties that cannot be assimilated with the relationships that certain species of animals or ultrasophisticated intelligent machines can have. On the whole, Baker's theory is inadequate to the task of accounting for the possibilities and the limits of the processes of hybridisation because it does not account for the relational constitution of identities and social forms.

The hybridisation of human relations, paradoxically, is due to the logic inherent in such relationships, which, by definition, must be open to the world and cannot remain closed in self-referentiality, as artificial machines do. In interacting repeatedly with machines (AI/robots), agents can incorporate certain aspects of the communication logic of the machines in their way of relating to others and the world. However, we cannot equate the relations between humans with those between humans and robots. How can one not see that the qualitative difference between these relations, for example, when someone says that the robot is 'my best friend', reduces the relationship to pure communication (i.e.

to communication and only to communication, as Luhmann 1995 states)? This is an unjustified and unjustifiable reduction because communication always takes place within a concrete relationship and takes its meaning, qualities, and properties from the kind of relationship in which it occurs, whereas the kind of relationship depends on the kind of subjects that are in communication.

The problems caused by DTM (ICTs and robotics) arise when sociability is entrusted to algorithms. The introduction of the sentient robot changes the context and the relationships between human subjects as well as the form of a social organisation.

Let us take the example of a social structure of a sports type like that of a football team and a football match. How does the behaviour of a football team that is equipped with a goalkeeper-robot capable of parrying all rolls change and if two teams agree to play a game in which each of them has a robot as a goalkeeper, what will be the consequences on human behaviour?

There cannot be a 'we believe' or a relational good between humans and robots for the simple fact that the supervenience human–robot relation is ontologically different from the supervenient relationship between human beings.[9]

The partial or total hybridisation cycle of relationships is selective and stratified on the basis of: (1) how the personal and relational reflexivity of agents operates, given the fact that reflexivity on these processes is necessary to get to know each other through one's own 'relational self-construal' (Cross and Morris 2003) and (2) how the reflectivity[10] of the network or organisational system in which the agents are inserted operates.

We can see this by using the morphogenetic scheme (Fig. 2), which describes how the hybridisation processes depend on the reflexivity that the subjects and their social network exert on the different forms of human enhancement.

Given an organisational context (or network) in which human persons must relate themselves to a DTM of sentient machines, social relationships can be hybridised in various ways, and with different intensity, depending on whether the human person adopts:

(a) A reflexivity that is *purely dependent* on the machine; in practice, the agent person relates to the machine by identifying herself with the digital code, which means

---

[8]For a critique of Baker's constitution view, see Houkes and Meijers (2006). As I have argued elsewhere (Donati 2020), the first-person perspective is not enough to define 'the essential' of the human, we have to consider also the second-person and third-person perspectives.

[9]A set of properties *A* supervenes upon another set *B* just in case no two things can differ with respect to *A*-properties without also differing with respect to their *B*-properties. In slogan form, 'there cannot be an *A*-difference without a *B*-difference' (Stanford Encyclopedia of Philosophy 2018). In our case, the human–robot relationship would be equivalent to the human–human relationship if the robot were human.

[10]I use the term reflexivity to indicate the reflexivity of networks, while I use the term reflexivity to indicate the reflexivity of human subjects (Donati 2014).

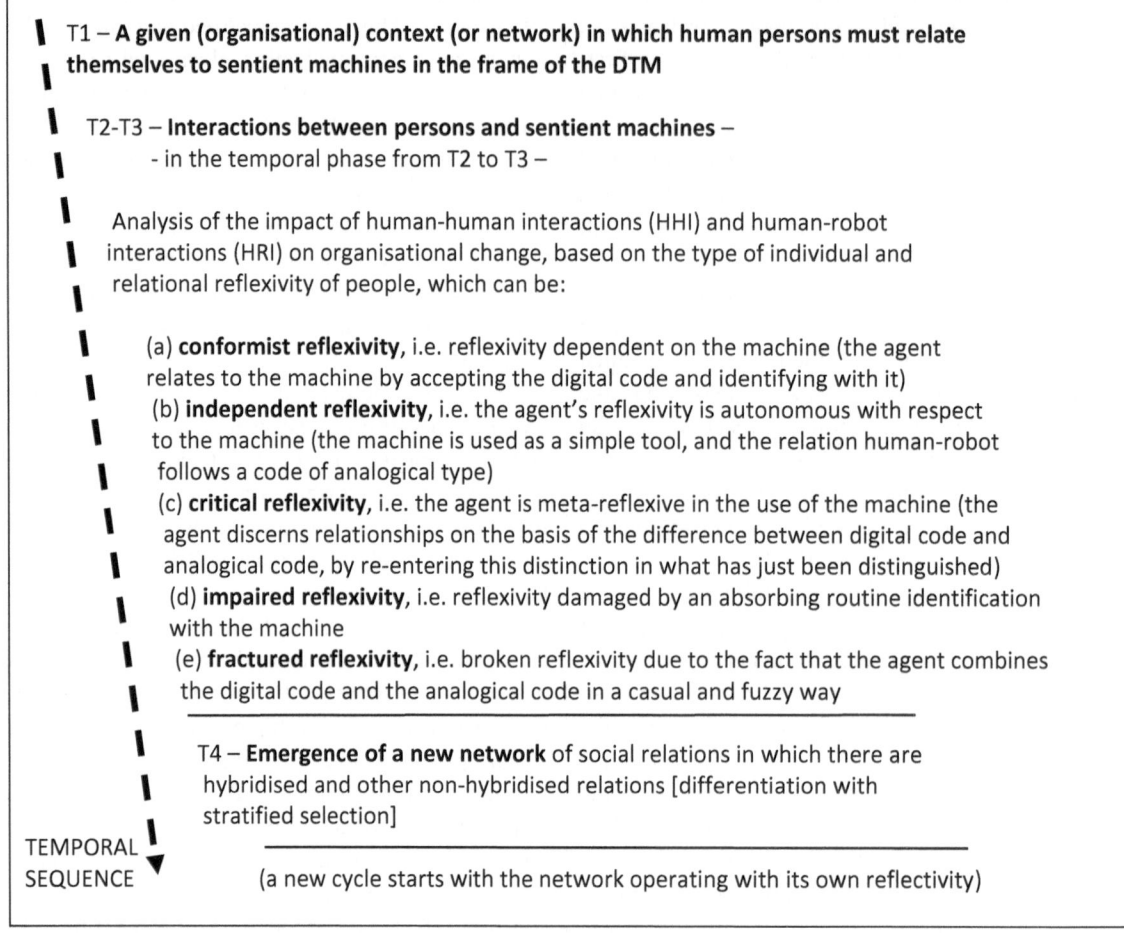

Fig. 2 The morphogenetic cycle through which identities and social relations are hybridised. Source: Designed by the author

that she 'connects' to the machine without establishing a real relationship (connection is not a social relationship); it happens, for instance, when people identify themselves with their Facebook profile;

(b) A reflexivity that is *autonomous* with respect to the machine; the machine is used as a simple tool, and the human–robot relationship follows basically an analogic code;

(c) A *critical* reflexivity (meta-reflexivity) as a use of the machine, a use that continuously redefines the purpose of interaction, in order to establish a more satisfying relationship; this mode of reflexivity reenters[11] the distinction digital/analogic into what has emerged in previous actions and their outcomes;

(d) An impeded reflexivity that is caused by an absorbing routine identification of the agent with the machine;

(e) A reflexivity that is *fractured* due to the fact that the agent combines the digital code and the analogical code in a casual and fuzzy way.

Take the case of domestic ICTs. They constitute reference points around which identity, gender, and intersubjectivity are articulated, constructed, negotiated, and contested. They are points through which people construct and express definitions of selves and other. As Lally further explained: 'The development of the human subject (individual and collective) takes place through a progressive series of processes of externalisation (or self-alienation) and sublation (reabsorption or reincorporation). Human subjects and the (material and immaterial) objects of their sociocultural environment form a subject-object relation which is mutually evolving, and through which they form a recursively defined, irreducible entity' (Lally 2002: 32).

These subject-object sociotechnical associations constitute the material culture of home life; they are bundled together with the affective flows of human relations and the emergence of a lifeworld in which images and emotions play a crucial role in the hybridisation of relationships because it is through them that the subjects identify with the logic of virtual/digital relationships (LaGrandeur 2015).

---

[11] The operation of *reentry* is here understood according to the logic of Spencer-Brown (1979).

Fig. 3 The hybridisation of the human person (her identity and social relations) due to the influence of DTM on the natural, practical, and social realities. Source: Designed by the author

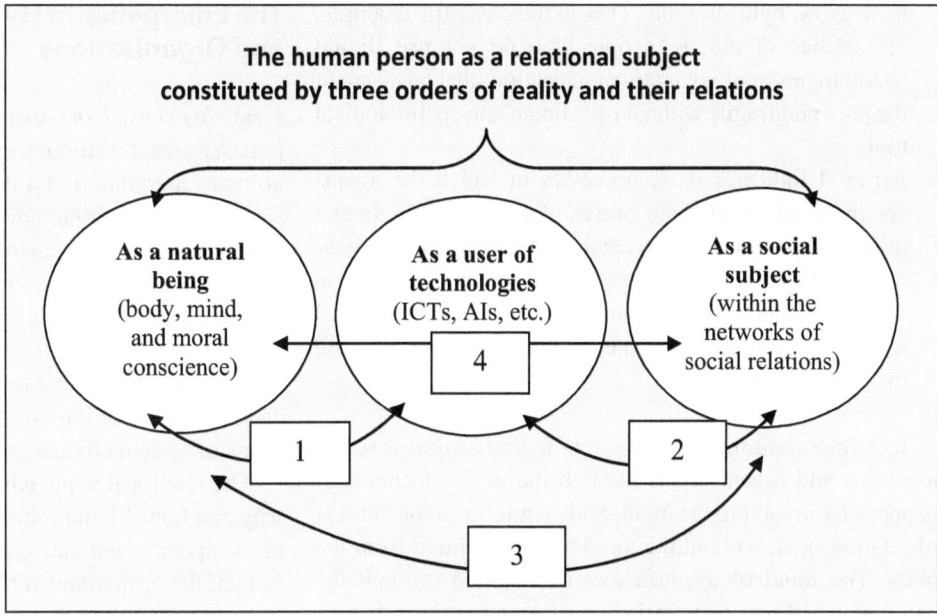

The Hybridisations of Identities and Social Relationships Are Interdependent

The hybridisations of human identity and social relations through DTM are connected to each other because the human being is a relational subject. The dynamics of mutual influence between identity and relations in the hybridisation processes are not linear, but proceed along very different paths.

We see the different hybridisation paths in Fig. 3.

1. Arrow 1 indicates the two-way interactions between the natural being and the use of technology (regardless of the social context). From a phenomenological point of view, the relational context is always there, but in many cases the agents do not take this into account, as often happens in medical settings and biological laboratories.

   Examples of type 1 hybridisations are those widely used in medicine, where they involve checking whether the technology helps repair damaged organs or cure diseases. Baker (2013) generally refers to this type, obviously considering the individual as natural beings. But even when she speaks of human enhancement in the proper sense, i.e. as making someone or something 'more than normal' by applying digital technological devices that are well beyond therapy, she does not consider the implications on the level of the social order. Since the relational order (context) is practically ignored in her theory, Baker's vision remains restricted to a mentalised interpretation of the hybridisation of the individual.

2. Arrow 2 indicates two-way interactions between the use of technology and sociality. In this case, the actor/agent is a virtual person who disregards her body-mind constitution. An example of a type 2 hybridisation is the phrase: 'I do not use the blog; *I am the blog*'. Hybridisation happens through the formation of a relationship in which Ego's identity is in his blog. There is an identification of subject and object: no longer, 'I make my blog in order to communicate', but, 'I am my communication' (Stichweh 2000). The logic of the blog becomes part of the logic of the Subject, who reflects according to the logic of what was once an instrumental object (a means) and now has become his way of relating (a norm), which reflects a value placed in the communication itself, in the presence of an undetermined goal.

   The dialogue between the '*I*' (i.e. the Self that dialogues with itself), its '*Me*' (i.e. what the 'I' has done in the past), and its '*You*' (i.e. what the 'I' is willing to do in its future)[12] takes place within an interactive time register, in which time has no duration (it is neither relational nor symbolic, but purely interactional) (Donati 2011). The temporal history of the body–mind complex, its stock of knowledge (Elster 2017) does not enter the process, except for the aspects in which it has been mentalised. The identification of the human person with the blog is the outcome of the process of mentalisation, which corresponds to the fact that an AI has become part of the Subject's Mind, forging its conscience.

3. Arrow 3 indicates the two-way interactions between the agent as natural being and sociality. In this case, the individual acts regardless of the technology, and, therefore,

---

[12]This sequence is elucidated by Archer (2013, figure 2: 4).

there is no hybridisation. This is the case, for example, of families of elderly people who do not use digital technologies or of voluntary organisations that take care of the poor and fragile without resorting to any technological tool.

4. Arrow 4 indicates those processes in which the agents are involved in all three orders of reality. Their body–mind complex and their sociality are mediated by technology, which is what happens 'normally', in fact, in any institution and organisation. The kind and degree of hybridisation depends on the network structure in which the agency is embedded.

It is important to emphasise that hybridisation of social identities and organisations through the use of technology happens by involving the mind–body complex in the context of relationships, without the mind being separated from the body. The mind of agents/actors is modified through the perception and activation of those social relationships that are possible only through the body.

Victoria Pitts-Taylor offers an account of how the mind works on the basis of a brain that is 'complex and multiple, rather than determined and determining' (Pitts-Taylor 2016: 8). Drawing on work by feminists, queer theorists, and disability theorists, she offers an understanding of bodies and cognition that can incorporate the cognitive differences that result from differences in embodiment and that can recognise the social shaping of both bodies and cognition. She notes that scientists have discovered that some neurons seem to respond both to our performance of an action and to our observation of others performing that action. These mirror neurons have been theorised to underlie social relationships, especially 'mind reading' and empathy. We need to do a better job of recognising the complexity and variety of social relationships. In particular, the assumption that the existence of mirror neurons shows that we are naturally in tune with others and empathetic to them leaves social neuroscientists unable to account for why our empathy for others is so often selective and so often fails. This is an example of how social neuroscience can take the embodiment of brain and social relationship seriously,[13] differently from those attempts to create digital identities as avatars that simulate the human without being human (for example, the *Soul Machines* company), thus producing a morphic sociability devoid of real humanity (Pitts-Taylor 2016).

Let me now move from the micro to the meso and macro levels of the social fabric, to consider the institutional and organisational hybridisation processes.

The Emergence of Hybridised Institutions and Organisations

I call *hybridised* organisations and institutions those that pursue a social configuration in which advanced digital technologies are conferred a certain degree (min/max) of decisional and operational autonomy. Technology (AI/robotics) takes on an autonomous and decisive role in managing the organisation of roles and relations between the members of the organisation. The digital logic with which hybrid organisations operate is that of increasing opportunities, conceived, however, not within the framework of their relational implications, but according to the maximum useful variability in terms of system efficiency.

The relational approach to social organisations can show why and how AI and robots cannot replace humans because of the specific generative character of interhuman relations. In fact, the utilitarianism of efficiency refers to the relationships between an actor/agent and 'things', while the search for relational goods (Donati 2019)—or the avoidance of relational evils—implies relationships between human beings, which, unlike algorithms, are generative of meta-reflexive solutions for problems of human relationships.[14]

Let me give an example of how an algorithm can generate relational evils. On 21 November 2017, the algorithm of the multinational company Ikea fires a worker of its megastore in a small town near Milan. Marica, a 39-year-old mother separated from her husband with two young children, one of whom is disabled, is fired because she does not observe the new work shift assigned to her by the algorithm. The algorithm has ordered her to show up at 7 a.m., and, instead, she arrives at 9 o'clock according to the old work shift because she has to look after the children, and, in particular, she has to take the disabled child to therapy. Previously, the woman had explained to the manager that she could not work that shift, and the manager said that he would consider her situation, but the algorithm worked on its own and fired her. The company did not review its decision, and, instead, continued to dismiss other workers on the grounds that they did not comply with the indications of the algorithm.

Undoubtedly, it cannot be said that the algorithm has not been proven to have its own 'personal' decision-making abil-

---

[13]On the embodied nature of human cognition see also Hayles (1999).

[14]By relying on the notion of *auctor* proposed by Donati (2015), Scott Eacott (2018: 81–94) observes that '*Auctor* provides the intellectual resource to overcome the structure/agency analytical dualism through recognition of relations and the generative rather than deterministic nature of activity without defaulting to a naïve form of autonomy (..) *Auctors* generate *organizing activity*. They are not separate from *spatio-temporal conditions* but simultaneously constitutive of and emergent from them. The substantialist basis of organisational theory in educational administration and leadership is broken down through relations. This is why, despite my (somewhat artificial) partitioning of *organizing activity*, *auctor*, and *spatio-temporal conditions* here, they work together, in relation, to generate an elaborated communication of activity'.

ity (some call it 'electronic person'), but it was certainly not a relational subject. Clearly, the algorithm simply followed its procedural rules, neglecting the needs of people and their relational context, which were unpredictable for him. If so, we can say that the algorithm is not a person, given the fact that a person is an 'individual-in-relation' (relationality being *constitutive* of the individual), although it is presented as such by management. The algorithm's personality is a convenient managerial fiction.

In the perspective of the future, in my opinion, in principle, it will be possible to build more sophisticated AI/robots that can take into account people's needs and their relationships. However, in addition to modifying the instructions provided to the algorithm accordingly, it will be always necessary to supervise its operations by a management that, by adopting a relational steering approach, must be able to deal with the individual problematic cases and the complexity and diversity of contingencies that the algorithm cannot handle. Without such relational steering, the hybridisation of organisational relationships due to increasing automation will only mean exploiting people and dismissing them by blaming the algorithm for making inhuman decisions and apologising for not being able to do otherwise.

From a more general view, as Teubner writes, 'Today, with the massive emergence of virtual enterprises, strategic networks, organisational hybrids, outsourcing and other forms of vertical disaggregation, franchising and just-in-time arrangements, intranets and extranets, the distinction of hierarchies and markets is apparently breaking down. The boundaries of formal organisations are blurring. This holds true for the boundaries of administration (hierarchies), of finance (assets and self-financing), of integration (organisational goals and shared norms and values) and of social relations (members and outside partners). In formal organisations, membership becomes ambiguous, geographical boundaries do not matter much anymore, hierarchies are flattened, functional differentiation and product lines are dissolved' (Teubner 2002: 311).

Hybrids raise problems of conflict between divergent normative arrangements. As a strategy to deal with these changes, Teubner recommends a 'polycontexturality which combines heterarchy with an overarching unity' (Teubner 2002: 331) assuming that this organisational solution would represent the new institutional logic capable of avoiding collisions between spheres ruled by incompatible norms.

I take Teubner's proposal as an example of an attempt to preserve the human (with its regulatory requirements) alongside the acceptance of the entry of new hybrids, through general rules that allow different norms to coexist in different domains (Teubner 2006b). *Unitas multiplex* is his keyword for preserving the integration of society, supposing that human beings and actants, human relations and artificial relations, could coexist within a neo-

functional differentiation architecture under the aegis of the DTM.

I have serious doubts about the tenability of this perspective (although it worked in the seventeenth century as the Hobbesian solution to the problem of social order). I think that it is an impracticable solution in a networked society governed by the DTM. In any case, it does not work for a 'society of the human'. As Hobbesian (Leibnizian) rationalism, Teubner's idea of a constitutionalisation of the private hybridised spheres does not address the issues of the relations between human and nonhuman, and the relations between divergent normative spheres. Teubner's perspective simply avoids the relational issue. In short, it has all the limitations and defects of a multicultural doctrine that ignores the reality of what lies *in between* opposite spheres that have incompatible normative orders. It avoids the issue of how far the hybridisation processes can go and to what extent they can affect the human.

The same difficulty is present in the Luhmannian theory that seems to do a good job of interpreting the hybridisation processes insofar as it places all cultures under the aegis of a conception of society as an 'operating system' (Clam 2000). According to Luhmann (1995), all systems, organisations, and interactions are forced to use a binary functional code that is precisely the one through which the DTM proceeds. Hybridisation follows a functional code. For this reason, he believes that the humanism of the old Europe is unsustainable and, therefore, sees no other way than that of considering the human (especially the human of the Western tradition) as a residual entity fluctuating in the environment of the DTM. Are there no alternatives? I think that there are. We must examine the possible scenarios.

Three Scenarios Dealing with the Processes of Hybridisation

The digital transformation of society is destined to produce different types of hybrids through different types of social morphogenesis. I would like to summarise them in three scenarios: adaptive morphogenesis, turbulent morphogenesis, and relationally steered morphogenesis.

1. *Adaptive MG producing hybrids by trial and error*: this is the scenario of a society that adapts itself to the hybridisation processes produced by DTM in an opportunistic way; it is configured as an afinalistic infosphere—without preestablished goals, as in Floridi (2015)—that tries to use the technologies knowing that they have an ambivalent character; they allow new opportunities but also involve new constraints and possible pathologies; therefore, it is essentially engaged in developing self-control tools to limit emerging harms (theories of harm reduction).

2. *Turbulent MG favouring mutations*: this is the scenario of a society that operates for the development of any form of hybridisation; it becomes 'normal' to anthropomorphise the robots, as well as to robotise the human body: in principal, it is run by anormativity and anomie (lack of presupposed moral norms) and openness to mutations, understood as positive factors of 'progress' (theory of singularity); it operates through ceaseless, unrepeatable, intermingling processes of relational flows with 'confluence of relating' (Shotter 2012).

3. *Relationally steered MG aiming to configure technologies in order to favour relational goods*: this is the scenario of a society that tries to guide the interactions between human subjects and technologies by distinguishing between humanising and non-humanising forms of hybridisation. The aim is to produce social forms in which the technologies are used reflexively in order to serve the creation of relational goods. This requires that producers and consumers of technologies work together interactively, that is, that they are co-producers and partners in the design and use of technologies, careful to ensure that technologies do not completely absorb or replace human social relations, but enrich them. This alternative is certainly much harder to pursue than harm reduction, but it is not impossible, and it is the one that leads to good life.[15]

Hybridisation cancels every dualism. In particular, it erases the dualism between the system and the lifeworld (theorised by Habermas) and replaces it with a complex relational system in which each action/communication must choose between different causal mechanisms (Elder-Vass 2018).

To be human, technological enhancement must be able not only to distinguish between the different causal mechanisms, but also to choose the most productive social relationships that generate relational goods. New technologies generate not only unemployment, as many claim. They also release energy for the development of many jobs in the field of virtual reality and make it possible to put human work into those activities that have a high content of care, such as education, social assistance, and health, or a high content of cultural creativity.

Laaser and Bolton (2017) have shown that the introduction of new technologies associated with the advance of performance management practices has eroded the ethics of the care approach in banking organisations. Under electronic performance management monitoring in bank branches, in particular, co-worker relationships have become increasingly objectified, resulting in disconnected and conflict-ridden forms of engagement. This research reveals the multilayered and necessarily complex nature of co-worker relationships

in a changing, technologically driven work environment and highlights the necessity for people to defend the capacity to care for others from the erosive tendencies of individualised processes. Within the relational approach, this entails assessing the way in which an organisation uses AI/robots to enhance human relations from the viewpoint of what I call 'ODG systems' aimed at the relational steering of digitised organisations.[16]

ODG systems are based on the sequence: Relational observation (O) → Relational diagnosis (D) → Relational guidance (G). The principles according to which the ODG systems operate are modalities that orient those who have the responsibility of guiding a network of relations among subjects to operate interactively on the basis of cooperative rules that allows the subjects, supported by AI/robots, to produce relational goods. The agency is made by all the parties, as in an orchestra or a sports team, where everyone follows a cooperative standard that is used to continually regenerate a non-hierarchical and non-individualistic social structure, and consequently modifies the behaviour of the individual subjects, who are driven to generate relational goods. The cooperative norm is obviously inserted as a basic standard in the AI/robot that supports the agents of the network.

Now let's see some details to explain the acronym ODG. (O) Relational observation aims to define the problem to which AI must respond in terms of a problem that depends on a certain relational context. Therefore, it favours the meso level (in which relational goods can be produced). (D) Relational diagnosis aims to define the satisfactory (or not satisfactory) conditions with respect to the effects produced by the way the AI works on the relational context (i.e. whether the AI contributes to produce relational goods instead of relational evils). (G) Relational guidance aims to modify the AI and its way of working to support a relational context that can be mastered by people in order to generate relational goods.

An example of product innovation can be that of systems that try to regulate the AIs of self-driving cars. The AI must be constructed in such a way as to take into account the main parameters of the relational context in which it operates. The AI must see objects and people around, and assess their relationships with respect to those who sit in the driver's seat to put him in a position to intervene on situations that present a high contingency. Relational observation implies the ability of AI to calculate the relationships given in a context and those possible in its short-medium-range evolution. Relational diagnosis concerns the ability to perceive possible clashes in case the relationships with objects andpeople can become dangerous while the car is on

---

[15]Practical examples can be found in the journal '*Relational Social Work*' (free access).

[16]For more details see P. Donati, *Teoria relazionale della società*, Milano: FrancoAngeli, 1991: 346–356.

the street. Relational guidance means the ability to regulate these relationships in order to make driving safer.

At the organisation level, we can consider any company that uses AI/robots in order to produce satisfactory goods for customers. The AI/robots that are used must have similar characteristics to those just mentioned for the self-driving car. It is up to those who organise the production, distribution, and sale of business products to configure the AI/robots so that they have the ability to contextually relate and evaluate the progress of relationships in various sectors within the company and in the context of sales and of consumption. It is not enough to improve the cognitive intelligence of the single AI/robot. It is necessary to build AI/robots that are able to regulate the progress of the network of relations between producers, distributors, and consumers of company products.

I do not know practical examples already in place, but the idea of 'distributed responsibility' among all the actors in the network that produces, distributes, and uses the goods produced goes in this line. It requires that the AI/robots be constructed and monitored within a project of (1) observation of their relational work, (2) with the ability to diagnose deviations from satisfactory procedures and results, and (3) the orientation of the management to operate according to a relational guidance program.

The Issue of Social Regulations

The process of hybridisation of human relations and the relationships between humans and the environment due to digital technologies proceed rapidly with the succession of generations. Everyone feels the need to control the undesirable and perverse effects of technologies on human beings and their identities. However, it is not just about checking. It is, above all, a matter of redefining the relationality of the human environment, that is, of integral ecology.

The EU high-level expert group on AI has proposed seven essentials for achieving trustworthy AI (EU high-level expert group on AI 2018). Trustworthy AI should respect all applicable laws and regulations, as well as a series of requirements; specific assessment lists aim to help verify the application of each of the key requirements:

- *Human agency and oversight*: AI systems should enable equitable societies by supporting human agency and fundamental rights, and not decrease, limit or misguide human autonomy.
- *Robustness and safety*: Trustworthy AI requires algorithms to be secure, reliable, and robust enough to deal with errors or inconsistencies during all life cycle phases of AI systems.

- *Privacy and data governance*: Citizens should have full control over their own data, while data concerning them will not be used to harm or discriminate against them.
- *Transparency*: The traceability of AI systems should be ensured.
- *Diversity, non-discrimination, and fairness*: AI systems should consider the whole range of human abilities, skills, and requirements, and ensure accessibility.
- *Societal and environmental well-being*: AI systems should be used to enhance positive social change and enhance sustainability and ecological responsibility.
- *Accountability*: Mechanisms should be put in place to ensure responsibility and accountability for AI systems and their outcomes.

As for the use of ICTs, the ethical principles proposed by the 'European Civil Law Rules in Robotics' (Nevejans 2016), which should guarantee greater security for humanity, have been stated in the following terms:

- Protect people from any possible damage caused by ICTs.
- Ensure that the person is always able to use the ICTs without being obliged to perform what they have requested.
- Protecting humanity from violations of privacy committed by ICTs.
- Maintain control over information captured and processed by ICTs.
- To avoid that for certain categories of people there can be a sense of alienation towards ICTs.
- Prevent the use of ICTs to promote the loss of social ties.
- Guaranteeing equal opportunities to access ICTs.
- Control the use of technologies that tend to modify the physical and mental characteristics of the human person.

The House of Lords (2018) has put forward five overarching principles for an AI (ethical) Code (of conduct):

- Artificial intelligence should be developed for the common good and benefit of humanity.
- Artificial intelligence should operate on principles of intelligibility and fairness.
- Artificial intelligence should not be used to diminish the data rights or privacy of individuals, families, or communities.
- All citizens have the right to be educated to enable them to flourish mentally, emotionally, and economically alongside artificial intelligence.
- The autonomous power to hurt, destroy, or deceive human beings should never be vested in artificial intelligence.

It is evident that all these indications, suggestions, and statements of principle are important, but they risk being just like wishful thinking. In my opinion, these general principles (1) are difficult to implement in the absence of a clear anthropology and (2) are not at all sufficient to prevent the negative and perverse effects of the DTM, because they do not sufficiently take into account the relational nature of the human person and the social formations in which her personality develops. To avoid the new forms of alienation generated by the hybridisation of the human we need a new relational thought that is able to re-enter the specific distinctions of the human, its qualities and properties, into people's identities and social relations.

Conclusions

With the fourth technological revolution, social identities, relations, and organisations are forced to take shape in the environment of a Digital Matrix that works through a symbolic code that tends to replace the ontological, ideal, moral, and theological matrices that have structured societies in the past. As a form of *Technological-Gnosis*, its peculiarity is that of making the boundaries between human and non-human labile and crossable in every way in order to foster hybrids. Hybrids, however, are not random and purely contingent entities. They stem from complex interactional networks in which social relations are mediated by the DTM. The processes of hybridisation are selective and stratified according to the ways in which the human/non-human distinction is thought and practised. Three possible scenarios of hybridisation can be outlined: adaptive, turbulent, and relationally steered.

As a criterion for evaluating hybridisation processes, I have proposed assessing how digital technologies mediate the transformations of people's mind-body identities with their sociality so as to assess when such mediations produce those relational goods that lead to a virtuous human fulfilment or, instead, relational evils.

The justification of this perspective is based on the fact that human beings are at the same time the creators of society and its product. They are the parents and the children of society. As a consequence, it is not the physical world (Anthropocene) and/or the artificial world (AI/robots) that can produce the human being as a human being, but society, which, from the point of view of my relational sociology, consists of relationships, i.e. 'is' (and not 'has') relationships. Therefore, the quality and causal property of what is properly *human* comes into existence, emerges and develops only in social life, that is, only in and through people's sociability, which is however embedded in the practical order and has roots in the natural order (see Fig. 3). In fact, only in sociality does nature exist for the human being as a bond with the other human being. The vital element of human reality lies in the relationality, good or bad, that connects Ego to Other and Other to Ego. The proper humanisation of the person is achieved only through the sociality that can be enjoyed by

generating those relational goods in which the naturalism of the human being and his technological enhancement complement each other.

References

Adams, M. (2006). Hybridizing habitus and reflexivity: Towards an understanding of contemporary identity? *Sociology, 40*(3), 511–528.
Archer, M. (1995). *Realist social theory: The morphogenetic approach.* Cambridge: Cambridge UP.
Archer, M. (2013). *Reflexivity.* Sociopedia.isa. https://doi.org/10.1177/205684601373.
Baker, L. R. (2000). *Persons and bodies: A constitution view.* Cambridge: Cambridge UP.
Baker, L. R. (2013). Technology and the future of persons. *The Monist, 96*(1), 37–53.
Barlas, Z. (2019). When robots tell you what to do: Sense of agency in human- and robot-guided actions. *Consciousness and Cognition, 75,* 102819. https://doi.org/10.1016/j.concog.2019.102819.
Bekoff, M., & Pierce, J. (2009). *Wild justice: The moral lives of animals.* Chicago: University of Chicago Press.
Bienvault, P. (2019). *Dans les Ehpad, les humanoids soulèvent un débat éthique.* Available via La Croix. Retrieved February 25, 2020, from https://www.la-croix.com/JournalV2/Ehpad-humanoides-soulevent-debat-ethique-2019-02-01-1100999447
Cappuccio, M., Peeters, A., & McDonald, W. (2019). Sympathy for Dolores: Moral consideration for robots based on virtue and recognition. *Philosophy and Technology, 34*(1), 129–136. https://doi.org/10.1007/s13347-019-0341-y.
Clam, J. (2000). System's sole constituent, the operation: Clarifying a central concept of Luhmannian theory. *Acta Sociologica, 43*(1), 63–79.
Cross, S., & Morris, M. (2003). Getting to know you: The relational self-construal, relational cognition, and well-being. *Personality and Social Psychology Bulletin, 29*(4), 512–523.
Donati, P. (2010). *La matrice teologica della società.* Soveria Mannelli: Rubbettino.
Donati, P. (2011). *Relational sociology: A new paradigm for the social sciences.* London: Routledge.
Donati, P. (2014). Morality of action, reflexivity and the relational subject. In D. K. Finn (Ed.), *Distant markets, distant harms: Economic complicity and Christian ethics* (pp. 54–88). Oxford: Oxford University Press.
Donati, P. (2015). Manifesto for a critical realist relational sociology. *International Review of Sociology, 25*(1), 86–109.
Donati, P. (2019). Discovering the relational goods: Their nature, genesis and effects. *International Review of Sociology, 29*(2), 238–259.
Donati, P. (2020). Being human in the digital matrix land. In M. Carrigan, D. Porpora, & C. Wight (Eds.), *The future of the human and social relations.* Abingdon: Routledge.
Donati, P., & Archer, M. (2015). *The relational subject.* Cambridge: Cambridge UP.
Eacott, S. (2018). Embedded and embodied auctors. In *Beyond leadership, a relational approach to organizational theory in education* (pp. 81–94). New York: Springer.
Elder-Vass, D. (2018). Lifeworld and systems in the digital economy. *European Journal of Sociology, 21*(2), 227–244.
Elster, J. (2017). The temporal dimension of reflexivity: Linking reflexive orientations to the stock of knowledge. *Distinktion, 18*(3), 274–293.
EU High-Level Expert Group on AI. (2018). *Policy and investment recommendations for trustworthy AI.* Brussels: European Commission.
Floridi, L. (Ed.). (2015). *The onlife manifesto: Being human in a hyperconnected era.* New York: Springer.

Floridi, L. (2019). *The logic of information: A theory of philosophy as conceptual design*. Oxford: Oxford UP.

Greenfield, S. (2009). *ID: The quest for identity in the 21st century*. London: Sceptre.

Greenfield, S. (2014). *Mind change: How digital technologies are leaving their mark on our brains*. New York: Random House.

Guardini, R. (1925). *Der Gegensatz*. Mainz: Matthias-Grunewald.

Harari, Y. N. (2017). *Homo Deus: A brief history of tomorrow*. New York: Harper Collins.

Hayles, K. (1999). *How we became posthuman. Virtual bodies in cybernetics, literature, and informatics*. Chicago: University of Chicago Press.

Houkes, W., & Meijers, A. (2006). The ontology of artefacts: The hard problem. *Studies in History and Philosophy of Science, 37*, 118–131.

House of Lords. (2018). *AI in the UK: Ready, willing and able?* Select Committee on Artificial Intelligence, HL Paper 100, London, 16 April 2018.

Laaser, K., & Bolton, S. (2017). Ethics of care and co-worker relationships in UK banks. *New Technology, Work and Employment, 32*(3), 213–227.

LaGrandeur, K. (2015). Emotion, artificial intelligence, and ethics. In J. Romportl, E. Zackova, & J. Kelemen (Eds.), *Beyond artificial intelligence: The disappearing human-machine divide* (pp. 97–110). Dordrecht: Springer.

Lally, E. (2002). *At home with computers*. Oxford/New York: Berg.

Lévy, P. (1997). *L'intelligence collective: Pour une anthropologie du cyberspace*. Paris: La Découverte/Poche.

Lindhard, T. (2019). Consciousness from the outside-in and inside-out perspective. *Journal of Consciousness Explorations & Research 10*(3), 1–15.

Luhmann, N. (1976). Generalized media and the problem of contingency. In J. Loubser et al. (Eds.), *Explorations in general theory in social science: Essays in honor of Talcott Parsons* (pp. 507–532). New York: Free Press.

Luhmann, N. (1995). *Social systems*. Palo Alto, CA: Stanford University Press.

Malabou, C. (2009). *Ontologie de l'accident: Essai sur la plasticité destructrice*. Paris: Editions Léo Scheer.

Nevejans, N. (2016). *European civil law rules in robotics*. Brussels: European Parliament: Policy Department Citizens' Rights and Constitutional Affairs.

Nørskov, M. (2015). Revisiting Ihde's fourfold "Technological Relationships": Application and modification. *Philosophy and Technology, 28*(2), 189–207.

Pana, L. (2006). Artificial intelligence and moral intelligence. *TripleC, 4*(2), 254–264.

Parsons, T. (1977). *Social systems and the evolution of action theory*. New York: Free Press.

Pitts-Taylor, V. (2016). *The brain's body: Neuroscience and corporeal politics*. Durham: Duke University Press.

Porpora, D. (2018). Materialism, emergentism, and social structure: A response to Wendt's Quantum Mind. *Journal for the Theory of Social Behaviour, 48*(2), 183–187.

Shotter, J. (2012). Gergen, confluence, and his turbulent, relational ontology: The constitution of our forms of life within ceaseless, unrepeatable, intermingling movements. *Psychological Studies, 57*(2), 134–141.

Spencer-Brown, G. (1979). *Laws of form*. New York: Dutton.

Stanford Encyclopedia of Philosophy. (2018). *Supervenience*. Stanford, CA: Stanford University. Retrieved from https://plato.stanford.edu/entries/supervenience/.

Stichweh, R. (2000). System theory as an alternative to action theory? The rise of 'communication' as a theoretical option. *Acta Sociologica, 43*(1), 5–13.

Teubner, G. (2002). Hybrid laws: Constitutionalizing private governance networks. In R. Kagan & K. Winston (Eds.), *Legality and community* (pp. 311–331). Berkley: Berkeley Public Policy Press.

Teubner, G. (2006a). The anonymous matrix: Human Rights violations by 'private' transnational actors. *Modern Law Review, 69*(3), 327–346.

Teubner, G. (2006b). Rights of non-humans? Electronic agents and animals as new actors in politics and law. *Journal of Law and Society, 33*(4), 497–521.

Wendt, A. (2015). *Quantum mind and social science: Unifying physical and social ontology*. Cambridge: Cambridge UP.

What Is It to Implement a Human-Robot Joint Action?

Aurelie Clodic and Rachid Alami

Contents

Abstract

Joint action in the sphere of human–human interrelations may be a model for human–robot interactions. Human–human interrelations are only possible when several prerequisites are met, inter alia: (1) that each agent has a representation within itself of its distinction from the other so that their respective tasks can be coordinated; (2) each agent attends to the same object, is aware of that fact, and the two sets of "attentions" are causally connected; and (3) each agent understands the other's action as intentional. The authors explain how human–robot interaction can benefit from the same threefold pattern. In this context, two key problems emerge. First, how can a robot be programed to recognize its distinction from a human subject in the same space, to detect when a human agent is attending to something, to produce signals which exhibit their internal state and make decisions about the goal-directedness of the other's actions such that the appropriate predictions can be made? Second, what must humans learn about robots so they are able to interact reliably with them in view of a shared goal? This dual process is here examined by reference to the laboratory case of a human and a robot who team up in building a stack with four blocks.

Keywords

Human · Robot interaction · Joint action

A. Clodic (✉) · R. Alami
LAAS-CNRS, Université de Toulouse, CNRS, Artificial and Natural Intelligence Toulouse Institute (ANITI), Toulouse, France
e-mail: aurelie.clodic@laas.fr

Introduction

In this chapter, we present what is it to implement a joint action between a human and a robot. Joint action is "a social interaction whereby two or more individuals coordinate their actions in space and time to bring about a change in the environment." (Sebanz et al. 2006: 70). We consider this implementation through a set of needed coordination processes to realize this joint action: Self-Other Distinction, Joint Attention, Understanding of Intentional Action, and Shared Task Representation. It is something that we have

J. von Braun et al. (eds.), *Robotics, AI, and Humanity*, https://doi.org/10.1007/978-3-030-54173-6_19

already talked about in Clodic et al. (2017) but we will focus here on one example. Moreover, we will speak here about several elements that are components of a more global architecture described in Lemaignan et al. (2017). We introduce a simple human-robot collaborative to illustrate our approach. This example has been used as a benchmark in a series of workshop "toward a Framework for Joint Action" (fja.sciencesconf.org) and is illustrated in Fig. 1. A human and a robot have the common goal to build a stack with four blocks. They should stack the blocks in a specific order (1, 2, 3, 4). Each agent participates in the task by placing his/its blocks on the stack. The actions available to each agent are the following: take a block on the table, put a block on the stack, remove a block from the stack, place a block on the table, and give a block to the other agent.

This presentation is a partial point of view regarding what is and can be done to implement a joint action between a robot and a human since it presents only one example and a set of software developed in our lab. It only intends to explain what we claim is needed to enable a robot to run such a simple scenario.

At this point, it has to be noticed that from a philosophical point of view, we have been taught that some philosophers such as Seibt (2017) stressed that the robotics intentionalist vocabulary that we use is considered as problematic especially when robots are placed in social interaction spaces. In the following, we will use this intentionalist vocabulary in order to describe the functionalities of the robot, such as "believe" and "answers," because this is the way we describe our work in robotics and AI communities. However, to accommodate the philosophical concern, we would like to note that this can be considered as shorthand for "the robot simulates the belief," "the robot simulates an answer," etc. Thus, whenever robotic behavior is described with a verb that normally characterizes a human action, these passages can be read as a reference to the robot's simulation of the relevant action.

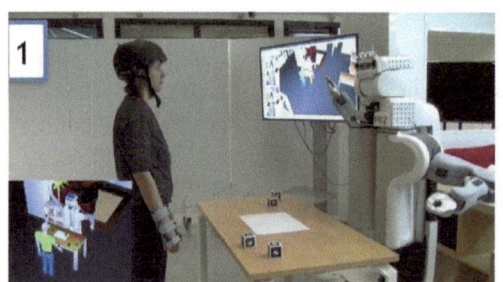
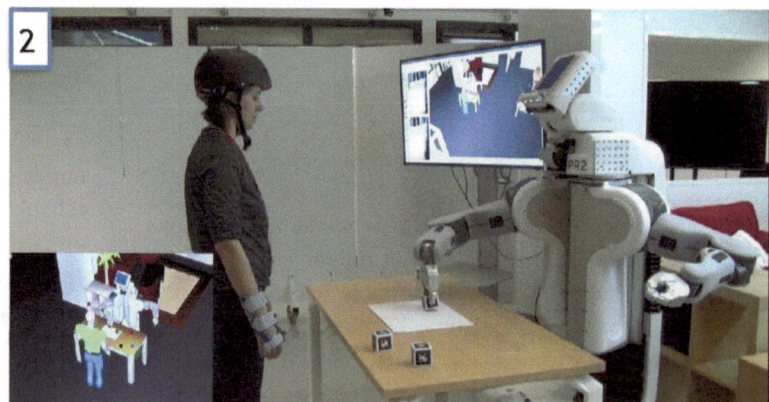
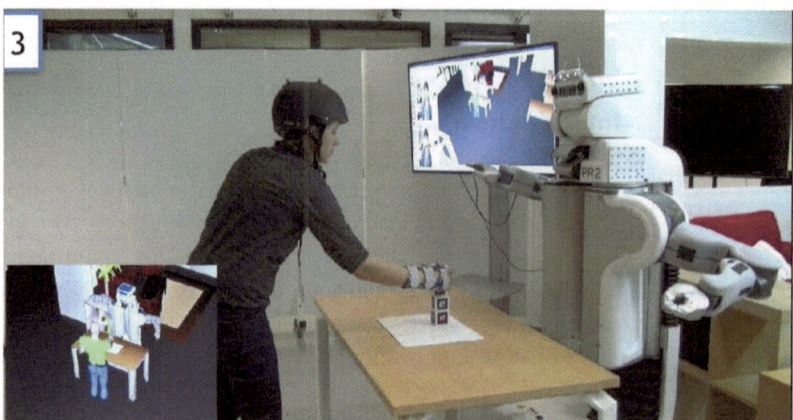
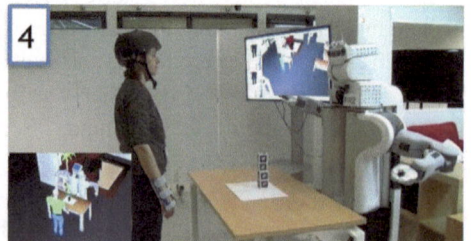

Fig. 1 A simple human–robot interaction scenario: A human and a robot have the common goal to build a stack with four blocks. They should stack the blocks in a specific order (1, 2, 3, 4). Each agent participates in the task by placing his/its blocks on the stack. The actions available to each agent are the following: take a block on the table, put a block on the stack, remove a block from the stack, place a block on the table, and give a block to the other agent. Also, the human and the robot observe one another. Copyright laas/cnrs https://homepages.laas.fr/aclodic

Self-Other Distinction

The first coordination process is Self-Other Distinction. It means that "for shared representations of actions and tasks to foster coordination rather than create confusion, it is important that agents also be able to keep apart representations of their own and other's actions and intentions" (Pacherie 2012: 359).

Regarding our example, it means that each agent should be able to create and maintain a representation of the world for its own but also from the point of view of the other agent. In the following, we will explain what the robot can do to build this kind of representation. The way a human (can) builds such representation for the robot agent (and on which basis) is still an open question.

Joint Attention

The second coordination process is Joint Attention. Attention is the mental activity by which we select among items in our

perceptual field, focusing on some rather than others (see Watzl 2017). In a joint action setting, we have to deal with joint attention, which is more than the addition of two persons' attention. "The phenomenon of joint attention involves more than just two people attending to the same object or event. At least two additional conditions must be obtained. First, there must be some causal connection between the two subjects' acts of attending (causal coordination). Second, each subject must be aware, in some sense, of the object as an object that is present to both; in other words, the fact that both are attending to the same object or event should be open or mutually manifest (mutual manifestness)" (Pacherie 2012: 355).

On the robot side, it means that the robot must be able to detect and represent what is present in the joint action space, i.e., the joint attention space. It needs to be equipped with situation assessment capabilities (Lemaignan et al. 2018; Milliez et al. 2014).

In our example, illustrated in Fig. 2, it means that the robot needs to get:

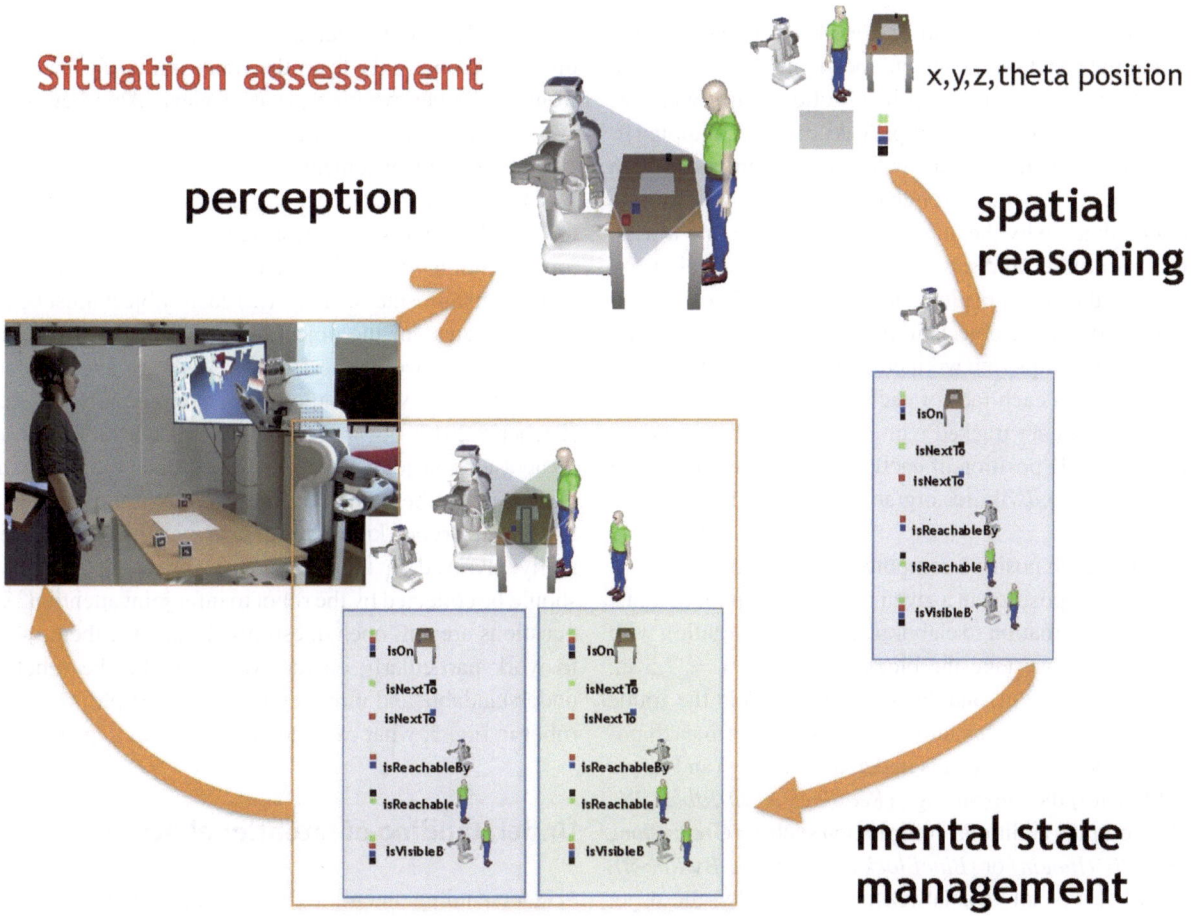

Fig. 2 Situation Assessment: the robot perceives its environment, builds a model of it, and computes facts through spatial reasoning to be able to share information with the human at a high level of abstrac-

tion and realizes mental state management to infer human knowledge. Copyright laas/cnrs https://homepages.laas.fr/aclodic

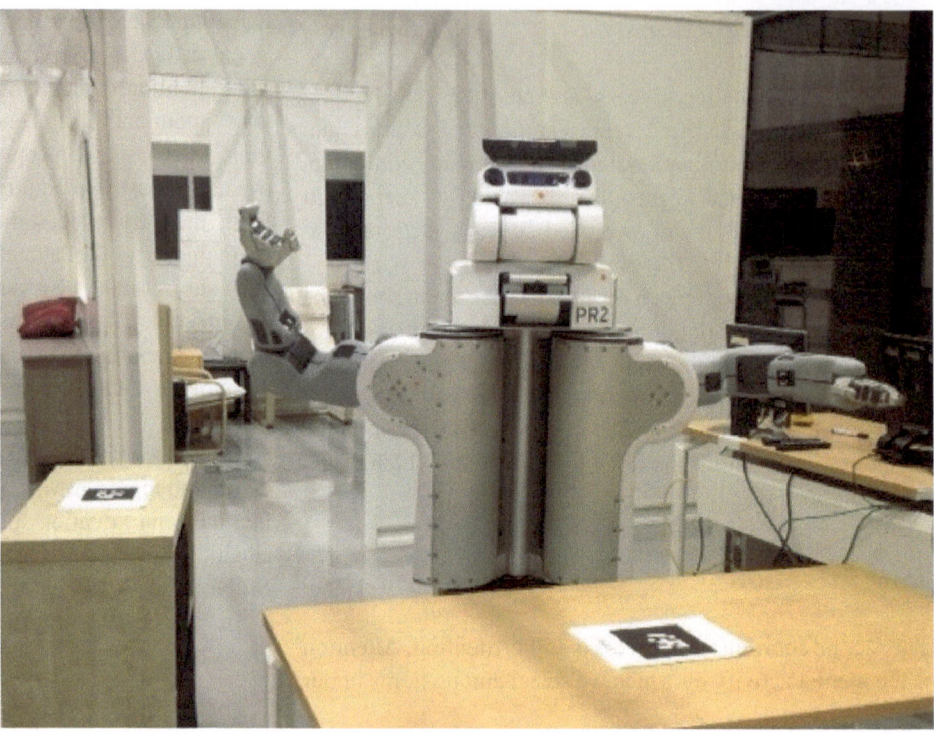

Fig. 3 What can we infer viewing this robot? There is no standard interface for the robot so it is difficult if not impossible to infer what this robot is able to do and what it is able to perceive (from its environment but also from the human it interacts with). Copyright laas/cnrs https:// homepages.laas.fr/aclodic

- its own position, that could be done for example by positioning the robot on a map and localizing it with the help of its laser (e.g., using amcl localization (http://wiki.ros.org/amcl) and gmapping (http://wiki.ros.org/gmapping))
- the position of the human with whom it interacts with (e.g., here it is tracked through the use of a motion capture system, that's why the human wears a helmet and a wrist brace. So more precisely, in this example, the robot has access to the head position and the right hand position)
- the position of the objects in the environment (e.g., here, a QR-code (https://en.wikipedia.org/wiki/QR_code) has been glued on each face of each block. These codes, and so, the blocks are tracked with one of the robot cameras. We get the 3D position of each block in the environment (e.g., with http://wiki.ros.org/ar_track_alvar))

However, each position computed by the robot is given as x, y, z, and theta position in a given frame. We cannot imagine to use such information to elaborate a verbal interaction with the human: "please take the block at position $x = 7.5$ m, $y = 3.0$ m, $Z = 1.0$ m, and theta $= 3.0$ radians in the frame map...". To overcome this limit, we must transform each position in an information that is understandable by (and hence shareable with) the human, e.g., *(RedBlock is On Table)*. We can also compute additional information such as *(GreenBlock is Visible By Human)* or *(BlueBlock is Reachable By Robot)*. This is what we call "spatial reasoning." Finally, the robot must also be aware that the information available to the human can be different from the one it has access to, e.g., an obstacle on the table can prevent her/him to see what is on

the table. To infer the human knowledge, we compute all the information not only from the robot point of view but also from the human position point of view (Alami et al. 2011; Warnier et al. 2012; Milliez et al. 2014), it is what we call "mental state management."

On the human side, we can infer that the human is able to have the same set of information from the situation. But joint attention is more than that. We have to take into account "mutual manifestness," i.e., "(...) each subject must be aware in some sense, of the object as an object that is present to both; in other words the fact that both are attending to the same object or event should be open or mutually manifest..." (Pacherie 2012: 355). It raises several questions. How can a robot exhibit joint attention? What cues the robot should exhibit to let the human to infer that joint attention is met? How can a robot know that the human it interacts with is really involved in the joint task? What are the cues that should be collected by the robot to infer joint attention? These questions are still open questions. To answer them, we have to work particularly on the way to make the robot more understandable and more legible. For example, viewing this robot in Fig. 3, what can one infer about its capabilities?

Understanding of Intentional Action

"Understanding intentions is foundational because it provides the interpretive matrix for deciding precisely what it is that someone is doing in the first place. Thus, the exact same physical movement may be seen as giving an object, sharing it, loaning it, moving it, getting rid of it, returning it, trading it, selling it, and on and on—depending on the goals and intentions of the actor" (Tomasello et al. 2005: 675). Understanding of intentional action could be seen as a building block of understanding intentions, it means that each agent should be able to read its partner's actions. To understand an intentional action, an agent should, when observing a partner's action or course of actions, be able to infer their partner's intention. Here, when we speak about partner's intention we mean its goal and its plan. It is linked to action-to-goal prediction (i.e., viewing and understanding the on-going action, you are able to infer the underlying goal) and goal-to-action prediction (i.e., knowing the goal you are able to infer what would be the action(s) needed to achieve it).

On the robot side, it means that it needs to be able to understand what the human is currently doing and to be able to predict the outcomes of the human's actions, e.g., it must be equipped with action recognition abilities. The difficulty here is to frame what should and can be recognized since the spectrum is vast regarding what the human is able to do. A way to do that is to choose to consider only a set of actions framed by a particular task.

On the other side, the human needs to be able to understand what the robot is doing, be able to infer the goal and to predict the outcomes of the robot's actions. It means, viewing a movement, the human should be able to infer what is the underlying action of the robot. That means the robot should perform movement that can be read by the human. Before doing a movement, the robot needs to compute it, it is motion planning. Motion planning takes as inputs an initial and a final configuration (for manipulation, it is the

position of the arms; for navigation, it is the position of the robot basis). Motion planning computes a path or a trajectory from the initial configuration to the final configuration. This path could be possible but not understandable and/or legible and/or predictable for the human. For example, in Fig. 4, on the left, you see a path which is possible but should be avoided if possible, the one on the right should be preferred.

In addition, some paths could be also dangerous and/or not comfortable for the human, as illustrated in Fig. 5. Human-aware motion planning (Sisbot et al. 2007; Kruse et al. 2013; Khambhaita and Alami 2017a, b) has been developed to enable the robot to handle the choice of a path that is acceptable, predictable, and comfortable to the human the robot interacts with.

Figure 6 shows an implementation of a human-aware motion planning algorithm (Sisbot et al. 2007, 2010; Sisbot and Alami 2012) which takes into account safety, visibility, and comfort of the human. In addition, this algorithm is able to compute a path for both the robot and the human, which can solve a situation where a human action is needed or can be used to balance effort between the two agents.

However, it is not sufficient. When a robot is equipped with something that looks like a head, for example, people tend to consider that it should act like a head because people anthropomorphize. It means that we need to consider the entire body of the robot and not only the base or the arms of the robot for the movement even if it is not needed to achieve the action (e.g., Gharbi et al. 2015; Khambhaita et al. 2016). This could be linked to the concept of coordination smoother which is "any kind of modulation of one's movements that reliably has the effect of simplifying coordination" (Vesper et al. 2010, p. 1001).

Fig. 4 Two final positions of the arm of the robot to get the object. The one at right is better from an interaction point of view since it is easily understandable by the human. However, from a computational point of view (and even from an efficiency if we just consider the robot action that needs to be performed) they are equivalent. Consequently, we need to take these features explicitly into account when planning robot motions. That is what human-aware motion planning aims to achieve. Copyright laas/cnrs https://homepages.laas.fr/aclodic

Human-Aware Motion Planning

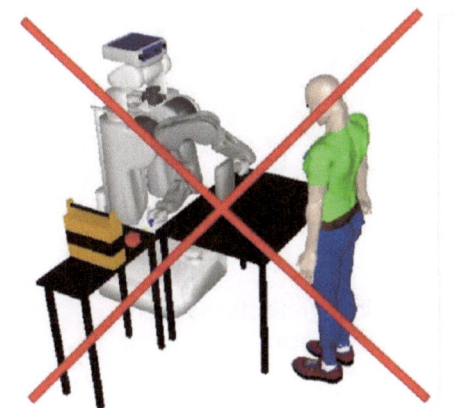

Fig. 5 Not "human-aware" positions of the robot. Several criteria should be taken into account, such as safety, comfort, and visibility. This is for the hand-over position but also for the overall robot position itself. Copyright laas/cnrs https://homepages.laas.fr/aclodic

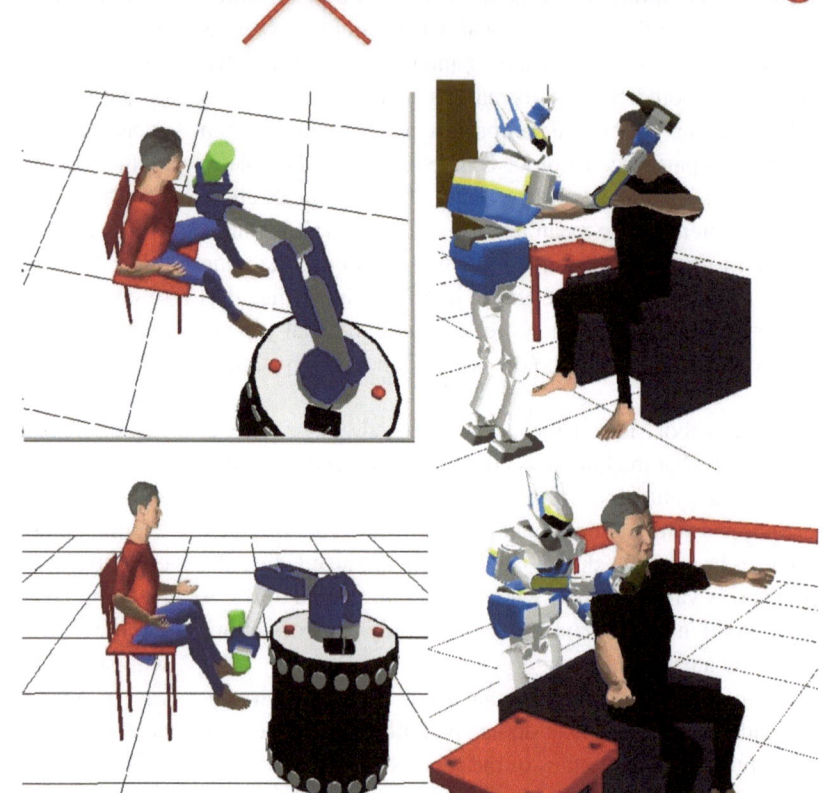

Human Aware Motion Planning

3 different HRI properties are defined and represented as 3D cost grids around the human

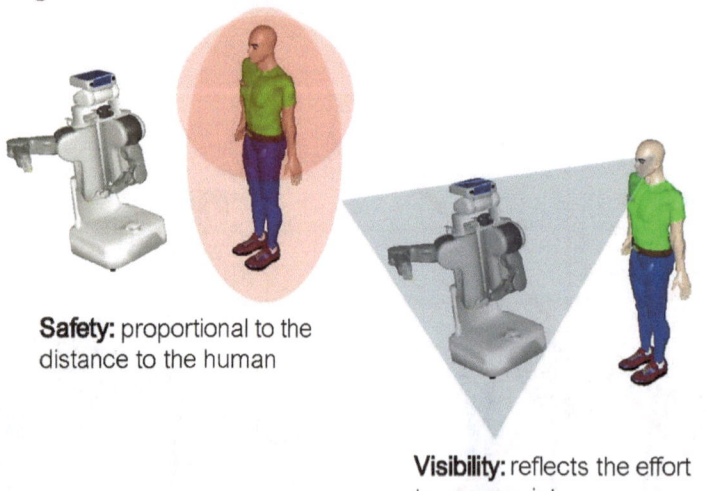

Safety: proportional to the distance to the human

Visibility: reflects the effort to see a point

Arm Comfort (left/right): combination of degree of freedom difference and potential energy

Fig. 6 An example of human-aware motion planning algorithm combining three criteria: safety of the human, visibility of the robot by the human, and comfort of the human. The three criteria can be weighed according to their importance with a given person, at a particular location or time of the task. Copyright laas/cnrs https://homepages.laas.fr/aclodic

Shared Task Representations

The last coordination process is shared task representations. As emphasized by Knoblich and colleagues (Knoblich et al. 2011), shared task representations play an important role in goal-directed coordination. Sharing representations can be considered as putting in perspective all the processes already described, e.g., knowing that the robot and the human track the same block in the interaction scene through joint attention and that the robot is currently moving this block in the direction of the stack by the help of intentional action understanding make sense in the context of the robot and the human building a stack together in the framework of a joint action.

To be able to share task representations, we need to have the same ones (or a way to understand them). We developed a Human-Aware Task Planner (HATP) based on Hierarchical Task Network (HTN) representation (Alami et al. 2006; Montreuil et al. 2007; Alili et al. 2009; Clodic et al. 2009; Lallement et al. 2014). The domain representation is illustrated in Fig. 7, it is composed of a set of actions (e.g., placeCube) and a set of tasks (e.g., buildStack) which combine action(s) and task(s). One of the advantages of such representation is that it is human readable. Here, placeCube

(Agent R, Cube C, Area A) means that for an Agent R, to place the Cube C in the Area A, the precondition is that R has in hand the Cube C and the effects of the action is that R has no more the Cube C in hand but the object C is on the stack of Area A. It is possible to add cost and duration to each action if we want to weigh the influence of each of the actions.

On the other hand, BuildStack is done by adding a cube (addCube) and then continue to build the stack (buildStack). Then each task is also refined until we get an action. HATP computes a plan both for the robot and the human (or humans) it interacts with as illustrated in Fig. 8. The workload could be balanced between the robot and the human; moreover, the system enables to postpone the choice of the actor at execution time (Devin et al. 2018). However, one of the drawbacks of such representation is that it is not expandable. Once the domain is written, you cannot modify it. One idea could be to use reinforcement learning. However, reinforcement learning is difficult to use "as is" in a human–robot interaction case. The reinforcement learning system needs to test any combination of actions to be able to learn the best one which could lead to nonsense behavior of the robot. This can be difficult to interpret for the human it interacts with and it will be difficult for him to interact with the robot,

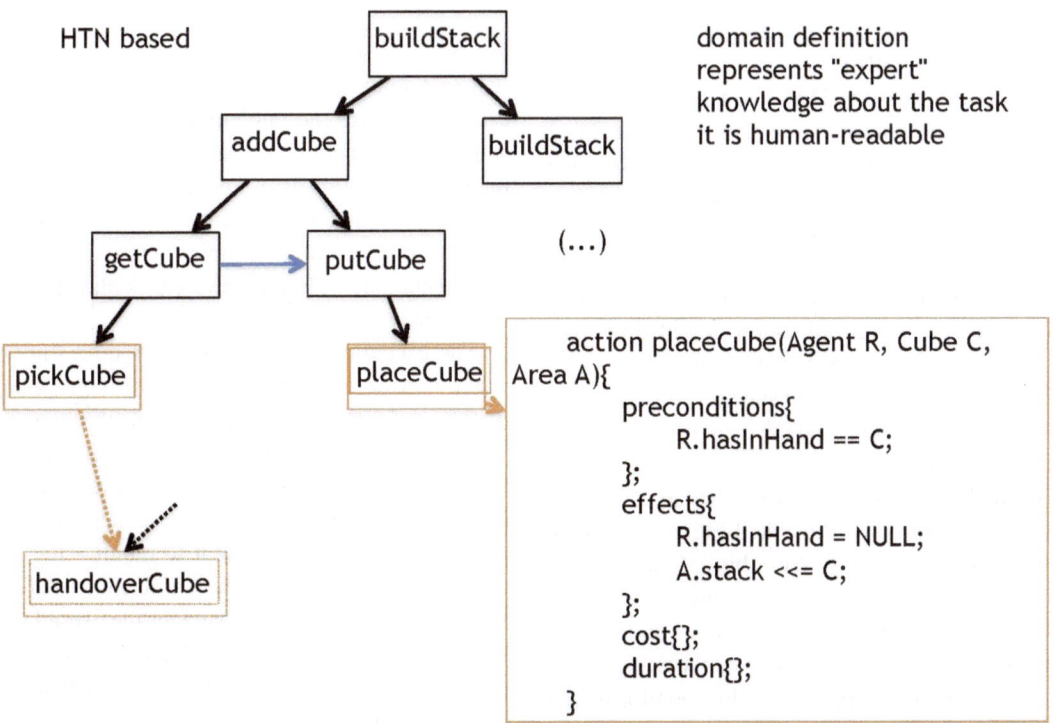

Fig. 7 HATP domain definition for the joint task buildStack and definition of the action placeCube: The action placeCube for an Agent R, a Cube C in an Area A, could be defined as follows. The precondition is that Agent R has the Cube C in hand before the action, the effect of the action is that Agent R does not have the Cube C anymore and the cube C is on the stack in Area A. Task buildStack combines addCube and buildStack. Task addCube combines getCube and putCube. Task getCube could be done either by picking the Cube or doing a handover. Copyright laas/cnrs https://homepages.laas.fr/aclodic

Human-Aware Task Planner

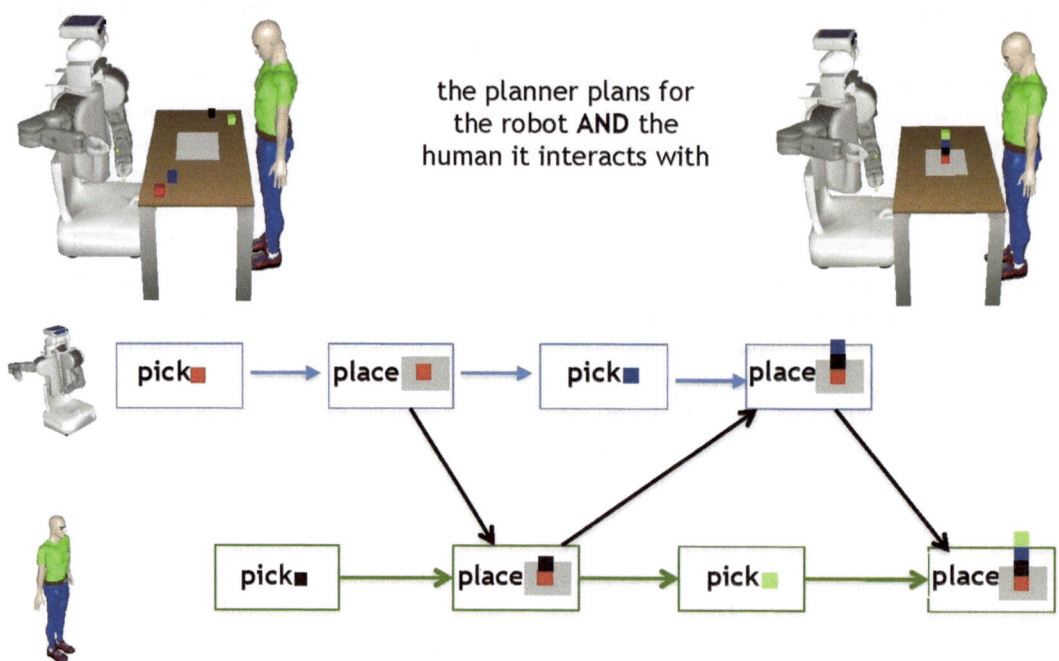

Fig. 8 HATP shared (human and robot) plan example for the stack of cubes example. Copyright laas/cnrs https://homepages.laas.fr/aclodic

and will lead to learning failure. To overcome this limitation, we have proposed to mix the two approaches by using HATP as a bootstrap for a reinforcement learning system (Renaudo et al. 2015; Chatila et al. 2018).

With a planning system as HATP, we have a plan for both the robot and the human it interacts with but this is not enough. If we follow Knoblich and colleagues (Knoblich et al. 2011) idea, shared task representations do not only specify in advance what the respective tasks of each of the coagents are, they also provide control structures that allow agents to monitor and predict what their partners are doing, thus enabling interpersonal coordination in real time. This means that the robot not only need the plan, but also ways to monitor this plan. Besides the world state (cf. Fig. 2 section regarding situation assessment) and the plan, we developed a monitoring system that enables the robot to infer plan status and action status both from its point of view and from the point of view of the human as illustrated Fig. 9 (Devin and Alami 2016; Devin et al. 2017). With this information, the robot is able to adapt its execution in real time. For example, there may be a mismatch between action status on the robot side and on the human side (e.g., the robot waiting for an action from the human). Equipped with this monitoring, the robot can detect the issue and warn. The issue can be at plan status level, e.g., the robot considering that the plan is no longer achievable while it detects that the human continues to act.

Conclusion

We have presented four coordination processes needed to realize a joint action. Taking these different processes into account requires the implementation of dedicated software: self-other distinction → mental state management; joint attention → situation assessment; understanding of intentional action → action recognition abilities as well as human-aware action (motion) planning and execution; shared task representations → human-aware task planning and execution as well as monitoring.

The execution of a joint action requires not only for the robot to be able to achieve its part of the task but to achieve it in a way that is understandable to the human it interacts with and to take into account the reaction of the human if any. Mixing execution and monitoring requires making some choices at some point, e.g., if the camera is needed to do an action, the robot cannot use it to monitor the human if it is not in the same field of view. These choices are made by the supervision system which manages the overall task execution from task planning to low-level action execution.

We talked a little bit about how the human was managing these different coordination processes in a human–robot interaction framework and about the fact that there was still some uncertainty about how he was managing things. We believe that it may be necessary in the long term to give the human the means to better understand the robot at first.

Shared Plan Management

Taking others mental state during execution

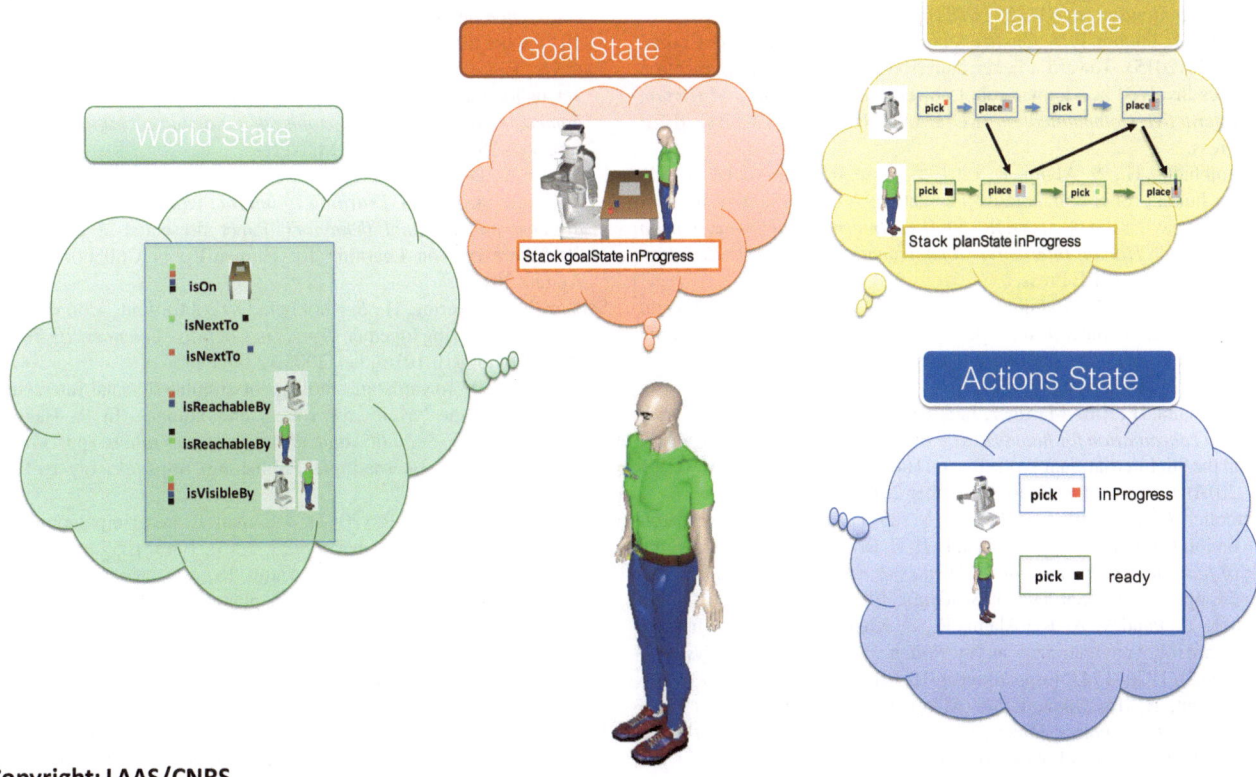

Copyright: LAAS/CNRS

Fig. 9 Monitoring the human side of the plan execution: besides the world state, the robot computes the state of the goals that need to be achieved, the status of the on-going plans as well of each action. It is done not only from its point of view but also from the point of view of the human. Copyright laas/cnrs https://homepages.laas.fr/aclodic

Finally, what has been presented in this chapter is partial for at least two reasons. First, we have chosen to present only work done in our lab but this work already covers the execution of an entire task and in an interesting variety of dimensions. Second, we make the choice to not mention the way to handle communication or dialog, to handle data management or memory, to handle negotiation or commitments management, to enable learning, to take into account social aspects (incl. privacy) or even emotional ones, etc. However, it gives a first intuition to understand what needs to be taken into account to make a human–robot interaction successful (even for a very simple task).

References

Alami, R., Clodic, A., Montreuil, V., Sisbot, E. A., & Chatila, R. (2006). Toward human-aware robot task planning. In *AAAI spring symposium: To boldly go where no human-robot team has gone before* (pp. 39–46). Palo Alto, CA.

Alami, R, Warnier, M., Guitton, J., Lemaignan, S., & Sisbo, E. A. (2011). When the robot considers the human. In *15th Int. Symposium on Robotics Research*.

Alili, S., Alami, R., & Montreuil, V. (2009). A task planner for an autonomous social robot. In H. Asama, H. Kurokawa, J. Ota, & K. Sekiyama (Eds.), *Distributed autonomous robotic systems* (Vol. 8, pp. 335–344). Berlin/Heidelberg: Springer.

Chatila, R., Renaudo, E., Andries, M., Chavez-Garcia, R. O., Luce-Vayrac, P., Gottstein, R., Alami, R., Clodic, A., Devin, S., Girard, B., & Khamassi, M. (2018). Toward self-aware robots. *Frontiers in Robotics and AI*, 51–20. https://doi.org/10.3389/frobt.2018.00088.

Clodic, A., Cao, H., Alili, S., Montreuil, V., Alami, R., & Chatila, R. (2009). Shary: A supervision system adapted to human-robot interaction. In R. Clodic, H. Cao, S. Alili, V. Montreuil, R. Alami, & R. Chatila (Eds.), *Experimental robotics* (pp. 229–238). Berlin/Heidelberg: Springer.

Clodic, A., Pacherie, E., Alami, R., & Chatila, R. (2017). Key elements for human-robot joint action. In R. Hakli & J. Seibt (Eds.), *Sociality and normativity for robots* (pp. 159–177). Cham: Springer International Publishing.

Devin, S., & Alami, R. (2016). An implemented theory of mind to improve human-robot shared plans execution. In *The Eleventh ACM/IEEE International Conference on Human Robot Interaction (HRI)* (pp. 319–326). Christchurch, New Zealand, March 7–10, 2016.201, IEEE Press.

Devin, S., Clodic, A., & Alami, R. (2017). About decisions during human-robot shared plan achievement: Who should act and how? In A. Kheddar, E. Yoshida, S. S. Ge, K. Suzuki, J. J. Cabibihan, F. Eyssel, & H. He (Eds.), *Social robotics* (Vol. 10652, pp. 453–463). Cham: Springer International Publishing.

Devin, S., Vrignaud, C., Belhassein, K., Clodic, A., Carreras, O., & Alami, R. (2018). Evaluating the pertinence of robot decisions in a human-robot joint action context: The PeRDITA questionnaire. In *27th IEEE International Symposium on Robot and Human Interactive Communication (RO-MAN) 2018* (pp. 144–151), IEEE Press.

Gharbi, M., Paubel, P. V., Clodic, A., Carreras, O., Alami, R., & Cellier, J. M. (2015). Toward a better understanding of the communication cues involved in a human-robot object transfer. In *24th and Human Interactive Communication (RO-MAN), 2015* (pp. 319–324). IEEE Press.

Khambhaita, H., & Alami, R. (2017a). Assessing the social criteria for human-robot collaborative navigation: A comparison of human-aware navigation planners. In *26th IEEE International Symposium on Robot and Human Interactive Communication (RO-MAN) 2017* (pp. 1140–1145), IEEE Press.

Khambhaita, H., & Alami, R. (2017b). *Viewing robot navigation in human environment as a cooperative activity*. Paper presented at the international symposium on robotics research (ISSR), Puerto Varas, Chile, 11–14 December 2017.

Khambhaita, H., Rios-Martinez, J., & Alami, R. (2016). *Head-body motion coordination for human aware robot navigation*. Paper presented at the 9th International Workshop on Human-Friendly Robotics (HFR 2016), Genoa, Italy, 29-30 September 2016.

Knoblich, G., Butterfill, S., & Sebanz, N. (2011). Psychological research on joint action: Theory and data. In B. H. Ross (Ed.), *The psychology of learning and motivation: Advances in research and theory* (Vol. 54, pp. 59–101). San Diego: Academic, Elsevier.

Kruse, T., Pandey, A. K., Alami, R., & Kirsch, A. (2013). Human-aware robot navigation: A survey. *Robotics and Autonomous Systems, 61*(12), 1726–1743. https://doi.org/10.1016/j.robot.2013.05.007.

Lallement, R., De Silva, L., & Alami, R. (2014). *HATP: An HTN planner for robotics*. Paper presented at the 2nd ICAPS Workshop on Planning and Robotics, Portsmouth, 22–23 June 2014.

Lemaignan, S., Warnier, M., Sisbot, E. A., Clodic, A., & Alami, R. (2017). Artificial cognition for social human–robot interaction: An implementation. *Artificial Intelligence, 247,* 45–69. https://doi.org/10.1016/j.artint.2016.07.002.

Lemaignan, S., Sallami, Y., Wallbridge, C., Clodic, A., Belpaeme, T., & Alami, R. (2018). *UNDERWORLDS: Cascading situation assessment for robots*. Paper presented at the IEEE/RSJ international conference on intelligent robots and systems (IROS 2018), Madrid, Spain, 1–5 October 2018.

Milliez, G., Warnier, M., Clodic, A., & Alami, R. (2014). A framework for endowing an interactive robot with reasoning capabilities about perspective-taking and belief management. In *The 23rd IEEE international symposium on robot and human interactive communication* (pp. 1103–1109). IEEE Press.

Montreuil, V., Clodic, A., & Alami, R. (2007). *Planning human centered robot activities*. Paper presented on the 2007 IEEE international conference on systems, man and cybernetics, Montreal, Canada, 7–10 October 2007.

Pacherie, E. (2012). The phenomenology of joint action: Self-agency versus joint agency. In A. Seeman (Ed.), *Joint attention: New developments in psychology, philosophy of mind, and social neuroscience* (pp. 343–389). Cambridge: MIT Press.

Renaudo, E., Devin, S., Girard, B., Chatila Alami, R., Khamassi, M., & Clodic, A. (2015). *Learning to interact with humans using goal-directed and habitual behaviors*. Paper presented at the RoMan 2015 Workshop on Learning for Human-Robot Collaboration, 31 August 2015.

Sebanz, N., Bekkering, H., & Knoblich, G. (2006). Joint action: Bodies and minds moving together. *Trends in Cognitive Sciences, 10*, 70–76. https://doi.org/10.1016/j.tics.2005.12.009.

Seibt, J. (2017). Towards an ontology of simulated social interaction: Varieties of the "as if" for robots and humans. In R. Hakli & J. Seibt (Eds.), *Sociality and normativity for robots* (pp. 11–39). Cham: Springer International Publishing. https://doi.org/10.1007/978-3-319-53133-5_2.

Sisbot, E. A., & Alami, R. (2012). A human-aware manipulation planner. *IEEE Transactions on Robotics, 28*(5), 1045–1057.

Sisbot, E. A., Marin-Urias, L. F., Alami, R., & Siméon, T. (2007). A human aware mobile robot motion planner. *IEEE Transactions on Robotics, 23*(5), 874–883.

Sisbot, E. A., Marin-Urias, L., Broquère, X., Sidobre, D., & Alami, R. (2010). Synthesizing robot motions adapted to human presence. *International Journal of Social Robotics, 2*(3), 329–343.

Tomasello, M., Carpenter, M., Call, J., Behne, T., & Moll, H. (2005). Understanding and sharing intentions: The origins of cultural cognition. *The Behavioral and Brain Sciences, 28*(5), 675–691. https://doi.org/10.1017/S0140525X05000129.

Vesper, C., Butterfill, S., Knoblich, G., & Sebanz, N. (2010). A minimal architecture for joint action. *Neural Networks, 23*(8–9), 998–1003. https://doi.org/10.1016/j.neunet.2010.06.002.

Warnier, M., Guitton, J., Lemaignan, S., & Alami, R. (2012). When the robot puts itself in your shoes. Explicit geometric management of position beliefs. In *The 21st IEEE international symposium on robot and human interactive communication (RO-MAN) 2012* (pp. 948–954), IEEE Press.

Watzl, S. (2017). *Structuring mind: The nature of attention and how it shapes consciousness*. Oxford: Oxford University Press.

Responsible Robotics and Responsibility Attribution

Aimee van Wynsberghe

Contents

Abstract

This paper stresses the centrality of human responsibility as the necessary foundation for establishing clear robotics policies and regulations; responsibility not on the part of a robot's hardware or software, but on the part of the humans behind the machines—those researching and developing robotics. Simply put, we need *responsible robotics*. Responsible robotics is a term that has recently 'come into vogue', yet an understanding of what *responsible robotics* means is still in development. In light of both the complexity of development (i.e. the many hands involved) and the newness of robot development (i.e. few regulatory boards established to ensure accountability), there is a need to establish procedures to assign future responsibilities among the actors involved in a robot's development and implementation. The three alternative laws of responsible robotics by Murphy and Wood play a formidable contribution to the discussion; however, they repeat the difficulty that Asimov introduced, that is, laws in general, whether they are for the robot or for the roboticist, are incomplete when put into practice. The proposal here is to extend the three alternative laws of responsible robotics into a more robust framework for responsibility

A. van Wynsberghe (✉)
Department of Philosophy, Technical University of Delft, Delft, The Netherlands
e-mail: aimeevanrobot@protonmail.com

© The Author(s) 2021
J. von Braun et al. (eds.), *Robotics, AI, and Humanity*, https://doi.org/10.1007/978-3-030-54173-6_20

attribution as part of the responsible robotics goal. This framework requires making explicit various factors: the type of robot, the stage of robot development, the intended sector of use, and the manner of robot acquisition. With this in mind, one must carefully consider the scope of the ethical issue in question and determine the kind of responsibility attributed to said actor(s).

Keywords

Responsibility · Robotics · Development · Implementation · Regulation

Introduction

The responsible development and use of robotics may have incredible benefits for humanity, from replacing humans in dangerous, life-threatening tasks with search and rescue robots (Murphy 2014) to the last mile delivery of lifesaving resources in humanitarian contexts (Gilman and Easton 2014; Chow 2012). Despite the success and efficiency that robots promise to bring, however, there are societal and ethical issues that need to be addressed. For the last 20 years, robot ethicists have flagged some of the ethical concerns related to robots, for example: the dehumanization and de-skilling of care workers, care receivers, and care practices when robots are used in care contexts (Sharkey 2014; Sparrow and Sparrow 2006; Vallor 2011; van Wynsberghe 2012); the loss of contextual learning necessary for understanding the detailed needs of others when robots replace humans in surgical or humanitarian care (van Wynsberghe and Comes 2019; van Wynsberghe and Gastmans 2008); and a risk of deceiving children when using robots in the classroom (Sharkey 2016), to name a few. To exacerbate these issues, there is growing concern regarding private organizations proving themselves unworthy of society's trust by demonstrating a lack of concern for safety, e.g., the fatal crashes of self-driving cars. If robotics is truly to succeed in making our world a better place, the public must be able to place their trust in the designers, developers, implementers, and regulators of robot technologies. To do this, the many hands in robot development must engage in responsible innovation and implementation, what I will refer to here as *responsible robotics*.

Responsible robotics is a term that has recently "come into vogue" just as similar terms like *responsible research and innovation* (European Commission 2012, 2014; van de Poel and Sand 2018; van den Hoven 2013), *value sensitive design* (Friedman et al. 2015; Friedman 1996; van den Hoven 2013), and other forms of innovation that take societal values, such as privacy, safety, and security, explicitly into account

in the design of a product. In recent years, research courses[1] and articles (Murphy and Woods 2009) have been dedicated to the topic, and a not-for-profit established with the aim of promoting responsible robotics.[2] Yet, the understanding of what *responsible robotics* means is still in development. In light of both the complexity (i.e., the many hands involved) and newness of robot development (i.e., few regulatory boards established to ensure accountability), there is a need to establish procedures to assign future responsibilities among the actors involved in a robot's development and implementation. This paper starts with an analysis of the three alternative laws of responsible robotics proposed by Murphy and Woods aimed at shifting the discussion away from robot responsibility (in Asimov's stories) to human responsibility. While acknowledging the incredible benefit that these alternative laws bring to the field, the paper presented here will introduce several shortcomings, namely, the need for: a more nuanced understanding of responsibility; recognition of the entire development process of a robot; and recognition of the robot's impact extending beyond the human–robot interaction alone. The paper proceeds by showing the complexity of the concept of "responsibility" and what it might mean in a discussion of responsible robotics. Finally, I suggest a preliminary responsibility attribution framework—a way in which robot applications should be broken down into the various stages of development, sectors, and patterns of acquisition (or procurement) so as to identify the individuals responsible for ensuring that responsibility to tackle ethical issues of prospective robots are addressed proactively.

The "Laws of Responsible Robotics"

In 1942, science fiction writer Isaac Asimov published the short story "Runaround" in which the three laws of robotics first appeared. These laws would become a "literary tool" to guide many of his future robot stories, illustrating the difficulty for robots to embody the same ability for situational judgment as humans. As such, "although the robots usually behaved 'logically', they often failed to do the 'right' thing" (Murphy and Woods 2009: 14). The three laws were formulated as follows:

"One, A robot may not injure a human being or, through inaction, allow a human being to come to harm…

Two, a robot must obey the orders given it by human beings except where such orders would conflict with the First Law…

---

[1] IDEA League summer school on responsible robotics, see http://idealeague.org/blog-summer-school-responsible-robotics-and-ai/ (retrieved Jan 14, 2020).

[2] See responsiblerobotics.org (retrieved Jan 8, 2010).

Three, a robot must protect its own existence as long as such protection does not conflict with the First or Second Laws." (Asimov 2004: 37).

In 2009 came the first scholarly work to outline the three *alternative* laws of "responsible robotics," an alternative to the three unworkable laws of robotics found in the Asimov stories (Murphy and Woods 2009). These "alternative three laws of responsible robotics" are stated as follows:

1. A human may not deploy a robot without the human–robot work system meeting the highest legal and professional standards of safety and ethics.
2. A robot must respond to a human as appropriate for their roles.
3. A robot must be endowed with sufficient situated autonomy to protect its own existence as long as such protection provides smooth transfer of control to other agents consistent with the first and second laws (Murphy and Woods 2009: 19).

To be sure, these alternative laws are monumental for moving robotics research forward in a responsible way. Particularly noteworthy is their emphasis on the centrality of the researcher's responsibility for following ethical and professional design protocols, on maintaining safety in the robot architecture, and on ensuring the robot is capable of smooth transfer of control between states of human-in-command (or human operated) and full autonomy (or acting without direct real time human interaction). In line with the pivotal role of human responsibility embedded in these alternative laws is a recognition of the difference between robot agency, as presented in Asimov's stories, and human agency, i.e., the type of agency presumed (and prescribed) for the alternative laws; this difference "illustrates why the robotics community should resist public pressure to frame current human–robot interaction in terms of Asimov's laws" (Murphy and Woods 2009: 19). Thus, the three alternative laws redirect the world's attention to the responsibilities of the human actors, e.g., the researchers, in terms of safe human–robot interactions.

Beyond the "Laws of Responsible Robotics"

Without diminishing the value of the contribution that these alternative laws provide, they do fall short in five critical ways. First, although the alternative laws place an emphasis on the centrality of human responsibility, the authors fall short of capturing the ethical nuance in the term "responsibility" outside of a colloquial usage. Responsibility may, in some instances, refer to individual accounts while in other instances may refer to collective accounts: the roboticist is

responsible for high-level planning, while the company is responsible for having a code of conduct for employees. Responsibility may be forward looking (for future consequences) or backward looking (for past consequences). Giving voice to the nuances of responsibility helps to shape a more robust account of responsible robotics: who is responsible for what in the development, deployment, and implementation of robots.

Second, while the alternative laws do refer to the "user" in the human–robot interaction, the focus does not acknowledge the size and scale of today's robot development process and the many actors involved in bringing a robot from idea to deployment. For some of the more complex technologies of today, there is a recent phenomenon used to describe the difficulty (or inability) to assign any particular person as "responsible" because of the number of actors involved. This is known as "the problem of many hands" (van de Poel et al. 2012). There are "many hands" (Johnson 2015; van de Poel et al. 2012) involved in the development chain of robots from robot developers to producers to implementers to regulators, each with differing roles and responsibilities. Understanding this network of actors, alongside the variety of kinds of responsibility, is also important for identifying who in this network is responsible for what and the form (or kind) of responsibility (e.g., legal vs. moral responsibility).

Third, each of the alternative laws makes reference to the workings of the robot within the human–robot interaction, yet it is important to acknowledge that the impact of the robot will extend far beyond the direct human–robot interaction (van Wynsberghe and Li 2019). Consider surgical robots in the healthcare system as an example; understanding the responsible use of these robots includes understanding that the robot does more than interact safely with surgeon and patient. These robots have changed: the manner of education for medical students (they must also be trained on a robot as well as conventional and/or laparoscopy) (van Koughnett et al. 2009), the allocation of funding and other resources within a hospital (surgical robots costs upwards of $1 million dollars), and the standard for care (e.g., surgical robots are thought to provide the highest standard of care in many instances and patients may demand this level of care). Responsible robotics should recognize that the introduction of a robot will be felt across an entire system, rather than strictly within the human–robot interaction.

Fourth, the alternative laws do not acknowledge the different stages and/or contexts of robot design, and as such, the various kinds of decision-making (and attributed responsibilities for said decision-making) that occur across these different stages. Being responsible for safe human–robot interactions in a healthcare context in which one must work together with FDA standards differs from ensuring safety for robots in edutainment contexts, where few standards exist regarding data privacy.[3]

Fifth, the formulation of the alternative laws hints at an assumption that it is already known what values a robot should be programmed for, e.g., safety. Also, the alternative laws imply that the obstacles facing roboticists are already known and understood. Yet it should be acknowledged that robotics, whether done in the academic space or made in a corporate space and sold to consumers, is very much an experiment, a social experiment when used in society (van de Poel 2013). Society has little experience with robots in personal and professional spaces so it should not be assumed that there is an understanding of the kinds of obstacles or ethical issues to take into account; these insights will only come with more experience of humans and robots interacting in various contexts.

Based on the these five elaborations on the three alternative laws of responsible robotics—a recognition of the many hands involved in robot development, the various kinds of responsibility attributed to these hands, the various stages of robot development and deployment, the experimental nature of robots in society, and the impact of the robot extending beyond the human—robot interaction alone—it becomes paramount to take a closer look at the three alternative laws in order to provide clear, complete guidelines for *responsible robotics*. What is needed for robot designers, developers, implementers, and regulators, is a tool to assign (a type of) responsibility to various (groups of) actors at the various stages of robot development and deployment in an academic (or any other) context. One may suggest an extension of these three alternative laws so as to incorporate various conceptions of responsibility or the various actors involved. But perhaps it is, instead, necessary to question the utility of strict laws or rules governing robotics in the first place. Given the complexity of robotics (e.g., different actors produce the hardware from the actors who create something with it and these again differ from the actors who will implement the robot), the experimental nature of robotics (whether the robot is being tested in an academic setting or in the wild) and the oftentimes lack of policy or regulations to guide the various actors (e.g., few Universities have ethical review boards dedicated to reviewing robotics experiments at Universities) perhaps a more flexible approach is needed to conceptualize responsible robotics. Specifically, what is missing from the alternative laws, and sorely needed, is a robust framework for responsibility attribution. Such a framework should be tasked with outlining the responsible actor or organization at the various stages in a robot's development. Even if an organization buys into the idea of *responsible robotics* in

theory, without designating a specific actor as responsible for these processes, they may never actually take place. The responsible actor(s) are tasked with the development process of the robot itself, but should also seek to go beyond these values through engaging with an ethical technology assessment (Palm and Hansson 2006), as an example, in order to capture the range of ethical issues in need of addressing.

With these thoughts in mind, the next section of this paper will take a closer look at the concept of responsibility to explore some of the distinctions in the term that can add help to shape a responsibility attribution framework for responsible robotics.

Responsibility

There is a wealth of literature on the concept of responsibility aimed at differentiating the different meanings of the term (Feinberg 1988; Johnson 2015; van de Poel and Sand 2018). Some works focus on distinctions between: normative and descriptive notions of responsibility (van de Poel et al. 2012; van de Poel and Sand 2018); senses of responsibility (Hart 2008); and/or between temporal accounts of responsibility, e.g., forward-looking and backward-looking responsibility (van de Poel et al. 2012; van de Poel and Sand 2018). Added to this, there are also heated discussions about the "traditional precondition of responsibility," namely that individuals who can be held responsible must meet the conditions of: "capacity, causality, knowledge, freedom, and wrong-doing" (van de Poel et al. 2012: 53). It would be impossible and unnecessary to cover all aspects of responsibility in this in this paper; however, for the purposes here—sketching a framework to prospectively assign responsibilities in the development of future robots—it is necessary to identify certain salient distinctions that help shape the concept of responsible robotics. The purpose of presenting certain distinctions in the next section is therefore twofold: first, to highlight the need for more granular discussions when using a phrase like "responsible robotics"; and, second, to identify the conceptualizations of responsibility necessary to establish a framework for assigning responsibility among the various actors in a robot's development and implementation.

Forward Looking–Backward Looking Responsibility

One of the first distinctions to address is that between forward-looking responsibility and backward-looking responsibility. Backward-looking responsibility relates to "things that have happened in the past and usually involves an evaluation of these actions and the attribution of blame or praise to the agent" (van de Poel and Sand 2018: 5; see also

---

[3]To be sure, reference to few privacy standards outside of a European Context where the new General Data Protection Regulation exists to ensure data privacy of consumers and citizens across Europe.

Smith 2007; Watson 2004). In robotics, backward-looking responsibility may refer to identifying who is responsible for a robot malfunctioning.

Forward-looking responsibility, on the other hand, refers to "things that have not yet occurred" (van de Poel et al. 2012: 51; van de Poel and Sand 2018). Forward-looking responsibility, therefore, can be understood as the responsibility to prevent harm from happening and/or possessing the character trait of being responsible (van de Poel and Sand 2018: 6). In robotics, forward-looking responsibility may refer to having an individual (or team) in a robotics company tasked with uncovering novel unintended consequences arising from new robot capabilities, and working towards their mitigation.

Of course, it is not possible to entirely separate backward- and forward-looking responsibility; if one establishes forward-looking responsibilities as moral obligations, then at a certain moment they could be breached, and as such, a discussion of backward-looking responsibility occurs. While it is necessary to consider both temporal dimensions in the assignment of responsibility, let us consider, for this paper, the prospective robots of the future, the robots that are currently in the stage of idea generation and will be designed, developed, and deployed over the next 5–10 years. For many of these robots there are no policy structures to guide researchers or developers. In such experimental situations of development, it is necessary to establish norms and expectations about who is responsible for what. One specific sense of forward-looking responsibility is understood as responsibility-as-obligation and refers to instances in which "one has to see to it that a certain desirable state-of-affairs is obtained, although one is free in how this state-of-affairs is to be brought about" (van de Poel et al. 2012: 52). Therefore, in this paper, let us consider what type of responsibility attribution framework is needed in the situations where we must establish prospective rules of engagement. Furthermore, in the creation of the responsibility attribution framework here, our goal is to identify the key players who should hold forward-looking responsibility-as-obligation to realize responsible robotics procedures and products.

Moral vs. Legal Responsibility

Another interesting distinction to be made in a discussion of responsibility is that between moral responsibility and legal responsibility. Moral responsibility has been defined as "responsibility that is attributed on moral grounds rather than on basis of the law or organizational rules" (van de Poel et al. 2012: 51). Legal responsibility may be considered more descriptive than moral. Considering the alternative law of responsible robotics #1 (i.e., A human may not deploy a robot without the human–robot work system meeting the highest legal and professional standards of safety and ethics), it may not be as simple as saying that humans "must meet the highest legal and ethical standards" given that the legal standards of the organization in which one works may be in conflict with the moral standards of the individual roboticist. One may be morally responsible to follow his/her own ethical principles, for instance voicing unethical practices (aka whistleblowing) within a company where he or she works even when he or she has a contract and maybe even an NDA (nondisclosure agreement) that imposes a legal responsibility for him/her not to speak out (Lenk and Maring 2001).

To be sure, keeping to one's contract is also a type of moral responsibility (in addition to a legal one) and creates a conflict when deciding which moral responsibility to prioritize above the other. There are many situations throughout history when we see employees speaking out against the company for which they work; for example, in the 1986 *Challenger* explosion, the project leader warned against the scheduled space travel to NASA and was ignored, resulting in the *Challenger* explosion killing seven astronauts (Lenk and Maring 2001).

Of late, we hear more and more stories of "whistleblowers," individuals breaking NDAs to speak out about ethically problematic treatment of employees or company practices in large tech companies. For example, there was a recent case in which Facebook content moderators broke their NDAs to speak about the psychological and physical suffering resulting from their jobs (Newton 2019). A roboticist working in a social robotics company may develop certain moral misgivings about their work after more sophisticated prototypes verge on deceiving the human users who interact with it into believing it can form emotional bonds. Roboticists may question their legal responsibility to their employer when it conflicts with their own moral responsibility to maintain expectations and/or be truthful with the public.

The distinction between moral and legal responsibility highlights once again the need for a more granular understanding of responsibility in a discussion of responsible robotics; perhaps following a code of conduct legally dictated by a company may impede following one's own moral responsibility. What such a distinction also points towards is a difference, and possible conflict, between individual and collective forms of responsibility.

Individual vs. Collective Responsibility

Not only are there temporal distinctions in a discussion of responsibility but also distinctions about the agents that can bear responsibility. For some, when discussing moral responsibility "only human beings can be held responsible and not collective entities" (Miller 2006). Moreover, "... social groups and organizations, have collective moral responsibility only in the sense that the individual human persons who constitute such entities have individual moral responsibility, either individually or jointly" (Miller 2006: 176).

To put this in more concrete terms, a robotics team may be collectively (or jointly) responsible for creating a robot prototype (i.e., the end), but the realization of this collective end results from the individual activities on team members in, for example, navigation, haptics, high-level planning, mechatronics, ethics, and so on. Thus, we might say that companies have a collective end of creating robots in a responsible way. However, it is not through a collective responsibility but through individual responsibilities for specific employees that acts of responsibility occur, including establishing an ethics review board to assess workflows, hire a privacy expert to ensure data privacy, or a sustainability expert to ensure selection of sustainable materials, and so on. Many organizations can have the same collective end of responsible robotics but the responsibilities of the individuals within said organization will differ depending on the type of organization, for example—a private company will have to answer to shareholders, while a University will have to answer to a public board. This distinction is important for understanding that it will not be possible to simply state that a company has the collective end of responsible robotics; rather, it will be necessary to identify what the individual responsibilities are within said company that are needed in order to realize the collective end.

Responsibility and Accountability

Thus far, the discussion has focused on responsibility—who is responsible for any ethical lapses that have occurred, or that may occur in the future, in relation to a particular robot in development. However, roboticists working in a company developing commercially available products may also feel *accountable* to the public at large for developing products that contribute to the well-being of individuals, or at the very least do not introduce new forms of harm. "Accountability-responsibility is embedded in relationships that involve norms and expectations . . . In accountability relationships those who are accountable believe they have an obligation to a forum, e.g., a community, the public, a particular individual or group of individuals. Members of the forum believe that they are owed an explanation; they expect that those who are accountable will answer (provide an account) when they fail to adhere to appropriate norms, i.e., fail to live up to expectations" (Johnson 2015: 713). Accountability differs from responsibility in that in the case of accountability, the responsible party must justify their decisions and actions to an outside entity. For example, many companies may feel accountable to the general public, who

may have certain expectations of robot makers (although it is worth nothing that these expectations may or may not be based in reality vs. taken from what they see in movies or hear from the press).

While the norms and expectations up for discussion here are usually known to companies attempting to meet those expectations (e.g., a University advisory board may provide written, public reports), they may be established in either a formal or an informal manner, according to Johnson (2015). They can be formally enshrined in legal obligations or codes of conduct of engineers or they may be informally held by a group based on experience and/or public communication. The general public, for example, may have certain expectations of robot capabilities based on stories they see in the press that giver certain expectations of robot capabilities—robots capable of falling in love (Levy 2008), of moral agency (Anderson and Anderson 2011), or of having consciousness (Himma 2009). It may be roboticists (i.e., academic and corporate) who are accountable for recalibrating expectations of the public.

Responsibility and Technology

The question concerning responsibility and robots is further complicated by the tangled relationship that responsibility shares with technology in general and robotics (or autonomous agents) in particular. An added concern for the field of robotics is the level of autonomy a robot may achieve if embedded with artificial intelligence that can allow the robot to learn and function in real time without direct human intervention. Such sophisticated levels of autonomy have led some scholars to raise the question of whether or not a responsibility gap will ensue: "we will not be able to control or predict how [highly specialized artificial agents] will behave" (Johnson 2015: 709).

There have been various rejections of the so-called responsibility gap. Some authors refer to accountability-responsibility relations and also perhaps the legal responsibilities of engineers and remind us that "engineers would be held responsible for the behavior of artificial agents even if they can't control them, on grounds of professional responsibility" (Nagenborg et al. 2008). Other scholars respond to the idea of a responsibility gap by suggesting that robots themselves could be held responsible. While once again bracketing whether or not a robot could meet the conditions of human responsibility the argument raised in such situations "rests finally on the tendency of humans to assign responsibility to computers and robots rather than something that would justify the attribution of responsibility" (Johnson 2015: 705).

Whereas Murphy and Woods address the impossibility of robots being responsible in their discussion of the three alternative laws by stressing the technical impossibility ["robots cannot infallibly recognize humans, perceive their intent, or reliably interpret contextualized scenes" (Murphy and Woods 2009: 15)], Johnson appeals to the wishes of society, that society would not flourish in a state in which no humans were accountable or responsible for the consequences of robot actions, or the malfunction of robot products; "that the human actors involved would decide to create, release, and accept technologies that are incomprehensible and out of the control of humans" (Johnson 2015: 712).

Let me suggest, in accordance with Johnson, Murphy, and Woods among others, that the concept of responsibility in the phrase *responsible robotics* should be a label attributed not to the robots themselves but to the humans acting to make, study, use, regulate, or take apart robot products and services. Therefore, responsible robotics ultimately needs to refer to the kinds of choices made by the humans involved in a robot's design, development, deployment, and regulation: how were decisions calculated, what other options were explored, what kinds of assessments were done to understand and minimize (or mitigate) negative consequences, and what kind of transparency developers and researchers provided to users/customers.

Responsible Robotics Through Responsibility Attribution

Let us suggest that the collective goal of an organization is to develop responsible robotics—to have procedures in place for establishing procedural trust, and for creating products that are considered to be responsibly developed. As discussed so far, we would not consider these organizations to be morally responsible as a whole, but we would expect the organization to designate responsible individuals, and we would then consider the individuals within said organization to be morally responsible for the collective end of developing responsible robotics.

At this moment, *responsible robotics* requires the execution of (at least) two steps or phases: a phase in which ethical issues are uncovered within an organization, and a second phase in which responsibility (in a forward-looking responsibility-as-obligation sense) for solving said issues is attributed to an individual or group of individuals.

For a first step, one could suggest that each organization involved in creating robots that wants to do it responsibly should be accountable for addressing ethical issues in the research and design (R & D) of their robot. An organization could rely on some of the more well-known ethical issues, e.g., privacy, sustainability, safety, and security, and translate these issues to their context and robot prototype. Meaning, designers, and implementers of a robot for a hospital context may design for the value of privacy and interpret privacy

as both corporeal privacy of patient bodies and privacy of personal (medical) data that the robot has access to.

While a substantive conversation about ethical issues is an important first step towards engaging in responsible robotics, there are formalized processes available to make sure that such conversations are as in-depth, comprehensive, and ultimately as effective as possible. Specifically, an organization could engage in a more in-depth assessment to uncover a greater range of the possible ethical issues they may encounter in R & D, for example, methods such as ethical Technology Assessment (eTA) (Palm and Hansson 2006), Care Centred Value Sensitive Design (CCVSD) (van Wynsberghe 2012, 2013), constructive technology assessment (Schot and Rip 1997), and/or ethicist as designer (van Wynsberghe and Robbins 2014), among others. Each of these approaches differs in scope, but is able to produce a list of ethical and societal concerns related to the development process and/or the resulting artifact. An eTA, for example, may provide information concerning privacy issues related to a certain robot application. A CCVSD approach alternatively will produce a list of ethical concerns related to the impact of a robot prototype on the expression of care values. Ideally, the choice of framework for identifying ethical issues should be informed by the context of use. A care robot to be designed for, or used in, a healthcare institution would benefit from a CCVSD approach whereas the developers of a personal robot assistant could benefit from a more generic eTA approach.

In either case, once an issue (past or future) has been identified, one must assign an agent responsible for mitigating or preventing said issue. However, in situations with so "many hands," how can this be done, and more importantly how can this be done in a systematic way to create a level playing field for all robotic companies, organizations, and institutions? This is precisely where we are in need of a framework to help solve the attribution of responsibility to individuals. Such a framework must be broad enough that it can capture the range of variables and stakeholders involved while at the same time specific enough that it allows one to appropriately assign responsibility at a more granular level (making distinctions in kinds of responsibility). As mentioned earlier, in practice this should be possible to do in both a forward and/or backward-looking sense, but for this paper we will consider a forward-looking sense predominantly.

Framework for Identifying Responsible Individuals

The components of the framework, when taken together, provide an analysis of a particular robot prototype or class of robots in various stages of development (Fig. 1). For example, the daVinci surgical robot is a prototype, while

Fig. 1 Framework for responsibility attribution. Source: created by the author

the surgical robots are a class of robots (and consist of a variety of commercial products including the daVinci, Zeus, etc.). To do this, we begin with identifying the type, or class, of robot. If speaking about one robot on the market, for example, a DGI drone, versus a class of robots on the market (e.g., drones or surgical robots), then the scope of responsible individuals is narrowed down. Moreover, this criterion is also about specifying the type of robot in terms of its tasks, goals, and capabilities. A greeter robot in hospitals without the capability to collect or store data on individuals in its path will raise a different set of privacy questions compared to a greeter robot in a store designed to collect and store data on customers it interacts with. Thus, the type of robot also alerts one to the range of ethical issues at stake (perhaps in addition to those identified in the eTA or CCVSD analysis).

Next, one must specify the moment in the robot's life cycle, for example, idea generation, first robot prototype, prototype following numerous iterations, large-scale development, distribution, implementation, or regulation. The idea behind this step is to introduce the understanding that the responsible individuals are not just the robot developers, but also the people who will be purchasing, implementing or even regulating, robots. Accordingly, if we consider the surgical robot daVinci, a robot that is produced on a large scale by a company called Intuitive Surgical but is sometimes distributed around the world by other intermediary companies, there will be certain responsibilities that the producers will have (e.g., choice of materials, meeting of ISO standards), while other responsibilities will fall on distributors (e.g., safe delivery and assistance with implementation), and even other responsibilities will fall on hospital administrators to ensure responsible deployment in the hospital (e.g., safety protocols, training of surgeons and other medical professionals).

Third, following identification of the maturity of the robot's development, the next step is to identify the context in which the robot is intended to be used, e.g., agriculture, defense, or healthcare. Such a specification also acts to define (and in some instances limit) the scope of actors able to take responsibility. For example, robots bought and used in a healthcare context may demand responsibility from individuals related to hospital administration, insurance companies, the FDA/EMA, and/or other related institutions. In another context of use—for example agriculture— different stakeholders, like farmers, may be introduced. If the robot is still early on in the experimental stages and is not yet commercially available, companies may be working closely with farmers to understand their specific needs and farmers may have a moral (not a legal) responsibility to assist in this process. Once the robot is commercially available, as in the case with robots for milking cows, feeding livestock, or cleaning stalls, farmers may no longer be responsible for providing insights into their daily needs, but they may be responsible for providing feedback on the efficacy, safety, or data security measures of the robot.

Fourth, one should identify the robot's mode of acquisition. Many robots used in an academic context are off-the-shelf products made and tested in industry, e.g., the Nao and Pepper platforms of Softbank. If one were to consider the ethical concern of e-waste, it seems beyond the limits of academics to be held responsible for this when they are not themselves making the robot and, even worse, have no environmentally sustainable alternative. In such instances, it would seem adequate to hold industry responsible for sustainability issues related to the robots made for study within academia. If, however, there were multiple robotics platforms available, one of which is a sustainably sourced one, then it seems appropriate to hold academics responsible for the purchasing of sustainable products. The same considerations hold for companies or NGOs who purchase off-the-shelf robots for their own use, e.g., grocery stores that are buying robots to implement in their warehouses and distribution centers. In the case of companies making and distributing their own robots to consumers/users, these companies have a responsibility to search for sustainable avenues of doing so (e.g., percentage of recycled plastics, minerals sourced for the batteries coming from certified mines).

In short, the framework presented here is meant to create a procedure for assigning responsibilities to the variety of actors involved in robot R & D, purchasing, implementation, and regulation of robots. At each stage of the framework one is called upon to list all possible responsible individuals for the ethical issue in question and to refine this list at each of the subsequent stages. By engaging in this framework in this way, one is then able to define the scope of individuals within organizations who bear responsibility for the consequences of the ethical issue at stake (whether in the design, development, use, implementation, and/or regulation). Depending on the ethical issue, the sector, context, and type of robot, the individual who should be assigned responsibility may change. In some instances, it may be developers of robots that bear responsibilities (e.g., companies making robots responsible for reducing e-waste concerns in their production or procurement officers responsible for providing alternatives to small start-up companies) whereas in other instances it may be the ones purchasing the robots that bear responsibilities (e.g., consumers or companies using robots in their retail processes may be responsible for purchasing sustainable robots). In other instances, it may be the policy makers responsible for developing governance tools to facilitate sustainable development of robots through the creation of subsidies to companies, laws to prevent the breaking of International e-waste treaties, and guidelines to show support for the responsible development of robots.

What's Missing in This Framework: Necessary But Not Sufficient for Labelling "Responsible Robotics"

There are two main criticisms of the work I have presented here: first, distinguishing between "should" questions and "how should" questions. The second being the need for broader policy and infrastructure questions. Neither one of these criticisms takes away from the content presented but each should be acknowledged nonetheless.

First, each of the alternative laws of responsible robotics, along with the responsibility attribution framework presented here, presume that the robot in question *should* be made and the challenge lies in teasing out *how* the robot should be made or in assigning a responsible individual for mitigating an ethical concern. However, responsible robotics must also be about questioning the very application of a robot. A robot for babysitting children may fit all ISO/FDA standards, it may achieve smooth transfer of autonomy, it may be designed with clear expectations of accountability and chains of responsibility to avoid the problem of many hands, and yet these criteria do not address the question of whether or not the robot *should* be made and/or used in the first place. Such a "babysitting" robot could erode bonds between parents and children over time or drastically change the socialization of children and for these reasons be considered unethical to develop and/or use no matter the production process. Thus, the framework presented here may be considered a necessary criterion of responsible robotics but by no means sufficient to claim responsible robotics.

Second, a big question missing here is: who does the assigning of responsible individuals? This is precisely where ethics meets policy as the responsibility infrastructure I am envisioning here will extend far outside an organization and into the realm of developer, implementer, and user. Who will determine the responsible individual (and the determinants of such a role) and who will make the responsible party do

the work of being responsible are questions inviting answers from the policy making space.

Concluding Remarks: Why We Need a Framework for Responsible Robotics

Robotics has come a long way in the last decades. Improvements in technology allow already innovative robots to do even more than they once could. For example, robots now work in factories alongside humans (i.e., cobots) rather than behind cages as they originally did. Financially speaking, robot sales reportedly increase every year. According to the International Federation of Robotics "in 2017, [industrial] robot sales increased by 30% to 381,335 units, a new peak for the fifth year in a row" (IFR n.d.-a: 13) and "the total number of professional service robots sold in 2017 rose considerably by 85% to 109,543 units up from 59,269 in 2016. The sales value increased by 39% to US$ 6.6bn" (IFR n.d.-b: 11). Globally speaking, "Robotics investments in December 2018 totaled at least $652.7 million worldwide with a total of 17 verified transactions" (Crowe 2019). With continued investments in robotics, it seems likely that such trends will continue. The question to ask is how to design and develop this technology in a way that pays tribute to societal values, resists the urge to exacerbate existing ethical problems (such as environmental sustainability), and proceeds in a resilient manner to navigate unknown ethical issues as they are revealed.

If robotics is truly to succeed in making our world a better place, the public must be able to place their trust in the designers, developers, implementers, and regulators of robot technologies. To do this, we must engage in the responsible research and innovation of robot development processes and the robots that result from these processes; we need *responsible robotics*. The three alternative laws of responsible robotics by Murphy and Wood play a formidable contribution to the discussion on responsible robotics; however, they repeat the difficulty that Asimov introduced, that is, laws in general in the robotics space, whether they are for the robot or for the roboticist (or any other actor in the design process), are incomplete when put into practice. The proposal here is to extend the three alternative laws of responsible robotics into a more robust framework for responsibility attribution as part of the responsible robotics goal. Such a framework is meant to draw attention to the network of actors involved in robot design and development, and to the differences in kinds of responsibility that each of these actors (either individuals or organizations) may have.

The responsibility attribution framework requires identification of various factors: the type of robot, the stage of robot development, the intended sector of use, and the manner of robot acquisition. Identifying these details in a step-by-step manner allows one to land on the stakeholder deserving of responsibility. With this in mind, one must carefully consider the kind of ethical issue (or societal value) in question and determine the kind of responsibility attributed to said actor(s). Such a framework rests on four starting assumptions related to the definition and concept of responsible robotics: (1) Responsibility of humans (and not robots) involved in the creation and use of robots; (2) Responsibility understood as a vast concept with various distinctions; (3) Robotics understood as a process with various stages of development for which different actors will bear different responsibilities; (4) An understanding of the impact of robots on systems rather than ending at the human–robot interaction.

The question of how to make robotics in a responsible way and what such products would look like is colossal—impossible to answer in one paper. This paper was meant to open the door for the discussion of a framework to encourage analysis allowing one to arrive at a decision concerning *who* is responsible for mitigating or solving a particular ethical/societal issue. In short, the phrase *responsible robotics* follows from the recognition of robotics as a social experiment and is meant to convey that the robotics experiment be done responsibly. It is directed at the people who design, develop, regulate, implement, and use the entire range of robotics products. It is, furthermore, about ensuring that those people are responsible for proactively assessing and taking actions, which ensures that robotics products respect important societal values.

References

Anderson, M., & Anderson, S. L. (2011). *Machine ethics*. Cambridge: Cambridge UP.

Asimov, I. (2004). Runaround. In I. Asimov (Ed.), *I, Robot* (pp. 25–45). New York: Random House.

Chow, J. C. (2012). *The case for humanitarian drones*. Available via OpenCanada. Retrieved February 28, 2020, from https://www.opencanada.org/features/the-case-for-humanitarian-drones/

Crowe, S. (2019). *Robotics investments recap: December 2018*. Available via The Robot Report. Retrieved May 7, 2019, from https://www.therobotreport.com/robotics-investments-recap-december/

European Commission. (2012). *Responsible research and innovation: Europe's ability to respond to societal challenges*. London: E Union.

European Commission. (2014). *Rome declaration on responsible research and innovation in Europe*. London: E Union. Available via European Commission. Retrieved February 28, 2020, from https://ec.europa.eu/research/swafs/pdf/rome_declaration_RRI_final_21_November.pdf.

Feinberg, J. (1988). Responsibility for the future. *Philosophy Research Archives, 14*, 93–113. https://doi.org/10.5840/pra1988/19891427.

Friedman, B. (1996). Value-sensitive design. *Interactions, 3*(6), 16–23. https://doi.org/10.1145/242485.242493.

Friedman, B., Hendry, D., Huldtgren, A., Jonker, C., van den Hoven, J., & van Wynsberghe, A. (2015). Charting the next decade for value sensitive design. *Aarhus Series on Human Centered Computing, 1*(1), 4. https://doi.org/10.7146/aahcc.v1i1.21619.

Gilman, D., & Easton, M. (2014). *Unmanned aerial vehicles in humanitarian response*. Occasional Policy Paper 010. Available via United Nations Office for the Coordination of Humanitarian Affairs. Retrieved February 28, 2020, from https://docs.unocha.org/sites/dms/Documents/Unmanned%20Aerial%20Vehicles%20in%20 Humanitarian%20Response%20OCHA%20July%202014.pdf

Hart, H. L. A. (2008). *Punishment and responsibility: Essays in the philosophy of law*. Oxford: Oxford UP.

Himma, K. E. (2009). Artificial agency, consciousness, and the criteria for moral agency: What properties must an artificial agent have to be a moral agent? *Ethics and Information Technology, 11*(1), 19–29. https://doi.org/10.1007/s10676-008-9167-5.

IFR. (n.d.-a). *Executive summary world robotics 2018 industrial robots*. Available via IFR. Retrieved May 7, 2019, from https://ifr.org/downloads/press2018/Executive_Summary_WR_2018_Industrial_Robots.pdf

IFR. (n.d.-b). *Executive summary world robotics 2018 service robots*. Available via IFR. Retrieved May 7, 2019, from https://ifr.org/downloads/press2018/Executive_Summary_WR_Service_Robots_2018.pdf

Johnson, D. G. (2015). Technology with no human responsibility? *Journal of Business Ethics, 127*(4), 707–715. https://doi.org/10.1007/s10551-014-2180-1.

Lenk, H., & Maring, M. (2001). Responsibility and technology. In A. E. Auhagen & H.-W. Bierhoff (Eds.), *Responsibility: The many faces of a social phenomenon* (pp. 93–108). London: Routledge.

Levy, D. (2008). *Love and sex with robots: The evolution of human-robot relationships*. New York: Harper Perennial.

Miller, S. (2006). Collective moral responsibility: An individualist account. *Midwest Studies in Philosophy, 30*(1), 176–193. https://doi.org/10.1111/j.1475-4975.2006.00134.x.

Murphy, R. (2014). *Disaster robotics*. Cambridge, MA: MIT Press.

Murphy, R., & Woods, D. (2009). Beyond Asimov: The three laws of responsible robotics. *IEEE Intelligent Systems, 24*(4), 14–20. https://doi.org/10.1109/MIS.2009.69.

Nagenborg, M., Capurro, R., Weber, J., & Pingel, C. (2008). Ethical regulations on robotics in Europe. *AI & Society, 22*(3), 349–366. https://doi.org/10.1007/s00146-007-0153-y.

Newton, C. (2019). *Bodies in Seats: At Facebook's worst-performing content moderation site in North America, one contractor has died, and others say they fear for their lives*. Available via The Verge. Retrieved April 30, 2020, from https://www.theverge.com/2019/6/19/18681845/facebook-moderator-interviews-video-trauma-ptsd-cognizant-tampa

Palm, E., & Hansson, S. O. (2006). The case for ethical technology assessment (eTA). *Technological Forecasting and Social Change, 73*(5), 543–558. https://doi.org/10.1016/j.techfore.2005.06.002.

Schot, J., & Rip, A. (1997). The past and future of constructive technology assessment. *Technological Forecasting and Social Change, 54*(2–3), 251–268. https://doi.org/10.1016/S0040-1625(96)00180-1.

Sharkey, A. (2014). Robots and human dignity: A consideration of the effects of robot care on the dignity of older people. *Ethics and Information Technology, 16*(1), 63–75. https://doi.org/10.1007/s10676-014-9338-5.

Sharkey, A. (2016). Should we welcome robot teachers? *Ethics and Information Technology, 18*, 1–15. https://doi.org/10.1007/s10676-016-9387-z.

Smith, A. M. (2007). On being responsible and holding responsible. *The Journal of Ethics, 11*(4), 465–484. https://doi.org/10.1007/s10892-005-7989-5.

Sparrow, R., & Sparrow, L. (2006). In the hands of machines? The future of aged care. *Minds and Machines, 16*(2), 141–161. https://doi.org/10.1007/s11023-006-9030-6.

Vallor, S. (2011). Carebots and caregivers: Sustaining the ethical ideal of care in the twenty-first century. *Philosophy and Technology, 24*(3), 251–268. https://doi.org/10.1007/s13347-011-0015-x.

van de Poel, I. (2013). Why new technologies should be conceived as social experiments. *Ethics, Policy & Environment, 16*(3), 352–355. https://doi.org/10.1080/21550085.2013.844575.

van de Poel, I., & Sand, M. (2018). Varieties of responsibility: Two problems of responsible innovation. *Synthese.* https://doi.org/10.1007/s11229-018-01951-7.

van de Poel, I., Nihlén Fahlquist, J., Doorn, N., Zwart, S., & Royakkers, L. (2012). The problem of many hands: Climate change as an example. *Science and Engineering Ethics, 18*(1), 49–67. https://doi.org/10.1007/s11948-011-9276-0.

van den Hoven, J. (2013). Value sensitive design and responsible innovation. In R. Owen, J. Bessant, & M. Heintz (Eds.), *Responsible innovation: Managing the responsible emergence of science and innovation in society* (pp. 75–84). London: Wiley.

van Koughnett, J., Jayaraman, S., Eagleson, R., Quan, D., van Wynsberghe, A., & Schlachta, C. (2009). Are there advantages to robotic-assisted surgery over laparoscopy from the surgeon's perspective? *Journal of Robotic Surgery, 3*(2), 79–82. https://doi.org/10.1007/s11701-009-0144-8.

van Wynsberghe, A. (2012). Designing robots for care: Care centered value-sensitive design. *Science and Engineering Ethics, 19*(2), 407–433. https://doi.org/10.1007/s11948-011-9343-6.

van Wynsberghe, A. (2013). A method for integrating ethics into the design of robots. *Industrial Robot, 40*(5), 433–440. https://doi.org/10.1108/IR-12-2012-451.

van Wynsberghe, A., & Comes, T. (2019). Drones in humanitarian contexts, robot ethics, and the human–robot interaction. *Ethics and Information Technology, 22,* 43. https://doi.org/10.1007/s10676-019-09514-1.

van Wynsberghe, A., & Gastmans, C. (2008). Telesurgery: An ethical appraisal. *Journal of Medical Ethics, 34*(10), e22. https://doi.org/10.1136/jme.2007.023952.

van Wynsberghe, A., & Li, S. (2019). A paradigm shift for robot ethics: From HRI to human–robot–system interaction (HRSI). *Medicolegal and Bioethics, 9,* 11–20. https://doi.org/10.2147/MB.S160348.

van Wynsberghe, A., & Robbins, S. (2014). Ethicist as designer: A pragmatic approach to ethics in the lab. *Science and Engineering Ethics, 20*(4), 947–961. https://doi.org/10.1007/s11948-013-9498-4.

Watson, G. (2004). Reasons and responsibility. In G. Watson (Ed.), *Agency and answerability: Selected essays* (pp. 289–317). Oxford: Oxford UP.

Regulating AI: Considerations that Apply Across Domains

Angela Kane

Contents

Abstract

Awareness that AI-based technologies have far outpaced the existing regulatory frameworks have raised challenging questions about how to set limits on the most dangerous developments (lethal autonomous weapons or surveillance bots, for instance). Under the assumption that the robotics industry cannot be relied on to regulate itself, calls for government intervention within the regulatory space—national and international—have multiplied. The various approaches to regulating AI fall into two main categories. A sectoral approach looks to identify the societal risks posed by individual technologies, so that preventive or mitigating strategies can be implemented, on the assumption that the rules applicable to AI, in say the financial industry, would be very different from those relevant to heath care providers. A cross-sectoral approach, by contrast, involves the formulation of rules (whether norms adopted by industrial consensus or laws set down by governmental authority) that, as the name implies, would have application to AI-based technologies in their generality. After surveying some domestic and international initiatives that typify the two approaches, the chapter concludes with a list of 15 recommendations to guide reflection on the promotion of societally beneficial AI.

A. Kane (✉)
Vienna Center for Disarmament and Non-Proliferation (VCDNP),
Vienna, Austria
e-mail: akane@vcdnp.org

Keywords

AI · Regulation · Domestic · International ·
Recommendations

J. von Braun et al. (eds.), *Robotics, AI, and Humanity*, https://doi.org/10.1007/978-3-030-54173-6_21

Introduction[1]

While we think of AI as a phenomenon that has rapidly arisen over the last few years, we should remember that it was already 80 years ago that Alan Turing laid down the mathematical basis of computation. ARPANET began in 1969, the Internet Protocol in 1974, and the World Wide Web 30 years ago, in 1989. Do any of you remember when we first started to access the internet, with modems? The distinctive whirring burpy sound they made when connecting—ever so slowly—to the web? This now seems very quaint, as the improvements in speed and performance, as well as the cost reductions in memory and information technology, have made possible the enormous expansion of data that now fuels the engine of global growth.

Harnessing AI has challenges and opportunities in many areas and domains: technical, ethical, political, social, and cultural. These are accompanied by the need for accountability, algorithmic explainability, and even legal liability. If we do not understand how a system works, then it blurs lines of who can or who should be responsible for the outcome or the process of the decision. Should that be the innovator? The regulator? The operator? And how can both policymakers and the public trust technology when it is not properly understood?

These are vexing questions that have been further compounded by the rise in disclosures of data and privacy leaks, of hacking into sites containing sensitive personal information, of spoofing, of selling consumer data without consent and, to make matters worse, of concealing or delaying disclosure of such egregious violations of privacy.

The debate about these issues has become louder and more polarized; it is pitting powerful companies against governments and consumers. Scientists are weighing in—as do employees of technology companies, as we have seen with Google. Until 2015, Google's motto was "Don't be evil" but it was then changed to "Do the right thing" within its corporate code of conduct. Swarms of bots, dark posts, and fake news websites inundate the web, ricochet around chatrooms, and overwhelm the legitimate media outlets.

Let us remember just a few recent events: in the US presidential elections in 2016, Russia supported one candidate (who subsequently won) by waging a campaign with paid advertisements and fake social media accounts that contained polarizing content. Concerns also abound in China about millions of cameras deployed with face recognition software which record streams of data about citizens. In India, it was

reported that the "fake news problem plagues several popular social networks" (Metha 2019) by spreading misinformation, doctored photos and videos which resulted in several cases of killing and even lynching.

More and more thoughtful questions about social platforms are being asked that do not lend themselves to easy answers. Technology companies are coming under increasing scrutiny, as they are seen to be operating without accountability. Facebook CEO Mark Zuckerberg testified in US Congress on efforts to address privacy issues and data sharing, but subsequent Facebook data leaks showed that his assurances to prevent a recurrence were hollow. Talking about regulation, he said: "My position is not that there should be no regulation. I think the real question is, as the internet becomes more important in people's lives, is what is the right regulation, not whether there should be or not" (Zuckerberg in Watson 2018).

In October 2019, responding to concerns that the social network has too much power to shape political and social issues, Zuckerberg pushed back against criticism that Facebook was not doing enough to combat hate speech, misinformation, and other offensive content, by opining that people in a democracy did not want a private company censoring the news (Wong 2019). Does free speech then allow the placing of ads with deliberate falsehoods? This question has taken added relevance, particularly in the run-up to the 2020 presidential elections in the USA where the use of social media is a prime factor in the campaign.

I will take stock of some of the efforts to address the attempts to regulate AI and technology, fully aware that the paper will outdate very quickly, as new initiatives and considerations are coming up quickly.

Curbing Lethal Autonomous Weapon Systems: An Early Effort at Regulating AI

In Wikipedia's definition, artificial general intelligence is the intelligence of a machine that can understand or learn any intellectual task that a human being can. Yet while the jury is still out whether AI will bring enormous benefits to humanity or bring possible calamity, the applications of AI abound in a variety of sectors. Deep learning algorithms are embedded already in our daily life; they are used in social media, medicine, surveillance, and determining government benefits, among others. By way of example, let me therefore look at one of the sectoral approaches, that of using AI in weapons.

In 2013, a report was published by the United Nations Special Rapporteur on extrajudicial, summary or arbitrary execution, Christof Heyns, on the use of lethal force through what he called "lethal autonomous robotics (LAR)." He approached the issue from the perspective of protection of

---

[1]This chapter is based on an earlier version presented at the Pontifical Academy of Sciences in May 2019 that was also placed on the website https://www.united-europe.eu/2019/05/angela-kane-regulating-ai-considerations-that-apply-across-domains/

life during war and peace and made a number of urgent recommendations to organizations, States, developers of robotic systems, and nongovernmental organizations.

Following its publication, 16 countries put the questions related to emerging—or "robotic"—technologies on the agenda of the Convention on Certain Conventional Weapons (CCW) in Geneva. The first meetings on these issues took place in 2014 and they showed that few countries had developed any policy on the matter. Thematic sessions, with significant input from AI scientists, academics, and activists, dealt with legal aspects, ethical and sociological aspects, meaningful human control over targeting and attack decisions, as well as operational and military aspects. And what Christof Heyns had called lethal autonomous robotics is now referred to as "lethal autonomous weapon systems" or LAWS. While there is no singularly accepted definition of LAWS, the term now generally covers a broad array of potential weapon systems, from fully autonomous weapons that can launch attacks without any human intervention to semiautonomous weapons that need human action to direct or execute a mission.

The first debates were conducted in an Open-Ended Working Group, in which any State could freely participate. Yet in 2016, governments decided to form a Group of Governmental Experts (GGE) to advance the issue. The crucial difference is that a GGE operates on a consensus basis, which essentially gives a veto right to any decisions or statements adopted by the GGE to any one participating State.

Twenty-nine States now openly call for a ban on these weapons. Austria, Brazil, and Chile have recently proposed a mandate to "negotiate a legally-binding instrument to ensure meaningful human control over the critical functions of weapon systems," but the prospects for such a move are slim. So far, no legally binding or political actions have been adopted by the Group due to the objections of about a dozen States: Australia, Belgium, France, Germany, Israel, Republic of Korea, Russian Federation, Spain, Sweden, Turkey, the United Kingdom, and the United States. These States argue that concrete action on LAWS is "premature" and that the Group could instead explore "potential benefits" of developing and using LAWS.

The opposing positions do not augur well for any legislative progress in the issue of LAWS. Yet the voices in favor of a total ban are getting louder and louder. Already in 2015, at one of the world's leading AI conferences, the International Joint Conference on Artificial Intelligence (IJCAI 15), an Open Letter from AI & Robotics Researchers—signed by nearly 4000 of the preeminent scientists such as Stuart Russell, Yann LeCun, Demis Hassabis, Noel Sharkey, and many many others—and over 22,000 endorsers including Stephen Hawking, Elon Musk, and Jaan Tallinn, to name just a few, warned against AI weapons development and posited that "most AI researchers have no interest in building AI

weapons, and do not want others to tarnish their field by doing so" (FLI 2015).

The decision by Google to end cooperation with the US Department of Defense on Project Maven—a minor contract in financial terms—was ended in 2018 due to strong opposition by Google employees who believed that Google should not be in the business of war. UN Secretary-General Guterres, former High Commissioner for Human Rights Zeid Ra'ad Al Hussein, and Pope Francis have weighed in, calling autonomous weapons "morally repugnant" and calling for a ban (UN 2018a).

There are also parliamentary initiatives in capitals. In April 2018, for example, the Lord's Select Committee on AI challenged the UK's futuristic definitions of autonomous weapon systems as "clearly out of step" with those of the rest of the world and demanded that the UK's position be changed to align these within a few months.

Yet the Government's response was limited to one paragraph which stated that the Ministry of Defense "has no plans to change the definition of an autonomous system" and notes that the UK will actively participate in future GGE meetings in Geneva, "trying to reach agreement (on the definition and characteristics of possible LAWS) at the earliest possible stage" (UK Parliament 2018, recommendations 60–61).

Interest in other European parliaments is also high, as awareness of the issue has grown exponentially. It is the hot topic of the day.

The European Commission issued a communication in April 2018 with a blueprint for "Artificial Intelligence for Europe" (European Commission 2018). While this does not specifically refer to LAWS, it demands an appropriate ethical and legal framework based on the EU's values and in line with the Charter of Fundamental Rights of the Union.

In July 2018, the European Parliament adopted a resolution that calls for the urgent negotiation of "*an international ban on weapon systems that lack human control over the use of force.*" The resolution calls on the European Council to work towards such a ban and "*urgently develop and adopt a common position on autonomous weapon systems*" (European Parliament 2018). In September 2018, EU High Representative Federica Mogherini told the EU Parliament that "*the use of force must always abide by international law, including international humanitarian law and human rights laws. (. . .) How governments should manage the rise of AI to ensure we harness the opportunities while also addressing the threats of the digital era is one of the major strands of open debate the EU has initiated together with tech leaders*" (EEAS 2018).

The issue of lethal autonomous weapons has clearly raised the profile of legislating AI. Advocacy by civil society, especially the Campaign to Stop Killer Robots, a coalition of NGOs seeking to pre-emptively ban lethal autonomous weapons, has been instrumental in keeping the issue prominent in the media, but this single-issue focus is not easily replicable in other AI-driven technologies.

Can We Ever Hope to Regulate and Govern AI?

Artificial intelligence is a universal subject that breaks down into many variations and applications. Rather than tackling AI as a whole, it is easier to address a sector-specific AI application—like LAWS—than general AI that is broad, adaptive, and advanced as a human being across a range of cognitive tasks.

We already have a myriad of automated decision systems that are being used by public agencies, in criminal justice systems, in predictive policing, in college admissions, in hiring decisions, and many more. Are these automated decision systems appropriate? Should they be used in particularly sensitive domains? How can we fully assess the impact of these systems? Whose interests do they serve? Are they sufficiently nuanced to take into account complex social and historical contexts? Do they cause unintended consequences?

The difficulty in finding answers to these questions is the lack of transparency and information. Many of these systems operate in a black box and thus outside the scope of understanding, scrutiny and accountability. Yet algorithms are endowed with a specific structuring function, as designed by individuals. The General Data Protection Regulation (GDPR) which the European Union adopted in 2018 includes an "explainability requirement" that applies to AI, but it is not clear exactly how much.

"Can You Sue an Algorithm for Malpractice?" was the headline of a magazine article in the USA in 2019 (Forbes 2019). Clearly, algorithms are being litigated, as a 2018 report by the AI Now Institute shows (AI Now 2018a) and which has resulted already in more study and scrutiny of the use of such systems across public agencies. Several lawsuits proved that decision-making formulas were corrupt due to data entry errors and biased historical data, while aimed to produce cost savings or to streamline work without assessment how they might harm vulnerable populations. While this showed the limits of AI use in public policy, it is clear that lawsuits set precedent in law but cannot establish regulations and the rule of law.

But if the litigation shows us anything, it is that AI-driven technology has become an important issue for people and for governments. In response, we are seeing two distinct trends:

- The AI and tech industry have become a hub for ethics advisory boards and related efforts to buff their credentials in what I would call "responsible AI".

- Private organizations have been established like Partnership for AI (mission: to benefit people and society), or Open AI (mission: to ensure that artificial general intelligence benefits all of humanity).
- Academic institutions—such as New York University—have set up institutes like AI Now, a research institute examining the social implications of AI; the Massachusetts Institute of Technology (MIT) conducts a project on AI Ethics and Governance to support people and institutions who are working to steer AI in ethically conscious directions.
- Workshops and conferences with a range of tech and non-tech stakeholders are being organized to debate the scope of the challenges as well as exploring solutions.

Governments Are Stepping Up

The second trend is the increasing focus by Governments on the disruption by artificial intelligence and the search for shaping the ethics of AI. Let me mention some statements by leaders.

When Russian President Putin in 2018 said to a group of school children that "whoever controls AI, will become the ruler of the world," it made headlines. China's blueprint—issued in 2017 and called the "New Generation Artificial Intelligence Development Plan"—outlined China's strategy to become the world player in AI by 2030. The Plan barely mentions information on laws, regulations, and ethical norms, since China's authoritarian approach is less restrained by attention to values and fundamental rights as well as ethical principles such as accountability and transparency. In the 3 years since its publication, China is already starting to overtake the USA as the leader in AI.

In Europe, French President Macron in 2018 called the technological revolution that comes with AI "in fact a political revolution," and said that in shaping how AI would affect us, you have to be involved at the design stage, *and set the rules* (italics added). He committed the French government to spend Euro 1.5 billion over 5 years to support research in the field, encourage startups, and collect data that can be used, and shared, by engineers.

A French data protection agency (Commission Nationale de l'Informatique et des Libertés, CNIL) issued a 75-page report in December 2017 about the results of a public debate about AI, algorithms, ethics, and how to regulate it. The report set out six areas, which predominate the ethical dilemmas:

1. Autonomous machines taking decisions
2. Tendencies, discrimination and exclusion which are programmed, intentionally or unintentionally
3. Algorithmic profiling of people

4. Preventing data collection for machine learning
5. Challenges in selecting data of quality, quantity, and relevance
6. Human identity in the age of artificial intelligence

Recommendations made in the report primarily focus on the individual by urging enhanced information and education but also request private industry to focus on ethics by establishing ethics committees and an ethics code of conduct or an ethics charter (CNIL 2017).

In the UK, the House of Lords Select Committee on Artificial Intelligence issued a report in April 2018 with the catchy title "AI in the UK: ready, willing and able?" The report was based on extensive consultations and contains an assessment of the current state of affairs as well as numerous recommendations on living with AI, and on shaping AI (House of Lords Artificial Intelligence Committee 2018).

The 183-page report has only two paragraphs on "regulation and regulators" which state that "Blanket AI-specific regulation, at this stage, would be inappropriate. We believe that existing sector-specific regulators are best placed to consider the impact on their sectors of any subsequent regulation *which may be needed*" (emphasis added). It also urges the Government Office for AI to "ensure that the existing regulators' expertise is utilized in informing any potential regulation that may be required in the future" and foresees that "the additional burden this could place on existing regulators could be substantial," recommending adequate and sustainable funding (House of Lords Artificial Intelligence Committee 2018: 386–387). In its final paragraphs, the report refers to the preparation of ethical codes of conduct for the use of AI by "many organizations" and recommends that a cross-sectoral ethical code of conduct—suitable for implementation across public and private sector organizations—be drawn up (. . .) with a sense of urgency. "In time, the AI code could provide the basis for statutory regulation, *if and when this is determined to be necessary*" (House of Lords Artificial Intelligence Committee 2018: 420, emphasis added).

In June 2018, the Government issued a 42-page response to the House of Lords' report. As to paragraph 386 (no blanket AI-specific regulation needed), the Government agreed with the recommendation. It stated its commitment to work with businesses to "develop an agile approach to regulation that promotes innovation and the growth of new sectors, while protecting citizens and the environment" (UK Parliament 2018). It further promises horizon-scanning and identifying the areas where regulation needs to adapt to support emerging technologies such as AI and the establishment of a Centre for Data Ethics and Innovation that "will help strengthen the existing governance landscape" (UK Parliament 2018: 108). Yet the Centre—established late last year—has only an advisory function, promoting best practices and

advising how Government should address potential gaps in the regulatory landscape.

Other European countries also addressed AI. Sweden published a report in May 2018 on its National Approach (a digestible 12 pages) which highlights the Government's goals to develop standards and principles for ethical, sustainable, and safe AI, and to improve digital infrastructure to leverage opportunities in AI. Finland was a bit ahead of the curve, issuing its first report on "Finland's Age of Artificial Intelligence" already in December 2017, but none of its eight proposals deal with rules and regulations.

Germany issued a 12-point strategy ("AI Made in Germany—a seal of excellence"), which focuses on making vast troves of data available to German researchers and developers, improves conditions for entrepreneurs, stops the brain drain of AI experts, and loosens or adapts regulation in certain areas. But it also heavily emphasizes the rights and advantages of AI for the citizens and underlines the ethical and legal anchoring of AI in Europe.

The European Union: "Placing the Power of AI at the Service of Human Progress"

Finally, let me focus on the European Union which in April 2018 issued "AI for Europe: Embracing Change" (European Commission 2018). This was the launch of a European Initiative on AI with the following aims:

1. Boost the EU's technological and industrial capacity and AI uptake across the economy
2. Prepare for socio-economic change
3. Ensure an appropriate ethical and legal framework

Under these three headings, ambitious plans were laid out, both in financial terms (stepping up investments) and in deliverables, with time lines until the end of 2020.

Let us not forget that the General Data Protection Regulation (GDPR) came into force the same year. While this regulation imposes a uniform data security law on all EU members, it is important to note that any company that markets good and services to EU residents, regardless of its location, is subject to the regulation. This means that GDPR is not limited to EU member states, but that it will have a global effect.

One of the deliverables was the setting up of an Independent High-Level Expert Group on Artificial Intelligence[2] which was asked to draft AI ethics guidelines and through an online framework called the European AI Alliance reached out to stakeholders and experts to contribute to this effort.

---

[2]Full disclosure: I was a reserve member of the High-Level Expert Group and participated in several of their meetings.

The draft ethics guidelines were issued in December 2018 and received over 500 comments, according to the EU. What resulted were the "Ethics Guidelines for Trustworthy AI," issued in April 2019, which defines trustworthy AI as follows: "*(It) has three components: (1) it should be lawful, ensuring compliance with all applicable laws and regulations (2) it should be ethical, demonstrating respect for, and ensure adherence to, ethical principles and values and (3) it should be robust, both from a technical and social perspective since, even with good intentions, AI systems can cause unintentional harm. Trustworthy AI concerns not only the trustworthiness of the AI system itself but also comprises the trustworthiness of all processes and actors that are part of the system's life cycle.*"

The Guidelines then list seven essentials for achieving trustworthy AI:

1. Human agency and oversight
2. Robustness and safety
3. Privacy and data governance
4. Transparency
5. Diversity, non-discrimination, and fairness
6. Societal and environmental well-being
7. Accountability

Again, the Guidelines are currently in a pilot phase for more time to receive feedback and to ensure that they can be issued by the end of 2019 and then implemented—which is expected in 2020 (European Commission 2019). At the same time, the EU Commission wants to bring their approach to AI ethics to the global stage: "because technologies, data and algorithms know no borders." Following the G7 summit in Canada in December 2018, where AI was prominently featured, the EU wants to strengthen cooperation with other "like-minded" countries like Canada, Japan, and Singapore, but also with international organizations and initiatives like the G20 to advance the AI ethics agenda.

Before we break out the champagne in celebration of the ethics guidelines, let me mention one dissenting voice from the High-Level Group: Thomas Metzinger, Professor of Theoretical Philosophy in Germany, wrote a scathing article entitled "Ethics washing made in Europe" in which he called the Trustworthy AI story "a marketing narrative invented by industry, a bedtime story for tomorrow's customers." The narrative, he claimed is "in reality, about developing future markets and using ethics debates as elegant public decorations for a large-scale investment strategy" (Metzinger 2019). Metzinger (2019) considers that "industry organizes and cultivates ethical debates to buy time—to distract the public and to prevent or at least delay effective regulation and policy-making. And politicians like to set up ethics committees because it gives them a course of action when, given the complexities of the issues, they simply don't know

what to do." Interestingly, he also mentions the use of lethal autonomous weapon systems as one of the "Red Lines," the non-negotiable ethical principles—which I outlined at the beginning of this paper.

Ethical AI—The New Corporate Buzz Phrase

I agree that the jury on the EU Ethics Guidelines is still out, but the criticism of major tech companies and academic ethics boards, especially in the USA, is very strong. Many tech companies have recently laid out ethical principles to guide their work on AI. Major companies like Microsoft, Facebook, and Axon (which makes stun guns and body cameras for police departments), all now have advisory boards on the issue. Amazon recently announced that it is helping fund research into "algorithmic fairness," and Salesforce employs an "architect" for ethical AI practice, as well as a "chief ethical and human use" officer. More examples could be cited.

Yet are these actions designed primarily to head off new government regulations? Is it a fig leaf or a positive step? "Ethical codes may deflect criticism by acknowledging that problems exist, without ceding any power to regulate or transform the way technology is developed and applied," wrote the AI Now Institute, a research group at New York University, in a 2018 report. "We have not seen strong oversight and accountability to backstop these ethical commitments" (AI Now 2018b).

The boards are also seen to mirror real-world inequality (mostly white men, very few women, few or no people of color or minorities) (see Levin 2019) or to have members who do not represent ethical values. The establishment of an ethics board by Google (actually called Advanced Technology External Advisory Council, ATEAC) lasted barely a week before it was disbanded amid great controversy.

The Google debate shows that discussing these issues in the public eye also invites public scrutiny. While I consider it positive that private industry is studying the issues and inviting views on company ethics, it is ultimately the CEO who gets to decide which suggestions on AI ethics would be incorporated into what are essentially business decisions. A company is clearly more concerned with the financial bottom line rather than sacrificing profit for ethical positions taken by an external advisory board, as there is no legal obligation to follow what are well-intentioned recommendations.

So the issue revolves around accountability, and in my view, government regulation will be needed to enforce it. Doteveryone, a UK organization (mission: Responsible Technology for a Fairer Future), issued a report entitled "Regulating for Responsible Technology" (Miller et al. 2018) which calls for a new independent regulatory body with three responsibilities:

1. Give regulators the capacity to hold technology to account.
2. Inform the public and policy-makers with robust evidence on the impacts of technology.
3. Support people to seek redress from technology-driven harms.

In addition to outlining that we currently have a "system in need of a steward," the organization also has a directory of regulation proposals in the UK to which it invites users to update (Doteveryone not dated). More surveys of such proposals might be very helpful in determining how best to go forward.

We should, however, also look at "soft law" which are substantive expectations that are not directly enforceable, as opposed to "hard law" which are legally enforceable requirements imposed by governments. As outlined by Wallach and Marchant, soft law includes voluntary programs, standards, codes of conduct, best practices, certification programs, guidelines, and statements of principles (Wallach and Marchant 2019). As an example of soft law being turned into hard law, they cite the Future of Life Institute Asilomar Principles (FLI 2017) adopted in 2017 as a soft law tool for AI governance, which have now been adopted by the State of California into its statutory law.

A Paradigm Shift Is Emerging

I believe one of the problems of the EU's High-Level Expert Group on AI is that it tries to be all-comprehensive and therefore tends towards more general and lofty declarations rather than be prescriptive in application. As I noted at the beginning of this paper, it is easier to address regulation in one aspect of AI rather than the entire gamut of applications. Let me focus on one such aspect that has started to capture attention in a major way: facial recognition and the pervasive use of cameras.

The Turing Award has been given to three preeminent computing scientists for their work on neural networks which has, inter alia, accelerated the development of face-recognition services. Yet they—together with some two dozen prominent AI researchers—have signed a letter to Amazon to stop selling its face-recognition technology (called "Rekognition") to law enforcement agencies because it is biased against women and people of color.

Facial recognition technology (FRT) has been used by government agencies, by retail industry, by Facebook with its millions of users posting photographs. In China, more than 176 million CCTV cameras are used for street monitoring and policing as well as in "cashless" stores and ATMs: where does consumer assistance start and surveillance begin?

Despite some positive aspects (reuniting missing children in India), there are major concerns about how to protect the privacy of those whose data is collected. With an industry quickly mushrooming to an estimated more than $10 billion in the next few years, alarms are beginning to sound about the lack of governmental oversight and the stealthy way it can be used to collect data on crowds of people—as we learned when it was revealed that the musician Taylor Swift had deployed FTR during her performances to root out stalkers. But is the technology only used for security?

Containing FTR is easier in Europe, where strict privacy laws are being enforced with the GDPR, but in other countries (and continents) no regulations exist. Yet even here in Europe people are warning against the "surveillance state." Looking at the increasing coverage and discussion of FTR, I am of the opinion that this will be one area of focus for regulation in the near future.

Could There Be a Role for International Organizations or Institutions?

UN Secretary-General Antonio Guterres weighed in on AI in July 2018, stating that "the scale, spread and speed of change made possible by digital technologies is unprecedented, but the current means and levels of international cooperation are unequal to the challenge (UN 2018b)." He set up a High-Level Panel on Digital Cooperation, with Melinda Gates and Jack Ma as Co-Chairs, and 18 additional members serving in their individual capacity. Their task was to submit a report by mid-2019—contributing to the broader public debate—which identified policy, research, and information gaps, and made proposals to strengthen international cooperation in the digital space.

The Panel has reached out and sought comments on their efforts from people all over the world, conducting a "global dialogue" to assist in reaching their final conclusions. Of course, it is important to bring this discussion to all member states, many of which do not have the capacity to harness new technology and lack a sophisticated understanding of the matter. It is also important for the Organization to embed this report in the universal UN values, and to consider practical ways to leverage digital technologies to achieve the Sustainable Development Goals.

The report—called "The Age of Digital Interdependence"—emphasizes the importance of fostering greater inclusivity and trust online and sets out recommendations for potential models of cooperation, yet the report is more of a summary overview of the current state of affairs rather than a model for implementation of ideas and suggestions (UN Secretary General 2019). It is vague how the report's wide-sweeping recommendations will be applied, and there appears no direct follow-up.

What is missing, in my opinion, is to take stock of existing—and emerging—normative, regulatory, and cooperative processes. I would not expect the UN to set rules and standards, but to have an inventory of the current state of affairs would be very valuable for national efforts to build on.

Past efforts by UN high-level panels have had mixed success. Despite the enormous work that goes into reports by high-ranking participants, their recommendations have at times been taken note of, politely debated—and then disappeared into a drawer without seeing implementation. Let us hope that the prominent co-chairs of this report will continue to contribute to a lively open debate and ensure that the recommendations will see further discussion and follow-up.

Summing Up: 15 Recommendations

Rapidly emerging technologies—AI and robotics in particular—present a singular challenge to regulation by governments. The technologies are owned by private industry, they advance in the blink of an eye, and they are not easily understood due to their complexity and may be obsolete by the time a government has agreed to regulate them.

This means that traditional models of government regulation cannot be applied. So if not regulation, what can be done? Here are my proposals:

1. Expand AI expertise so that it is not confined to a small number of countries or a narrow segment of the population.
2. Accept that the right decisions on AI technology will not be taken without strong input from the technologists themselves.
3. Find therefore a common language for government officials, policy-makers, and technical experts.
4. Begin dialogue so that (a) policies are informed by technical possibilities and (b) technologists/experts appreciate the requirements for policy accountability.
5. Discuss how to build a social license for AI, including new incentive structures to encourage governments and private industry to align the development and deployment of AI technologies with the public interest.
6. Focus on outcome, not process: principles, privacy protection, digital policy convergence, and differences in legal and regulatory systems and cultures between the USA, EU, and China.
7. Establish some "Red Lines"—no-go areas for AI technology, such as lethal autonomous weapon systems, AI-supported assessment of citizens by the government ("social scoring").

8. Use the strategy of "soft law" to overcome limitations and challenges of traditional government regulation for AI and robotics.
9. Discuss the challenges, costs, reliability, and limitations of the current state of art.
10. Develop strong working relationships, particularly in the defense sector, between public and private AI developers.
11. Ensure that developers and regulators pay particular attention to the question of human-machine interface.
12. Understand how different domains raise different challenges.
13. Compile a list of guidelines that already exist and see where there are gaps that need to be filled to offer more guidance on transparency, accountability and fairness of AI tools.
14. Learn from adjacent communities (cyber security, biotech, aviation) about efforts to improve safety and robustness.
15. Governments, foundations, and corporations should allocate funding to develop and deploy AI systems with humanitarian goals.

I encourage others to add to the list. What is really important here is that we come to a common understanding of what needs to be done. How do we develop international protocols on how to develop and deploy AI systems? The more people ask that question, the more debate we have on it, the closer we will get to a common approach. This is what is needed more than ever today.

References

AI Now. (2018a). *Litigating algorithms: Challenging government use of algorithmic decision systems.* AI Now Institute. Retrieved from https://ainowinstitute.org/litigatingalgorithms.pdf
AI Now. (2018b). *AI Now report 2018.* AI Now Institute. Retrieved from https://ainowinstitute.org/AI_Now_2018_Report.pdf
CNIL. (2017). *How can humans keep the upper hand? Report on the ethical matters raised by algorithms and artificial intelligence.* Retrieved from https://www.cnil.fr/en/how-can-humans-keep-upper-hand-report-ethical-matters-raised-algorithms-and-artificial-intelligence
European External Action Services (EEAS). (2018). *Autonomous weapons must remain under human control, Mogherini says at European Parliament.* Retrieved from https://eeas.europa.eu/topics/economic-relations-connectivity-innovation/50465/autonomous-weapons-must-remain-under-human-control-mogherini-says-european-parliament_en
European Commission. (2018). *Artificial intelligence for Europe.* Communication from the commission to the European Parliament, the European council, the council, the European economic and social committee and the committee of the regions, Brussel, 25 April 2018. Retrieved from https://ec.europa.eu/newsroom/dae/document.cfm?doc_id=51625

European Commission. (2019). *Ethics guidelines for trustworthy AI.* Retrieved from https://ec.europa.eu/digital-single-market/en/news/ethics-guidelines-trustworthy-ai

European Parliament. (2018). *Resolution of 12 September 2018 on autonomous weapon systems (2018/2752(RSP)).* Retrieved from https://www.europarl.europa.eu/doceo/document/TA-8-2018-0341_EN.html

The Future of Life Institute (FLI). (2015). *Autonomous weapons: An open letter from AI & robotics researchers.* Retrieved from https://futureoflife.org/open-letter-autonomous-weapons

The Future of Life Institute (FLI). (2017). *Asilomar AI principles.* Retrieved from https://futureoflife.org/ai-principles/

Forbes. (2019). *Can you sue an algorithm for malpractice?* Interview with W. Nicholson Price. Forbes insights, 11 February 2019. Retrieved from https://www.forbes.com/sites/insights-intelai/2019/02/11/can-you-sue-an-algorithm-for-malpractice/

House of Lords Artificial Intelligence Committee. (2018). *AI in the UK: Ready, willing and able.* Retrieved from https://publications.parliament.uk/pa/ld201719/ldselect/ldai/100/100.pdf

Levin, S. (2019). *'Bias deep inside the code': The problem with AI 'ethics' in Silicon Valley.* The Guardian, 29 March 2019. Retrieved from https://www.theguardian.com/technology/2019/mar/28/big-tech-ai-ethics-boards-prejudice

Metha, I. (2019). *It's not just WhatsApp—India's fake news problem plagues several popular social networks.* The Next Web, 29 January 2019. Retrieved from https://thenextweb.com/in/2019/01/29/its-not-just-whatsapp-indias-fake-news-problem-plagues-several-popular-social-networks/

Metzinger, T. (2019). *Ethics washing made in Europe.* Der Tagesspiegel, 8 April 2019. Retrieved from https://www.tagesspiegel.de/politik/eu-guidelines-ethics-washing-made-in-europe/24195496.html

Miller, C., Ohrvik-Stott, J., & Coldicutt, R. (2018). *Regulating for responsible technology: Capacity, evidence and redress: A new system for a fairer future.* London: Doteveryone. Retrieved from https://doteveryone.org.uk/project/regulating-for-responsible-technology/.

UK Parliament. (2018). *Government response to house of lords artificial intelligence select committee's report on AI in the UK: Ready, willing and able?* Retrieved from https://www.parliament.uk/documents/lords-committees/Artificial-Intelligence/AI-Government-Response.pdf

United Nations (UN). (2018a). *Machines with power, discretion to take human life politically unacceptable, morally repugnant, secretary-general tells Lisbon 'Web Summit'.* Press release, 5 November 2018. Retrieved from https://www.un.org/press/en/2018/sgsm19332.doc.htm

United Nations (UN). (2018b). *Secretary-general appoints high-level panel on digital cooperation.* Press release, 12 July 2018. Retrieved from https://www.un.org/press/en/2018/sga1817.doc.htm

UN Secretary General. (2019). *The age of digital interdependence—Report of the high-level panel on digital cooperation.* Retrieved from https://digitalcooperation.org/report

Wallach, W., & Marchant, G. (2019). Toward the agile and comprehensive international governance of AI and robotics. *Proceedings of the IEEE, 107*(3), 505–508.

Watson, C. (2018). *The key moments from Mark Zuckerberg's testimony to Congress.* The Guardian, 11 April 2018. Retrieved from https://www.theguardian.com/technology/2018/apr/11/mark-zuckerbergs-testimony-to-congress-the-key-moments

Wong, Q. (2019). *Facebook CEO Mark Zuckerberg pushes back against claims of anti-conservative censorship.* Cnet, 18 October 2019. Retrieved from https://www.cnet.com/news/facebook-ceo-mark-zuckerberg-pushes-back-against-claims-of-conservative-censorship/

A Human Blueprint for AI Coexistence

Kai-Fu Lee

Contents

Abstract

The positive coexistence of humans and AI is possible and needs to be designed as a system that provides for all members of society, but one that also uses the wealth generated by AI to build a society that is more compassionate, loving, and ultimately human. It is incumbent on us to use the economic abundance of the AI age to foster the values of volunteers who devote their time and energy toward making their communities more caring. As a practical measure, to protect against AI/robotics' labor saving and job displacement effects, a "social investment stipend" should be explored. The stipend would be given to those who invest their time and energy in those activities that promote a kind, compassionate, and creative society, i.e., care work, community service, and education. It would put the economic bounty generated by AI to work in building a better society, rather than just numbing the pain of AI-induced job losses.

Keywords

Artificial intelligence · Robotics · Employment · Income · Social investment · Compassion

K.-F. Lee (✉)
Sinovation Ventures, Beijing, China
e-mail: kfl@chuangxin.com

J. von Braun et al. (eds.), *Robotics, AI, and Humanity*, https://doi.org/10.1007/978-3-030-54173-6_22

What Does the Future Look Like?[1]

Artificial intelligence is a technology that sparks the human imagination. What will our future look like as we come to share the earth with intelligent machines? Utopians believe that once AI far surpasses human intelligence, it will provide us with near-magical tools for alleviating suffering and realizing human potential. In this vision, super-intelligent AI systems will so deeply understand the universe that they will act as omnipotent oracles, answering humanity's most vexing questions and conjuring brilliant solutions to problems such as disease and climate change.

But not everyone is so optimistic. The best-known member of the dystopian camp is the technology entrepreneur Elon Musk, who has called super-intelligent AI systems "the biggest risk we face as a civilization," comparing their creation to "summoning the demon" (Kumparak 2014). This group warns that when humans create self-improving AI programs whose intellect dwarfs our own, we will lose the ability to understand or control them. Which vision to accept? I'd say neither.

They simply aren't possible based on the technology we have today or any breakthroughs that might be around the corner. Both scenarios would require "artificial general intelligence"—that is, AI systems that can handle the incredible diversity of tasks done by the human brain. Making this jump would require several fundamental scientific breakthroughs, each of which may take many decades.

AI Revolution: Advantages and Limits

The AI revolution will be of the magnitude of the Industrial Revolution—but probably larger and definitely faster. Where the steam engine only took over physical labor, AI can perform both intellectual and physical labor. And where the Industrial Revolution took centuries to spread beyond Europe and the USA, AI applications are already being adopted simultaneously all across the world.

AI's main advantage over humans lies in its ability to detect incredibly subtle patterns within large quantities of data and to learn from them. While a human mortgage officer will look at only a few relatively crude measures when deciding whether to grant you a loan (your credit score, income and age), an AI algorithm will learn from thousands of lesser variables (what web browser you use, how often you buy groceries, etc.). Taken alone, the predictive power of each of these is minuscule, but added together, they yield a

far more accurate prediction than the most discerning people are capable of.

For cognitive tasks, this ability to learn means that computers are no longer limited to simply carrying out a rote set of instructions written by humans. Instead, they can continuously learn from new data and perform better than their human programmers. For physical tasks, robots are no longer limited to repeating one set of actions (automation) but instead can chart new paths based on the visual and sensor data they take in (autonomy).

Together, this allows AI to take over countless tasks across society: driving a car, diagnosing a disease, or providing customer support. AI's superhuman performance of these tasks will lead to massive increases in productivity. According to a June 2017 study by the consulting firm PwC, AI's advance will generate $15.7 trillion in additional wealth for the world by 2030 (PwC 2017). This is great news for those with access to large amounts of capital and data. It's very bad news for anyone who earns their living doing soon-to-be-replaced jobs.

There are, however, limits to the abilities of today's AI, and those limits hint at a hopeful path forward. While AI is great at optimizing for a highly narrow objective, it is unable to choose its own goals or to think creatively. And while AI is superhuman in the coldblooded world of numbers and data, it lacks social skills or empathy—the ability to make another person feel understood and cared for. Analogously, in the world of robotics, AI is able to handle many crude tasks like stocking goods or driving cars, but it lacks the delicate dexterity needed to care for an elderly person or infant.

Jobs at Risk

What does that mean for workers who fear being replaced? Jobs that are asocial and repetitive, such as fast-food preparers or insurance adjusters, are likely to be taken over in their entirety. For jobs that are repetitive but social, such as bartenders and doctors, many of the core tasks will be done by AI, but there remains an interactive component that people will continue to perform. The jobs that will be safe, at least for now, are those well beyond the reach of AI's capabilities in terms of creativity, strategy, and sociability, from social workers to CEOs.

Even where AI doesn't destroy jobs outright, however, it will exacerbate inequality. AI is inherently monopolistic: A company with more data and better algorithms will gain ever more users and data. This self-reinforcing cycle will lead to winner-take-all markets, with one company making massive profits while its rivals languish.

A similar consolidation will occur across professions. The jobs that will remain relatively insulated from AI fall on opposite ends of the income spectrum. CEOs, home care nurses, attorneys, and hairstylists are all in "safe" professions, but the people in some of these professions will be swimming in the riches of the AI revolution while others compete against a vast pool of desperate fellow workers.

We can't know the precise shape and speed of AI's impact on jobs, but the broader picture is clear. This will not be the normal churn of capitalism's creative destruction, a process that inevitably arrives at a new equilibrium of more jobs, higher wages, and better quality of life for all.

Many of the free market's self-correcting mechanisms will break down in an AI economy. The twenty-first century may bring a new caste system, split into a plutocratic AI elite and the powerless struggling masses.

Recent history has shown us just how fragile our political institutions and social fabric can be in the face of disruptive change. If we allow AI economics to run their natural course, the geopolitical tumult of recent years will look like child's play.

On a personal and psychological level, the wounds could be even deeper. Society has trained most of us to tie our personal worth to the pursuit of work and success. In the coming years, people will watch algorithms and robots easily outmaneuver them at tasks they've spent a lifetime mastering. I fear that this will lead to a crushing feeling of futility and obsolescence. At worst, it will lead people to question their own worth and what it means to be human.

Compassion and Love

But in developing a blueprint for human coexistence with AI, we need to remember that intelligent machines will increasingly be able to do our jobs and meet our material needs, disrupting industries and displacing workers in the process. But there remains one thing that only human beings are able to create and share with one another: love.

With all of the advances in machine learning, the truth remains that we are still nowhere near creating AI machines that feel any emotions at all. Can you imagine the elation that comes from beating a world champion at the game you've devoted your whole life to mastering? AlphaGo, the first computer Go program to beat a human professional Go player, did just that, but it took no pleasure in its success, felt no happiness from winning, and had no desire to hug a loved one after its victory. Despite what science-fiction films like Her—in which a man and his artificially intelligent computer operating system fall in love—portray, AI has no ability or desire to love or be loved.

It is in this uniquely human potential for growth, compassion, and love where I see hope. I believe we must forge a new synergy between artificial intelligence and the human heart, and look for ways to use the forthcoming material abundance generated by artificial intelligence to foster love and compassion in our societies.

If we can do these things, I believe there is a path toward a future of both economic prosperity and spiritual flourishing. Navigating that path will be tricky, but if we are able to unite behind this common goal, I believe humans will not just survive in the age of AI. We will thrive like never before.

The New Social Contract

The challenges before us remain immense. Within 15 years, I predict that we will technically be able to automate 40–50% of all jobs in the United States. That does not mean all of those jobs will disappear overnight, but if the markets are left to their own devices, we will begin to see massive pressure on working people. China and other developing countries may differ slightly in the timing of those impacts, lagging or leading in job losses depending on the structures of their economies. But the overarching trend remains the same: rising unemployment and widening inequality.

Techno-optimists will point to history, citing the Industrial Revolution and the nineteenth-century textile industry as "proof" that things always work out for the best. But as we've seen, this argument stands on increasingly shaky ground. The coming scale, pace, and skill-bias of the AI revolution mean that we face a new and historically unique challenge. Even if the most dire predictions of unemployment do not materialize, AI will take the growing wealth inequality of the internet age and accelerate it tremendously.

We are already witnessing the way that stagnant wages and growing inequality can lead to political instability and even violence. As AI rolls out across our economies and societies, we risk aggravating and quickening these trends. Labor markets have a way of balancing themselves out in the long run, but getting to that promised long run requires we first pass through a trial by fire of job losses and growing inequality that threaten to derail the process. Meeting these challenges means we cannot afford to passively react. We must proactively seize the opportunity that the material wealth of AI will grant us and use it to reconstruct our economies and rewrite our social contracts. The epiphanies that emerged from my experience with cancer were deeply personal, but I believe they also gave me a new clarity and vision for how we can approach these problems together.

Building societies that thrive in the age of AI will require substantial changes to our economy but also a shift in culture and values. Centuries of living within the industrial economy have conditioned many of us to believe that our primary role in society (and even our identity) is found in productive, wage-earning work. Take that away and you have broken one of the strongest bonds between a person and his or her community. As we transition from the industrial age to the

AI age, we will need to move away from a mindset that equates work with life or treats humans as variables in a grand productivity optimization algorithm. Instead, we must move toward a new culture that values human love, service, and compassion more than ever before.

No economic or social policy can "brute force" a change in our hearts. But in choosing different policies, we can reward different behaviors and start to nudge our culture in different directions. We can choose a purely technocratic approach—one that sees each of us as a set of financial and material needs to be satisfied—and simply transfer enough cash to all people so that they don't starve or go homeless. In fact, this notion of universal basic income (UBI) seems to be becoming more and more popular these days (Ito 2018). But in making that choice I believe we would both devalue our own humanity and miss out on an unparalleled opportunity. Instead, I want to lay out proposals for how we can use the economic bounty created by AI to double-down on what makes us human. Doing this will require rewriting our fundamental social contracts and restructuring economic incentives to reward socially productive activities in the same way that the industrial economy rewarded economically productive activities.

Market Symbiosis

The private sector is leading the AI revolution, and, in my mind, it must also take the lead in creating the new, more humanistic jobs that power it. Some of these will emerge through the natural functioning of the free market, while others will require conscious efforts by those motivated to make a difference.

Many of the jobs created by the free market will grow out of a natural symbiosis between humans and machines. While AI handles the routine optimization tasks, human beings will bring the personal, creative, and compassionate touch. This will involve the redefinition of existing occupations or the creation of entirely new professions in which people team up with machines to deliver services that are both highly efficient and eminently human. AI will do the analytical thinking, while humans will wrap that analysis in warmth and compassion.

A clear example of human-AI symbiosis for the upper-left-hand quadrant can be found in the field of medicine. I have little doubt that AI algorithms will eventually far surpass human doctors in their ability to diagnose disease and recommend treatments. Legacy institutions—medical schools, professional associations, and hospitals—may slow down the adoption of these diagnostic tools, using them only in narrow fields or strictly as reference tools. But in a matter of a few decades, I'm confident that the accuracy and efficiency

gains will be so great that AI-driven diagnoses will take over eventually.

One response to this would be to get rid of doctors entirely, replacing them with machines that take in symptoms and spit out diagnoses. But patients don't want to be treated by a machine, a black box of medical knowledge that delivers a cold pronouncement: "You have fourth-stage lymphoma and a 70 percent likelihood of dying within five years." Instead, patients will desire—and I believe the market will create—a more humanistic approach to medicine.

Traditional doctors could instead evolve into a new profession, one that I'll call a "compassionate caregiver." These medical professionals would combine the skills of a nurse, medical technician, social worker, and even psychologist. Compassionate caregivers would be trained not just in operating and understanding the diagnostic tools but also in communicating with patients, consoling them in times of trauma, and emotionally supporting them throughout their treatment. Instead of simply informing patients of their objectively optimized chances of survival, they could share encouraging stories, saying "Kai-Fu had the same lymphoma as you and he survived, so I believe you can too."

These compassionate caregivers would not compete with machines in their ability to memorize facts or optimize treatment regimens. In the long run that's a losing battle. Compassionate caregivers would be well trained, but in activities requiring more emotional intelligence, not as mere vessels for the canon of medical knowledge. They would form a perfect complement to the machine, giving patients unparalleled accuracy in their diagnoses as well as the human touch that is so often missing from our hospitals today. In this human-machine symbiosis created by the free market, we would inch our society ahead in a direction of being a little kinder and a little more loving.

Best of all, the emergence of compassionate caregivers would dramatically increase both the number of jobs and the total amount of medical care given. Today, the scarcity of trained doctors drives up the cost of healthcare and drives down the amount of quality care delivered around the world. Under current conditions of supply and demand, it's simply not cost-feasible to increase the number of doctors. As a result, we strictly ration the care they deliver. No one wants to go wait in line for hours just to have a few minutes with a doctor, meaning that most people only go to hospitals when they feel it's absolutely necessary. While compassionate caregivers will be well trained, they can be drawn from a larger pool of workers than doctors and won't need to undergo the years of rote memorization that is required of doctors today. As a result, society will be able to cost-effectively support far more compassionate caregivers than there are doctors, and we would receive far more and better care.

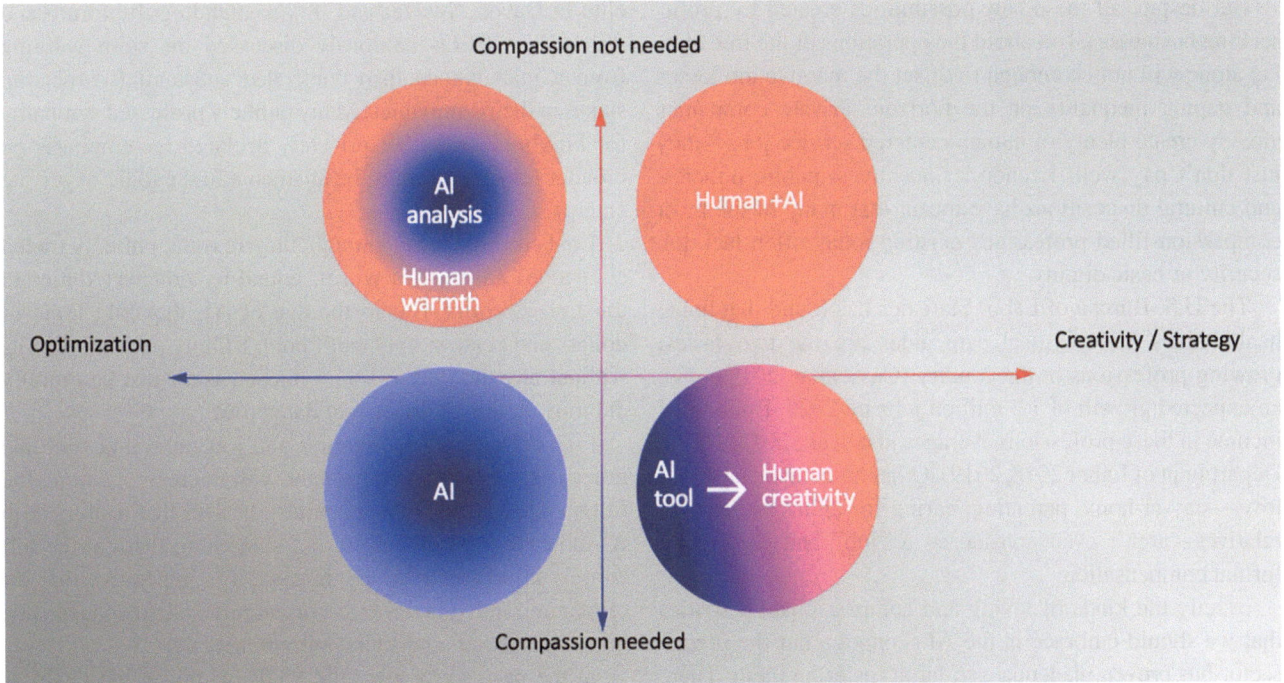

Human-AI coexistence in the labor market. Source: Lee (2018b: 211)

Similar synergies will emerge in many other fields: teaching, law, event planning, and high-end retail. Paralegals at law firms could hand their routine research tasks off to algorithms and instead focus on communicating more with clients and making them feel cared for. AI-powered supermarkets like the Amazon Go store may not need cashiers anymore, so they could greatly upgrade the customer experience by hiring friendly concierges.

For those in professional sectors, it will be imperative that they adopt and learn to leverage AI tools as they arrive. As with any technological revolution, many workers will find the new tools both imperfect in their uses and potentially threatening in their implications. But these tools will only improve with time, and those who seek to compete against AI on its own terms will lose out. In the long run, resistance may be futile, but symbiosis will be rewarded.

Finally, the internet-enabled sharing economy will contribute significantly to alleviating job losses and redefining work for the AI age. We'll see more people step out of traditional careers that are being taken over by algorithms, instead using new platforms that apply the "Uber model" to a variety of services. We see this already in Care.com,[2] an online platform for connecting caregivers and customers, and I believe we will see a blossoming of analogous models in education and other fields. Many mass-market goods and services will be captured by data and optimized by algorithms, but some of the more piecemeal or personalized work within the sharing economy will remain the exclusive domain of humans.

In the past, this type of work was constrained by the bureaucratic costs of running a vertical company that attracted customers, dispatched workers, and kept everyone on the payroll even when there wasn't work to be done. The platformatization of these industries dramatically increases their efficiency, increasing total demand and take-home pay for the service workers themselves. Adding AI to the equation—as ride-hailing companies like Didi and Uber have already done—will only further boost efficiency and attract more workers.

New Service Jobs

Beyond the established roles in the sharing economy, I'm confident we will see entirely new service jobs emerge that we can hardly imagine today. Explain to someone in the 1950s what a "life coach" was and they'd probably think you were goofy. Likewise, as AI frees up our time, creative entrepreneurs and ordinary people will leverage these platforms to create new kinds of jobs. Perhaps people will hire "season changers" who redecorate their closets every few months, scenting them with flowers and aromas that match the mood of the season. Or environmentally conscious families will hire "home sustainability consultants" to meet with the family and explore creative and fun ways for the household to reduce its environmental footprint.

[2]https://en.wikipedia.org/wiki/Care.com

But despite all these new possibilities created by profit-seeking businesses, I'm afraid the operations of the free market alone will not be enough to offset the massive job losses and gaping inequality on the horizon. Private companies already create plenty of human-centered service jobs—they just don't pay well. Economic incentives, public policies, and cultural dispositions have meant that many of the most compassion-filled professions existing today often lack job security or basic dignity.

The U.S. Bureau of Labor Statistics has found that home health aides and personal care aides are the two fastest growing professions in the country (Casselman 2017), with an expected growth of 1.2 million jobs by 2026. But annual income in these professions averages just over $20,000 (U.S. Department of Labor 2018, 2019). Other humanistic labors of love—stay-at-home parenting, caring for aging or disabled relatives—aren't even considered a "job" and receive no formal compensation.

exactly the kinds of loving and compassionate activities that we should embrace in the AI economy, but the private sector has proven inadequate so far at fostering them. There may come a day when we enjoy such material abundance that economic incentives are no longer needed. But in our present economic and cultural moment, money still talks. Orchestrating a true shift in culture will require not just creating these jobs but turning them into true careers with respectable pay and greater dignity.

Encouraging and rewarding these prosocial activities means going beyond the market symbiosis of the private sector. We will need to re-energize these industries through service sector impact investing and government policies that nudge forward a broader shift in cultural values.

Fink's Letter

When a man overseeing $5.7 trillion speaks, the global business community tends to listen. So when BlackRock founder Larry Fink, head of the world's largest asset management company, posted a letter to CEOs demanding greater attention to social impact, it sent shockwaves through corporations around the globe. In the letter, titled "A Sense of Purpose," Fink (2018) wrote, "We . . . see many governments failing to prepare for the future, on issues ranging from retirement and infrastructure to automation and worker retraining. As a result, society increasingly is turning to the private sector and asking that companies respond to broader societal challenges. . . . Society is demanding that companies, both public and private, serve a social purpose. . . . Companies must benefit all of their stakeholders, including shareholders, employees, customers, and the communities in which they operate." Fink's letter dropped just days before the 2018 World Economic Forum, an annual gathering of the global financial

elite in Davos, Switzerland. I was attending the forum and watched as CEOs anxiously discussed the stern warning from a man whose firm controlled substantial ownership stakes in their companies. Many publicly professed sympathy for Fink's message but privately declared his emphasis on broader social welfare to be anathema to the logic of private enterprise.

Looked at narrowly enough, they're right: publicly traded companies are in it to win it, bound by fiduciary duties to maximize profits. But in the age of AI, this cold logic of dollars and cents simply can't hold. Blindly pursuing profits without any thought to social impact won't just be morally dubious; it will be downright dangerous.

Fink referenced automation and job retraining multiple times in his letter. As an investor with interests spanning the full breadth of the global economy, he sees that dealing with AI-induced displacement is not something that can be left entirely up to free markets. Instead, it is imperative that we reimagine and reinvigorate corporate social responsibility, impact investing, and social entrepreneurship.

In the past, these were the kinds of things that business people merely dabbled in when they had time and money to spare. Sure, they think, why not throw some money into a microfinance startup or buy some corporate carbon offsets so we can put out a happy press release touting it. But in the age of AI, we will need to seriously deepen our commitment to—and broaden our definition of—these activities. Whereas these have previously focused on feel-good philanthropic issues like environmental protection and poverty alleviation, social impact in the age of AI must also take on a new dimension: the creation of large numbers of service jobs for displaced workers.

New Impact Investing

As a venture-capital investor, I see a particularly strong role for a new kind of impact investing. I foresee a venture ecosystem emerging that views the creation of humanistic service-sector jobs as a good in and of itself. It will steer money into human-focused service projects that can scale up and hire large numbers of people: lactation consultants for postnatal care, trained coaches for youth sports, gatherers of family oral histories, nature guides at national parks, or conversation partners for the elderly. Jobs like these can be meaningful on both a societal and personal level, and many of them have the potential to generate real revenue—just not the 10,000% returns that come from investing in a unicorn technology startup.

Kick-starting this ecosystem will require a shift in mentality for venture capitalists (VCs) who participate. The very idea of venture capital has been built around high risks and exponential returns. When an investor puts money into ten startups, they know full well that nine of them most likely will fail. But if that one success story turns into a billion-dollar company, the exponential returns on that one investment make the fund a huge success. Driving those exponential returns are the unique economics of the internet. Digital products can be scaled up infinitely with near-zero marginal costs, meaning the most successful companies achieve astronomical profits.

Service-focused impact investing, however, will need to be different. It will need to accept linear returns when coupled with meaningful job creation. That's because human-driven service jobs simply cannot achieve these exponential returns on investment. When someone builds a great company around human care work, they cannot digitally replicate these services and blast them out across the globe. Instead, the business must be built piece by piece, worker by worker. The truth is, traditional VCs wouldn't bother with these kinds of linear companies, but these companies will be a key pillar in building an AI economy that creates new jobs and fosters human connections.

There will of course be failures, and returns will never match pure technology VC funds. But that should be fine with those involved. The ecosystem will likely be staffed by older VC executives who are looking to make a difference, or possibly by younger VC types who are taking a "sabbatical" or doing "pro bono" work. They will bring along their keen instincts for picking entrepreneurs and building companies, and will put them to work on these linear service companies. The money behind the funds will likely come from governments looking to efficiently generate new jobs, as well as companies doing corporate social responsibility.

Together, these players will create a unique ecosystem that is much more jobs-focused than pure philanthropy, much more impact-focused than pure venture capital. If we can pull together these different strands of socially conscious business, I believe we'll be able to weave a new kind of employment safety net, all while building communities that foster love and compassion.

In this respect I am most excited about the impact of AI on healthcare and education. These two sectors are ready for AI disruption and can deploy AI for good. For example, we have invested in a company that uses AI and big data to optimize supply chains, reducing medication shortages for over 150 million people living in rural China. We also have drug discovery companies combining deep learning and generative chemistry to shorten drug discovery time by a factor of three to four. In education, we see companies developing AI solutions to improve English pronunciation, grade exams, and homework, and personalize and gamify math learning. This will free teachers from routine tasks, and allow them to spend time building inspirational and stimulating connections and with our next generations.

Big Changes and Big Government

And yet, for all the power of the private market and the good intentions of social entrepreneurs, many people will still fall through the cracks. We need look no further than the gaping inequality and destitute poverty in so much of the world today to recognize that markets and moral imperatives are not enough. Orchestrating a fundamental change in economic structures often requires the full force of governmental power. If we hope to write a new social contract for the age of AI, we will need to pull on the levers of public policy.

There are some in Silicon Valley who see this as the point where UBI comes into play. Faced with inadequate job growth, the government must provide a blanket guarantee of economic security, a cash transfer that can save displaced workers from destitution and which will also save the tech elite from having to do anything else about it.

The unconditional nature of the transfer fits with the highly individualistic, live-and-let-live libertarianism that undergirds much of Silicon Valley. Who is the government, UBI proponents ask, to tell people how to spend their time? Just give them the money and let them figure it out on their own. It's an approach that matches how the tech elite tend to view society as a whole. Looking outward from Silicon Valley, they often see the world in terms of "users" rather than citizens, customers rather than members of a community. I have a different vision. I don't want to live in a society divided into technological castes, where the AI elite live in a cloistered world of almost unimaginable wealth, relying on minimal handouts to keep the unemployed masses sedate in their place. I want to create a system that provides for all members of society, but one that also uses the wealth generated by AI to build a society that is more compassionate, loving, and ultimately human.

The Social Investment Stipend

Just as those volunteers devoted their time and energy toward making their communities a little bit more loving, I believe it is incumbent on us to use the economic abundance of the AI age to foster the values of volunteers who devote their time and energy toward making their communities a little bit more loving and encourage this same kind of activity. To do this, I propose we explore the creation not of a UBI but of what I call a social investment stipend. The stipend would be a decent government salary given to those who invest their time and energy in those activities that promote a kind, compassionate, and creative society. These would include three broad categories: care work, community service, and education.

These would form the pillars of a new social contract, one that valued and rewarded socially beneficial activities in the same way we currently reward economically productive activities. The stipend would not substitute for a social safety net—the traditional welfare, healthcare, or unemployment benefits to meet basic needs—but would offer a respectable income to those who choose to invest energy in these socially productive activities. Today, social status is still largely tied to income and career advancement. Endowing these professions with respect will require paying them a respectable salary and offering the opportunity for advancement like a normal career. If executed well, the social investment stipend would nudge our culture in a more compassionate direction. It would put the economic bounty of AI to work in building a better society, rather than just numbing the pain of AI-induced job losses.

Each of the three recognized categories—care, service, and education—would encompass a wide range of activities, with different levels of compensation for full- and part-time participation. Care work could include parenting of young children, attending to an aging parent, assisting a friend or family member dealing with illness, or helping someone with mental or physical disabilities live life to the fullest. This category would create a veritable army of people—loved ones, friends, or even strangers—who could assist those in need, offering them what my entrepreneur friend's touchscreen device for the elderly never could: human warmth. Service work would be similarly broadly defined, encompassing much of the current work of nonprofit groups as well as volunteers. Tasks could include performing environmental remediation, leading afterschool programs, guiding tours at national parks, or collecting oral histories from elders in our communities. Participants in these programs would register with an established group and commit to a certain number of hours of service work to meet the requirements of the stipend.

Finally, education could range from professional training for the jobs of the AI age to taking classes that could transform a hobby into a career. Some recipients of the stipend will use that financial freedom to pursue a degree in machine learning and use it to find a high-paying job. Others will use that same freedom to take acting classes or study digital marketing.

Bear in mind that requiring participation in one of the above activities is not something designed to dictate the daily activities of each person receiving the stipend. That is, the beauty of human beings lies in our diversity, the way we each bring different backgrounds, skills, interests, and eccentricities. I don't seek to smother that diversity with a command-and-control system of redistribution that rewards only a narrow range of socially approved activities. But by requiring some social contribution in order to receive the stipend, we would foster a far different ideology than the laissez-faire individualism of a UBI. Providing a stipend in exchange for participation in prosocial activities reinforces a clear message: It took efforts from people all across society to help us reach this point of economic abundance. We are now collectively using that abundance to recommit ourselves to one another, reinforcing the bonds of compassion and love that make us human.

Looking across all the activities above, I believe there will be a wide enough range of choices to offer something suitable to all workers who have been displaced by AI. The more people-oriented may opt for care work, the more ambitious can enroll in job-training programs, and those inspired by a social cause may take up service or advocacy jobs. In an age in which intelligent machines have supplanted us as the cogs and gears in the engine of our economy, I hope that we will value all of these pursuits—care, service, and personal cultivation—as part of our collective social project of building a more human society.

Open Questions

Implementing a social investment stipend will of course raise new questions and frictions: How much should the stipend be? Should we reward people differently based on their performance in these activities? How do we know if someone is dutifully performing their "care" work? And what kinds of activities should count as "service" work? These are admittedly difficult questions, ones for which there are no clear-cut answers. Administering a social investment stipend in countries with hundreds of millions of people will involve lots of paperwork and legwork by governments and the organizations that create these new roles.

But these challenges are far from insurmountable. Governments in developed societies already attend to a dizzying array of bureaucratic tasks just to maintain public services, education systems, and social safety nets. Our governments already do the work of inspecting buildings, accrediting schools, offering unemployment benefits, monitoring sanitary conditions at hundreds of thousands of restaurants, and providing health insurance to tens of millions of people. Operating a social investment stipend would add to this workload, but I believe it would be more than manageable. Given the huge human upside to providing such a stipend, I believe the added organizational challenges will be well worth the rewards to our communities. But what about affordability? Offering a living salary to people performing all of the above tasks would require massive amounts of revenue, totals that today appear unworkable in many heavily indebted countries. AI will certainly increase productivity across society, but can it really generate the huge sums necessary to finance such dramatic expansion in government expenditures?

This too remains an open question, one that will only be settled once the AI technologies themselves proliferate across our economies. If AI meets or exceeds predictions for productivity gains and wealth creation, I believe we could fund these types of programs through super taxes on super profits. Yes, it would somewhat cut into economic incentives to advance AI, but given the dizzying profits that will accrue to the winners in the AI age, I don't see this as a substantial impediment to innovation. But it will take years to get to that place of astronomical profits, years during which working people will be hurting. To smooth the transition, I propose a slow ratcheting up of assistance. While leaping straight into the full social investment stipend described above likely won't work, I do think we will be able to implement incremental policies along the way. These piecemeal policies could both counteract job displacement as it happens and move us toward the new social contract articulated above.

We could start by greatly increasing government support for new parents so that they have the choice to remain at home or send their child to full-time daycare. For parents who choose to home-school their kids, the government could offer subsidies equivalent to a teacher's pay for those who attain certain certifications. In the public school systems, the number of teachers could also be greatly expanded—potentially by a factor as high as ten—with each teacher tasked with a smaller number of students that they can teach in concert with AI education programs. Government subsidies and stipends could also go to workers undergoing job retraining and people caring for aging parents. These simple programs would allow us to put in place the first building blocks of a stipend, beginning the work of shifting the culture and laying the groundwork for further expansion.

As AI continues to generate both economic value and worker displacement, we could slowly expand the purview of these subsidies to activities beyond care work or job training. And once the full impact of AI—very good for productivity, very bad for employment—becomes clear, we should be able to muster the resources and public will to implement programs akin to the social investment stipend. When we do, I hope that this will not just alleviate the economic, social, and psychological suffering of the AI age. Rather, I hope that it will further empower us to live in a way that honors our humanity and empowers us to do what no machine can: share our love with those around us.

References

Casselman, B. (2017). *A peek at future jobs reveals growing economic divides.* Available via New York Times. Retrieved February 27, 2020, from https://www.nytimes.com/2017/10/24/business/economy/future-jobs.html

Fink, L. (2018). *Larry Fink's annual letter to CEOs: A sense of purpose.* Available via BlackRock. Retrieved February 27, 2020, from https://www.blackrock.com/corporate/en-us/investor-relations/larry-fink-ceo-letter

Ito, J. (2018). *The paradox of universal basic income.* Available via WIRED. Retrieved February 27, 2020, from https://www.wired.com/story/the-paradox-of-universal-basic-income/

Kumparak, G. (2014). *Elon Musk compares building artificial intelligence to "summoning the demon".* Available via TC. Retrieved February 27, 2020, from https://techcrunch.com/2014/10/26/elon-musk-compares-building-artificial-intelligence-to-summoning-the-demon/

Lee, K.-F. (2018a). *AI superpowers: China, Silicon Valley, and the new world order.* New York: Houghton Mifflin Harcourt.

Lee, K.-F. (2018b, September 14). The human promise of the AI revolution. *The Wall Street Journal.* https://www.wsj.com/articles/the-human-promise-of-the-ai-revolution-1536935115

PwC. (2017). *Sizing the price: PwC's global artificial intelligence study: Exploiting the AI revolution.* Available via pwc. Retrieved February 27, 2020, from https://www.pwc.com/gx/en/issues/data-and-analytics/publications/artificial-intelligence-study.html

U.S. Department of Labor, Bureau of Labor Statistics, Occupational Employment Statistics. (2018). *Personal care aides.* Available via U.S. Bureau of Labor Statistics. Retrieved February 27, 2020, from https://www.bls.gov/oes/current/oes399021.htm

U.S. Department of Labor, Bureau of Labor Statistics, Occupational Employment Statistics. (2019). *Home health aides and personal care aides.* Available via U.S. Bureau of Labor Statistics. Retrieved February 27, 2020, from https://www.bls.gov/ooh/healthcare/home-health-aides-and-personal-care-aides.htm